ALGEBRAIC RECURSION THEORY

Mathematics and its Applications
Series Editor: G. M. BELL, Professor of Mathematics, King's College (KQC), University of London

Statistics and Operational Research
Editor: B. W. CONOLLY, Professor of Operational Research, Queen Mary College, University of London

Mathematics and its applications are now awe-inspiring in their scope, variety and depth. Not only is there rapid growth in pure mathematics and its applications to the traditional fields of the physical sciences, engineering and statistics, but new fields of application are emerging in biology, ecology and social organisation. The user of mathematics must assimilate subtle new techniques and also learn to handle the great power of the computer efficiently and economically.

The need of clear, concise and authoritative texts is thus greater than ever and our series will endeavour to supply this need. It aims to be comprehensive and yet flexible. Works surveying recent research will introduce new areas and up-to-date mathematical methods. Undergraduate texts on established topics will stimulate student interest by including applications relevant at the present day. The series will also include selected volumes of lecture notes which will enable certain important topics to be presented earlier than would otherwise be possible.

In all these ways it is hoped to render a valuable service to those who learn, teach, develop and use mathematics.

For complete series list see back of book

ALGEBRAIC RECURSION THEORY

L. L. IVANOV
Institute of Mathematics, Bulgarian Academy of Sciences
and Faculty of Mathematics, University of Sofia

Editor:
Dr. J. L. BELL
Department of Mathematics
London School of Economics and Political Science

ELLIS HORWOOD LIMITED
Publishers · Chichester

Halsted Press: a division of
JOHN WILEY & SONS
New York · Chichester · Brisbane · Toronto

First published in 1986 by
ELLIS HORWOOD LIMITED
Market Cross House, Cooper Street,
Chichester, West Sussex, PO19 1EB, England
The publisher's colophon is reproduced from James Gillison's drawing of the ancient Market Cross, Chichester.

Distributors:
Australia and New Zealand:
JACARANDA WILEY LIMITED
GPO Box 859, Brisbane, Queensland 4001, Australia
Canada:
JOHN WILEY & SONS CANADA LIMITED
22 Worcester Road, Rexdale, Ontario, Canada
Europe and Africa:
JOHN WILEY & SONS LIMITED
Baffins Lane, Chichester, West Sussex, England
North and South America and the rest of the world:
Halsted Press: a division of
JOHN WILEY & SONS
605 Third Avenue, New York, NY 10158, USA

British Library Cataloguing in Publication Data
Ivanov, L.L.
Algebraic recursion theory. —
(Mathematics and its applications)
1. Recursive functions
I. Title II. Bell, J.L. III. Series
511.3 QA9.615

Library of Congress Card No. 86–20923

ISBN 0–7458–0102–1 (Ellis Horwood Limited)
ISBN 0–470–20725–6 (Halsted Press)

Printed in Great Britain by The Camelot Press, Southampton

Preface

This monograph on algebraic recursion theory has its origins in a disertation submitted to Sofia University in 1980. Numerous recent developments are included and, as is often the case, the very process of working on the manuscript added new episodes to the story.

Undoubtedly, the fundamental works of D. Skordev on algebraic recursion theory have made this book possible. Skordev's mathematical ideas and style of thinking have had a significant effect on the whole of my work and I am deeply indebted to him for his invaluable help and encouragement.

The papers of S. C. Kleene on higher recursion theory are of particular importance for me since the present approach was inspired by the idea of embracing Kleene's notion of recursiveness in higher types. The works of Y. N. Moschovakis on inductive definability and recursion in higher types should also be mentioned on acccount of their beneficial influence.

Acknowledgements:
I should like to acknowledge my colleagues D. Vakarelov, N. Georgieva, J. Zashev. I. Soskov, S. Passay, T. Tinchev, S. Radev, A. Buda, P. Petkov, G. Gargov and V. Sotirov for their helpful comments and discussions.

I thank Dr. J. L. Bell of London School of Economics who polished up the 'Bulgarian English' of this book; and also Mr Ellis Horwood, Mrs Patricia Morrison and Mrs Hazel Kelly, who made my work with Ellis Horwood Publishers a real pleasure.

The final version of the book was prepared during my stay at Oslo University during the spring term 1985. I am indebted to my Norwegian colleagues and especially to D. Normann for stimulating discussions.

Sofia, September 1986 L. L. Ivanov

To Pepa and Borislava

Table of contents

PART A

Introduction

CHAPTER 1

Preliminaries

This book aims to provide a straightforward introduction to some fundamentals of modern Recursion Theory and to establish a basis for further work in this and related areas. The mathematical core of the subject is the theory of an algebraic system which we call an *iterative operative space*, IOS for short, which is a partially ordered set with four basic operations satisfying several axioms.

The algebraic system in question has been specially designed to support a unified yet selective axiomatic study of the concept of effective computability arising in a wide range of disciplines, from Computer Science to Generalized Recursion Theory. Some reasons for our choice are given in the introductory chapter 2, while a further discussion follows in the Epilogue.

By way of illustration two standard spaces are presented in chapter 3. They are essentially different, so the axiomatic system of IOS has no single standard model to reflect adequately its intuitive aspects. Nonetheless, one could start with some particular spaces and only later pass to the general theory. It seems however that the detailed preliminary or parallel study of examples is not a very good introduction, for the key to the theory lies in the μ-induction technique which is equally clear in particular spaces and *in abstract*. Perhaps the best way to familiarize oneself with IOS is to practise proving properties of its initial operations as done in chapters 4, 6. Finally, passing from the general to the particular is in closer conformity with our traditional Bulgarian way of thinking; for instance, we address a letter to South Georgia, Grytviken, Scotia Arc College, not Scotia Arc College, Grytviken, South Georgia, as is the custom in other countries. These are the chief reasons for arranging the exposition as we have.

The book consists of seven parts, part A being the Introduction. In part B a general theory of recursion is developed. Abstract notions of primitive recursiveness and recursiveness are introduced, a number of properties of the IOS-operations are established and used to obtain some basic statements including a Pull Back Theorem, a Transition Theorem, Representation Theorems for primitive recursive and partial recursive number theoretic functions, Normal Form Theorems, Enumeration Theorems, a Kleene First Recursion Theorem, a Second Recursion Theorem, abstract Rice and Rogers Theorems.

Part C broadens the argument with recursion theory on so called consecutive spaces, spaces with additional operations and hierarchies of spaces. The hierarchies in question are transfinite sequences of consecutive IOS each of which is a proper subspace of its successors. In particular, abstract analogues to Platek's First Recursion Theorem are established. Thus parts B, C introduce the reader straightaway to the central topics of the book. To keep fiction in touch with reality certain examples of IOS are regularly used as sources of illustration, mainly in the exercises.

Part D deals with some constructions yielding IOS, consecutive IOS and hierarchies of IOS.

In part E a series of spaces is used to express particular notions of effective computability, e.g. the relative μ-recursiveness and partial recursiveness of Ordinary Recursion Theory, the stack recursiveness of Germano and Maggiolo–Schettini [1976], the prime and search computability of Moschovakis [1969], as well as computabilities by abstract programs. It is not our intention to present or study all known spaces and notions of effective computability but rather to suggest typical patterns of modelling applicable to other similar situations as well.

Connections with certain related theories are established in part F. The system of de Bakker and Scott, better known as Scott's μ-calculus, and those of Skordev [1982, 1976] are considered in chapters 26, 27. In chapters 28–30 appropriate spaces provide a framework for the theory of recursive functionals of Kechris and Moschovakis [1977], Higher Recursion Theory and Inductive Definability Theory. While parts E, F clarify the way in which the general axiomatic theory produces particular recursion theories, few details about these are given, for this is beyond the scope of the book. Works listed in the References offer further introduction and bibliography; especially useful is Barwise [1977], part C.

Part G comprises some final remarks on the axiomatization of Recursion Theory with special attention paid to the present approach.

The exposition is necessarily detailed since no comprehensive presentation of the subject can be found elsewhere. Nonetheless, the author has tried to clear a middle course between an encyclopaedic reference book and a readable text-book. A number of exercises have been provided, sometimes contributing to the main text.

A lecture course based on the book may include a minimum of general theory and Ordinary Recursion Theory (selected parts of chapters 4–9, 18, 19, 22) as applied to a more specific area, say the Mathematical Theory of Programs (chapters 23, 26) or Computable Functions over Arbitrary Domains (chapters 10, 21, 24), Recursion in Higher Types (chapters 28, 29). Inductive Definability (chapter 30), a corresponding introduction to Recursion on Set Functions or Ordinal Functions etc. If the field of further interest is to be Algebraic Recursion Theory itself, then substantial parts of chapters 10, 12–16, 20, 27 should be included instead.

Because of its axiomatic character the present work is essentially self-contained. The reader is assumed to possess the ordinary mathematical sophistication of an undergraduate student and a slender acquaintance with ordinals and transfinite induction.

Some standard notations are used throughout the book. 'Iff' is, as usual, written for 'if, and only if'. The letter ω denotes the set of all natural numbers (nonnegative integers), as well as the first infinite cardinal. Letters m, n, i, j, k, l stand for natural numbers and ξ, η, ζ for ordinals.

If M, N are sets, then $f{:}M \longrightarrow N$ means that f is a partial *single-valued function* from M to N, i.e. $f \subseteq M \times N$ and whenever (s,t), $(s,r) \in f$, then $r = t$. Symbols $\downarrow$ and $\uparrow$ are written respectively for 'is defined' and 'is not defined'. The equality sign is also used as a sign of conventional equality. For instance, if $f, g{:}M \longrightarrow N$, then $f(s) = g(s)$ means that either $f(s)$, $g(s)$ are defined and identical, or $f(s)\uparrow, g(s)\uparrow$ (the usual notation is $f(s) \simeq g(s)$). If f is a partial *multiple-valued function* from M to N, $f{:}M \longrightarrow 2^N \setminus \{\varnothing\}$, then we write also $f{:}M \to 2^N$ or $f \subseteq M \times N$, meaning $f(s)\uparrow$ if $f(s) = \varnothing$ and identifying f with its graph. A countable sequence $\varphi_0, \varphi_1, \ldots, \varphi_n, \ldots$ is written $\{\varphi_n\}_{n\in\omega}$ or just $\{\varphi_n\}$, and similarly $\{\varphi_\xi\}$ is a transfinite sequence.

If $\mathscr{F}$ is a set, $\mathscr{B} \subseteq \mathscr{F}$ and $\mathscr{B}'$ is a set of operations over $\mathscr{F}$ (i.e. mappings $\Gamma{:}\mathscr{F}^n \to \mathscr{F}, n > 0$), then $cl(\mathscr{B}/\mathscr{B}')$ stands for the *closure* of $\mathscr{B}$ under the operations in $\mathscr{B}'$. Formally, this is the least set $\mathscr{B}^*$ satisfying the following two conditions:

1. $\mathscr{B} \subseteq \mathscr{B}^*$.
2. If $\Gamma{:}\mathscr{F}^n \to \mathscr{F} \in \mathscr{B}'$, then $\Gamma(\mathscr{B}^{*n}) \subseteq \mathscr{B}^*$.

In other words, $cl(\mathscr{B}/\mathscr{B}')$ consists of those members of $\mathscr{F}$ which can be obtained from members of $\mathscr{B}$ by means of operations from $\mathscr{B}'$. Notice that whenever $\varphi \in cl(\mathscr{B}/\mathscr{B}')$, then there is a finite subset $\mathscr{B}_0$ of $\mathscr{B}$ such that $\varphi \in cl(\mathscr{B}_0/\mathscr{B}')$.

If $\mathscr{F}$ is a partially ordered set and $\Gamma{:}\mathscr{F} \to \mathscr{F}$, we write $\mu\theta.\Gamma(\theta)$ for the least solution of the inequality $\Gamma(\theta) \leq \theta$, if it exists. If Γ is monotonic, then $\theta_0 = \mu\theta.\Gamma(\theta)$ is also the *least fixed point* of Γ. Though the argument is immediate and well known, here it is. Indeed, it suffices to show that θ_0 is a fixed point of Γ. The inequality $\Gamma(\theta_0) \leq \theta_0$ implies $\Gamma(\Gamma(\theta_0)) \leq \Gamma(\theta_0)$ by the monotonicity of Γ; hence $\Gamma(\theta_0)$ is another solution to the inequality $\Gamma(\theta) \leq \theta$. This gives $\theta_0 \leq \Gamma(\theta_0)$, hence $\Gamma(\theta_0) = \theta_0$.

All the statements in the book are called propositions and enumerated uniformly. For instance, 5.11 refers to the eleventh statement of chapter 5. The exercises and examples are enumerated separately.

CHAPTER 2

Effective computability

What is *effective computability*?

Broadly speaking, a class of elements $\mathscr{F}$ is given, where the elements may be functions, functionals, operators, or any other kind of 'data processing' objects. Some elements are taken as initial because of their 'obvious effective computability'. Certain initial 'effective' operations over $\mathscr{F}$ are also chosen, assuming they preserve 'effective computability'. New 'effectively computable' elements are generated from initial ones by means of initial operations and their properties are studied.

A classical example of a class of effectively computable objects is the effectively computable number theoretic functions studied in Ordinary Recursion Theory. According to Church's thesis (which is widely accepted) these are exactly the *partial recursive functions*, which can be defined inductively as follows.

1. The functions $\lambda s.0$, $\lambda s.s+1$ and $\lambda s_1 \ldots s_n.s_i$, $1 \leq i \leq n$, are partial recursive.
2. If $f:\omega^m \dashrightarrow \omega$ and $g_1,\ldots,g_m:\omega^n \dashrightarrow \omega$ are partial recursive, then so is

$$h = \lambda s_1 \ldots s_n.f(g_1(s_1,\ldots,s_n),\ldots,g_m(s_1,\ldots,s_n)).$$

3. If $f:\omega \dashrightarrow \omega$ and $g:\omega^3 \dashrightarrow \omega$ are partial recursive, then so is $h:\omega^2 \dashrightarrow \omega$, where

$$h(s,0) = f(s),$$
$$h(s,t+1) = g(s,t,h(s,t)).$$

4. If $f:\omega^{n+1} \dashrightarrow \omega$ is partial recursive, then so is

$$h = \lambda s_1 \ldots s_n.\mu t(f(s_1,\ldots,s_n,t) = 0),$$

where $h(s_1,\ldots,s_n)$ is the least t such that $f(s_1,\ldots,s_n,t)=0$ and $f(s_1,\ldots,s_n,r)\downarrow$ for all $r<t$, while $h(s_1,\ldots,s_n)\uparrow$ if such t does not exist.

The respective operations in clauses 2–4 are called *composition, primitive recursion* and the *least number operator*. One gets the *primitive recursive functions* by skipping clause 4.

We shall modify the above definitions of primitive recursive and partial recursive functions to refer only to unary functions by making use of the fact that ω admits a primitive recursive pair coding, say $J = \lambda st.2^s(2t+1)-1$.

Take the class $\mathscr{F}_0 = \{\varphi/\varphi:\omega \dashrightarrow \omega\}$ and fix initial elements $\psi_0 = \lambda J(s,t).s$, $\psi_1 = \lambda J(s,t).t$, $\psi_2 = \lambda s.s$, $\psi_3 = \lambda s.0$, $\psi_4 = \lambda s.s+1$ and initial operations $\circ$,

$\Pi^*, Pr: \mathscr{F}_0^2 \to \mathscr{F}_0$ and $Ln: \mathscr{F}_0 \to \mathscr{F}_0$ such that

$$\varphi \circ \psi = \varphi\psi = \lambda s.\psi(\varphi(s)),$$
$$\Pi^*(\varphi,\psi) = \lambda s.J(\varphi(s),\psi(s)),$$
$$Pr(\varphi,\psi)(s) = \varphi(\psi_0(s)), \quad \text{if } \psi_1(s) = 0,$$
$$Pr(\varphi,\psi)(s) = \psi(J(\psi_0(s), J(\psi_1(s)-1, Pr(\varphi,\psi)(J(\psi_0(s), \psi_1(s)-1))))), \text{ if } \psi_1(s) > 0,$$
$$Ln(\varphi) = \lambda s.\mu t(\varphi(J(s,t)) = 0).$$

Then it can be shown that

$$\mathscr{U}_0 = cl(\psi_0,\ldots,\psi_4/\circ, \Pi^*, Pr)$$

is the set of all unary primitive recursive functions, while

$$\mathscr{U} = cl(\psi_0,\ldots,\psi_4/\circ, \Pi^*, Pr, Ln)$$

is the set of all unary partial recursive functions. Other initial elements and operations which also generate $\mathscr{U}$ will be suggested in chapter 3.

It is instructive to notice that two out of the above four operations have a least fixed point nature. Namely, let $\varphi \le \psi$ iff ψ is an extension of φ, i.e. iff $\varphi \subseteq \psi$. Making use of the function $\psi_5 = \lambda s.s \dot{-} 1(\psi_5(0) = 0, \psi_5(s+1) = s)$ and the branching operation

$$(\chi \to \varphi,\psi)(s) = \begin{cases} \varphi(s), & \text{if } \chi(s) = 0, \\ \psi(s), & \text{if } \chi(s) > 0, \\ \uparrow, & \text{if } \chi(s)\uparrow, \end{cases}$$

it follows that $Pr(\varphi,\psi)$ is the least (and unique) fixed point of the mapping

$$\lambda\theta.(\psi_1 \to \psi_0\varphi, \Pi^*(\psi_0, \psi_1\psi_5)\Pi^*(\psi_0, \Pi^*(\psi_1,\theta))\psi)$$

while, as shown in Skordev [1980], $Ln(\varphi)$ equals $\Pi^*(\psi_2,\psi_3)$ multiplied by the least fixed point of the mapping

$$\lambda\theta.(\varphi \to \psi_1, \Pi^*(\psi_0, \psi_1\psi_4)\theta).$$

Of course, the class $\mathscr{F}_0$ is very special and will not be our only starting point. Here is a rather different one consisting of *monotonic functionals*:

$$\mathscr{F}_1 = \{\phi/\phi:\omega \times \mathscr{F}_0 \dashrightarrow \omega \ \&\ \phi \text{ is monotonic}\},$$

where ϕ is monotonic iff whenever $\phi(s,\varphi) = t$ and $\varphi \le \psi$, then $\phi(s,\psi) = t$. Alternatively, one may consider *operators* instead of functionals, the class

$$\mathscr{F}_2 = \{\Gamma/\Gamma:\mathscr{F}_0 \to \mathscr{F}_0 \ \&\ \Gamma \text{ is monotonic}\},$$

where Γ is monotonic iff whenever $\varphi \le \psi$, then $\Gamma(\varphi) \le \Gamma(\psi)$. To each functional $\phi \in \mathscr{F}_1$ there corresponds an operator $\lambda\theta.\lambda s.\phi(s,\theta) \in \mathscr{F}_2$ and vice versa, a functional $\lambda s\theta.\Gamma(\theta)(s) \in \mathscr{F}_1$ corresponds to each operator $\Gamma \in \mathscr{F}_2$.

Intending as we do to develop a general theory of effective computability, we make it our immediate task to choose abstract initial elements and operations to fit the above two classes as well as other similar ones. This is a key step which in effect determines the shape of the theory. Now, what kind of initial operations are we looking for? Initially, they should be few in number, simple and convenient. Secondly, they must be sufficiently non-trivial to support a rich general theory.

We shall extract four initial operations. The first one is fundamental and inherent in the concept of effective computability. (And so are the others, but this is less obvious at this stage.) The second operation is needed for technical reasons, while the last two are added to ensure that our collection of operations is in a certain sense complete. Let us be more specific. Letters $\varphi, \psi, \chi, \theta, \tau$ will stand for arbitrary members of $\mathscr{F}$.

Suppose first that an 'effective' *multiplication operation* $\circ : \mathscr{F}^2 \rightarrow \mathscr{F}$ and an element $I \in \mathscr{F}$ can be specified in such a way that a semigroup with a unit I is obtained. It is surely difficult to imagine operations more natural for $\mathscr{F}_0, \mathscr{F}_2$ than composition of functions or operators. Moreover composition is associative. So take $\varphi\psi = \lambda s.\psi(\varphi(s))$, $I = \lambda s.s$, respectively $\Gamma_1\Gamma_2 = \lambda\theta.\Gamma_1(\Gamma_2(\theta))$, $I = \lambda\theta.\theta$. Passing from operators to functionals, one gets $\phi\Psi = \lambda s\varphi.\phi(s, \lambda t.\Psi(t, \varphi))$, $I = \lambda s\varphi.\varphi(s)$ in $\mathscr{F}_1$.

Furthermore, we need a *pairing operation* for the members of $\mathscr{F}$, which is why we assume that there is an 'effective' operation $\Pi : \mathscr{F}^2 \rightarrow \mathscr{F}$ and 'effectively computable' elements $L, R \in \mathscr{F}$ such that $L(\varphi, \psi) = \varphi$, $R(\varphi, \psi) = \psi$ for all φ, ψ, writing for short (φ, ψ) for $\Pi(\varphi, \psi)$. To realize this in our two control classes, just take $(\varphi, \psi)(2s) = \varphi(s)$, $(\varphi, \psi)(2s+1) = \psi(s)$, $L = \lambda s.2s$, $R = \lambda s.2s+1$ in $\mathscr{F}_0$ and $(\phi, \Psi)(2s, \varphi) = \phi(s, \varphi)$, $(\phi, \Psi)(2s+1, \varphi) = \Psi(s, \varphi)$, $L = \lambda s\varphi.\varphi(2s)$, $R = \lambda s\varphi.\varphi(2s+1)$ in $\mathscr{F}_1$. (More details about the classes $\mathscr{F}_0$, $\mathscr{F}_1$ and their initial operations will be given in the next chapter.)

The operations $\circ$ and Π are naturally connected by the distributive law $(\varphi, \psi)\chi = (\varphi\chi, \psi\chi)$ which can be easily substantiated in both $\mathscr{F}_0$ and $\mathscr{F}_1$.

We obtain our remaining initial operations from multiplication and pairing by making use of least fixed points. Experience from Ordinary Recursion Theory suggests that it is reasonable to suppose that with respect to a certain partial ordering all 'effective' operations $\Gamma : \mathscr{F} \rightarrow \mathscr{F}$ have least fixed points which are 'effectively computable' elements. Assuming this and examining the least fixed points of some simple mappings constructed by means of $\circ$, Π, we show that two of them behave very nicely. The first is the least fixed point of $\lambda\theta.(\varphi L, \theta R)$, denoted by $\langle\varphi\rangle$, while the second is the least fixed point of $\lambda\theta.(I, \varphi\theta)$, denoted by $[\varphi]$. We call the operations $\lambda\varphi.\langle\varphi\rangle$, $\lambda\varphi.[\varphi]$ respectively *translation* and *iteration*, the names deriving from the syntax or semantics of the operations concerned.

A partial order has already been introduced in $\mathscr{F}_0$. As for $\mathscr{F}_1$, take $\phi \leq \Psi$ iff Ψ is an extension of ϕ. The existence of operations $\langle\ \rangle$, $[\ ]$ in both classes will be established in the next chapter.

The elements I, L, R and the operations $\circ, \Pi, \langle\ \rangle, [\ ]$ are all we need to start developing a general recursion theory. As a matter of fact, the above remarks reflect the very route by which the author arrived at the present approach in May 1979, with $\mathscr{F}$ being the class of all monotonic unary mappings over an arbitrary Skordev combinatory space.

It is interesting to see how $\langle\ \rangle$, $[\ ]$ look in $\mathscr{F}_0$ or $\mathscr{F}_1$; however this yields an intuition which is not so helpful in the general theory. In compensation, the following intuitive algebraic interpretation proves very suggestive.

Consider informally $\langle\varphi\rangle$ as $(\varphi\bar{0},(\varphi\bar{1},(\varphi\bar{2},\ldots)))$, where $\bar{n}=LR^n$, and $[\varphi]$ as $(I,\varphi(I,\varphi(I,\ldots)))$. Now, for instance, the equality $[\varphi]\psi=R[(\psi L,\varphi R)]$ is easy to deduce, but how does one guess what $[\varphi]\psi$ is equal to in advance? To see this, we agree informally to get

$$\begin{aligned}[\varphi]\psi &= (I,\varphi(I,\ldots))\psi=(\psi,\varphi(\psi,\ldots))=(\psi L,\varphi R)(I,(\psi L,\varphi R)(I,\ldots))\\ &= (\psi L,\varphi R)[(\psi L,\varphi R)]=R[(\psi L,\varphi R)].\end{aligned}$$

In another example,

$$\begin{aligned}\langle\varphi\rangle\langle\psi\rangle &= (\varphi\bar{0},(\varphi\bar{1},\ldots))\langle\psi\rangle=(\varphi\bar{0}\langle\psi\rangle,(\varphi\bar{1}\langle\psi\rangle,\ldots))\\ &= (\varphi\psi\bar{0},(\varphi\psi\bar{1},\ldots))=\langle\varphi\psi\rangle\end{aligned}$$

suggests that $\langle\varphi\rangle\langle\psi\rangle=\langle\varphi\psi\rangle$. Similarly in other cases, some of which are far more complicated.

Now that the above two equalities have been suggested, how does one prove them formally?

Take the equality $(I,\varphi[\varphi])=[\varphi]$. Multiplying by ψ, one gets $(\psi,\varphi[\varphi]\psi)=[\varphi]\psi$; hence $[\varphi]\psi$ is a fixed point of the mapping $\lambda\theta.(\psi,\varphi\theta)$. In order to prove the equality $[\varphi]\psi=R[(\psi L,\varphi R)]$, however, it is not sufficient to know that $[\varphi]$ is the least fixed point of $\lambda\theta.(I,\varphi\theta)$. One should be given a little more, namely that $[\varphi]\psi$ is the least fixed point of $\lambda\theta.(\psi,\varphi\theta)$. (The latter implies the former by taking $\psi=I$.) That granted, the proof is easy.

To begin with,

$$(I,(\psi L,\varphi R)(I,[\varphi]\psi))=(I,(\psi,\varphi[\varphi]\psi))=(I,(I,\varphi[\varphi])\psi)=(I,[\varphi]\psi);$$

hence $[(\psi L,\varphi R)]\le(I,[\varphi]\psi)$. Multiplying on the left by R, we get $R[(\psi L,\varphi R)]\le[\varphi]\psi$. On the other hand,

$$(\psi,\varphi R[(\psi L,\varphi R)])=(\psi L,\varphi R)[(\psi L,\varphi R)]=R[(\psi L,\varphi R)];$$

hence $[\varphi]\psi\le R[(\psi L,\varphi R)]$. Therefore, $[\varphi]\psi=R[(\psi L,\varphi R)]$;

In order to prove $\langle\varphi\rangle\langle\psi\rangle=\langle\varphi\psi\rangle$ it suffices to be given that, whenever $R\chi=\chi\chi_1$, then $\langle\varphi\rangle\chi$ is the least fixed point of $\lambda\theta.(\varphi L\chi,\theta\chi_1)$. Actually, it follows that

$$(\varphi\psi L,\langle\varphi\rangle\langle\psi\rangle R)=(\varphi L\langle\psi\rangle,\langle\varphi\rangle R\langle\psi\rangle)=(\varphi L,\langle\varphi\rangle R)\langle\psi\rangle=\langle\varphi\rangle\langle\psi\rangle;$$

hence $\langle\varphi\psi\rangle\le\langle\varphi\rangle\langle\psi\rangle$. On the other hand, $R\langle\psi\rangle=\langle\psi\rangle R$ (here $\chi=\langle\psi\rangle$, $\chi_1=R$) and

$$(\varphi L\langle\psi\rangle,\langle\varphi\psi\rangle R)=(\varphi\psi L,\langle\varphi\psi\rangle R)=\langle\varphi\psi\rangle;$$

hence $\langle\varphi\rangle\langle\psi\rangle\le\langle\varphi\psi\rangle$. This completes the proof.

We collect all the necessary basic properties of $\langle\ \ \rangle$ and $[\ \ ]$ in an axiom called *μ-induction axiom*, *μ-axiom* for short. Several versions of this axiom are studied in this book. In essence, the properties assumed in the above two proofs compose the weakest axiom μA_0:

$$\mu A_0 \begin{cases} (\pounds) & (\varphi L, \langle \varphi \rangle R) \leq \langle \varphi \rangle, \\ & R\psi \leq \psi\psi_1 \,\&\, (\varphi L\psi, \tau\psi_1) \leq \tau \Rightarrow \langle \varphi \rangle \psi \leq \tau. \\ (\pounds\pounds) & (I, \varphi[\varphi]) \leq [\varphi], \\ & (\psi, \varphi\tau) \leq \tau \Rightarrow [\varphi]\psi \leq \tau. \end{cases}$$

This simple axiom turns out to engender almost all important statements of the general theory.

Thus we have designed an algebraic system intended to provide a framework for studying the concept of effective computability. We call this system an *operative space* since its elements are capable of *operating* in one way or another. An *operative space* (OS) is a 5-tuple $\mathscr{S} = (\mathscr{F}, I, \Pi, L, R)$ where $\mathscr{F}$ is a partially ordered semigroup with unit I, Π is a monotonic binary operation over $\mathscr{F}$ and L, R are distinct members of $\mathscr{F}$ such that $(\varphi, \psi)\chi = (\varphi\chi, \psi\chi)$, $L(\varphi, \psi) = \varphi$ and $R(\varphi, \psi) = \psi$. An *iterative operative space* (IOS) is an OS satisfying a certain μ-axiom, and we write IOS = OS + μA.

Our basic abstract notion of effective computability is that of *relative recursiveness* defined as follows. Given a subset $\mathscr{B}$ of $\mathscr{F}$, an element φ is *recursive in* $\mathscr{B}$ iff

$$\varphi \in cl(\{L, R\} \cup \mathscr{B} / \circ, \Pi, \langle \quad \rangle, [\quad]),$$

i.e. iff φ can be constructed from L, R and members of $\mathscr{B}$ by means of $\circ, \Pi, \langle \quad \rangle, [\quad]$. (The element I is not taken to be initial since $I = L[L]$.) This definition goes beyond what one might expect from the preceding discussion, for it introduces a notion of relative rather than absolute effective computability—an important step first made by Turing.

Starting with $\circ$ and Π, infinitely many 'effective' operations can be obtained via least fixed points. Such an operation is, say, the passage from φ, ψ to the least fixed point of $\lambda\theta.\varphi(I, \theta^2\psi)$. However, our choosing only two of these operations ($\langle \quad \rangle$ and $[\quad]$) leads to no loss of generality since all the others can be expressed in terms of $\circ, \Pi, \langle \quad \rangle, [\quad]$, as we will show later.

Least fixed points have always played an important role in Recursion Theory; indeed the least fixed point phenomenon gave the theory its name. Least fixed point machinery has been widely exploited recently in Recursion Theory (e.g. Platek [1966], Moschovakis [1977, 1984], Feferman [1977], Kleene [1978], Skordev [1980]), in Inductive Definability Theory (Moschovakis [1974]), as well as in Computer Science (Scott [1971], de Bakker [1971], Manna [1974]). However, what is characteristic of the iterative spaces studied in the present book is the axiomatic treatment of recursion. Similar μ-induction principles are used in the Skordev combinatory spaces and the system of de Bakker and Scott. In fact, we shall see in chapter 26 that the axiom μA_0 is a particular instance of Scott's μ-induction rule.

CHAPTER 3

Two examples

The axiomatic system of IOS outlined in the previous chapter has an enormous scope provided by its abundance of models. These are of two sorts; let us call them *first-order* and *higher order models*. The carrier of a first order model (such as $\mathscr{F}_0$) consists of function-like elements, while that of a higher order model (such as $\mathscr{F}_1$) consists of operator-like ones. As an illustration we present here two standard examples of IOS which will be studied in greater detail and in a more general form respectively in chapters 22, 28. It is convenient to use binary code, so that in this chapter $s0$ equals twice s (in particular, $00 = 0$), while $s1$ equals twice s plus one.

The first example corresponds to example 4 in Skordev [1980], chapter 2, and to example 1 in Skordev [1982].

Proposition 3.1 (Example 3.1). Let $\mathscr{F} = \{\varphi/\varphi{:}\omega \dashrightarrow \omega\}$ (the class $\mathscr{F}_0$ from the previous chapter), $\varphi \leq \psi$ iff $\varphi \subseteq \psi$, $\varphi\psi = \lambda s.\psi(\varphi(s))$, $(\varphi, \psi)(s0) = \varphi(s)$, $(\varphi, \psi)(s1) = \psi(s)$, $I = \lambda s.s$, $L = \lambda s.s0$ and $R = \lambda s.s1$. Then $\mathscr{S} = (\mathscr{F}, I, \Pi, L, R)$ is an IOS.

Proof. Let us first check the axioms of OS. We have

$$\varphi(\psi\chi)(s) = (\psi\chi)(\varphi(s)) = \chi(\psi(\varphi(s))) = \chi((\varphi\psi)(s)) = (\varphi\psi)\chi(s);$$

hence $\varphi(\psi\chi) = (\varphi\psi)\chi$. We also get

$$\varphi I(s) = I(\varphi(s)) = \varphi(s), \quad I\varphi(s) = \varphi(I(s)) = \varphi(s);$$

hence $\varphi I = I\varphi = \varphi$. Therefore, $\mathscr{F}$ is a semigroup with unit I.

The monotonicity of $\circ, \Pi$ is immediate. The equalities

$$(\varphi, \psi)\chi(s0) = \chi(\varphi(s)) = \varphi\chi(s) = (\varphi\chi, \psi\chi)(s0),$$
$$(\varphi, \psi)\chi(s1) = \chi(\psi(s)) = \psi\chi(s) = (\varphi\chi, \psi\chi)(s1)$$

imply $(\varphi, \psi)\chi = (\varphi\chi, \psi\chi)$. Finally,

$$L(\varphi, \psi)(s) = (\varphi, \psi)(s0) = \varphi(s),$$
$$R(\varphi, \psi)(s) = (\varphi, \psi)(s1) = \psi(s)$$

give $L(\varphi, \psi) = \varphi$, $R(\varphi, \psi) = \psi$; hence $\mathscr{S}$ is an OS.

If $\{\varphi_n\}$ is an increasing sequence in $\mathscr{F}$, then $\varphi_0 \subseteq \varphi_1 \subseteq \ldots$, hence $\varphi = \bigcup_n \varphi_n \in \mathscr{F}$ and $\varphi = \sup_n \varphi_n$. Moreover, we have $\varphi\psi = \sup_n(\varphi_n\psi)$, $\psi\varphi = \sup_n \psi\varphi_n$

and $(\psi, \varphi) = \sup_n(\psi, \varphi_n)$ for all ψ. If $O = \lambda s.\uparrow$, then $O \leq \psi$ and $O\psi = O$ for all ψ. Using these properties, we are going to establish the validity of the axiom μA_0.

Take $\theta_0 = O$ and $\theta_{n+1} = (\varphi L, \theta_n R)$. The monotonicity of the mapping $\lambda\theta.(\varphi L, \theta R)$ implies by induction $\theta_n \leq \theta_{n+1}$ for all n. Define $\langle\varphi\rangle = \sup_n \theta_n$. Then

$$(\varphi L, \langle\varphi\rangle R) = \left(\varphi L, \left(\sup_n \theta_n\right) R\right) = \left(\varphi L, \sup_n(\theta_n R)\right) = \sup_n(\varphi L, \theta_n R)$$
$$= \sup_n \theta_{n+1} = \langle\varphi\rangle.$$

Suppose that $R\psi \leq \psi\psi_1$ and $(\varphi L\psi, \tau\psi_1) \leq \tau$. Then $\theta_0\psi = O\psi = O \leq \tau$ and whenever $\theta_n\psi \leq \tau$, then

$$\theta_{n+1}\psi = (\varphi L, \theta_n R)\psi = (\varphi L\psi, \theta_n R\psi) \leq (\varphi L\psi, \theta_n\psi\psi_1) \leq (\varphi L\psi, \tau\psi_1) \leq \tau,$$

hence

$$\langle\varphi\rangle\psi = \left(\sup_n \theta_n\right)\psi = \sup_n(\theta_n\psi) \leq \tau.$$

Therefore, (£) holds.

In the case of iteration take $\theta_0 = O$, $\theta_{n+1} = (I, \varphi\theta_n)$ and $[\varphi] = \sup_n \theta_n$. Then

$$(I, \varphi[\varphi]) = \left(I, \varphi \sup_n \theta_n\right) = \left(I, \sup_n \varphi\theta_n\right) = \sup_n(I, \varphi\theta_n) = \sup_n \theta_{n+1} = [\varphi].$$

Suppose that $(\psi, \varphi\tau) \leq \tau$. Then $\theta_0\psi = O \leq \tau$ and if $\theta_n\psi \leq \tau$, then

$$\theta_{n+1}\psi = (I, \varphi\theta_n)\psi = (\psi, \varphi\theta_n\psi) \leq (\psi, \varphi\tau) \leq \tau.$$

Therefore,

$$[\varphi]\psi = \left(\sup_n \theta_n\right)\psi = \sup_n(\theta_n\psi) \leq \tau$$

and (££) holds. The proof is complete.

The next statement establishes explicit characterizations of the operations $\langle\ \rangle, [\ ]$. We begin with two remarks.

Take $L_1 = L^{-1} = \lambda s0.s$ and $R_1 = R^{-1} = \lambda s1.s$. (In other words, $L_1 = (I, O)$, $R_1 = (O, I)$.) Then it follows that $LL_1 = RR_1 = I$, $LR_1 = RL_1 = O$ and $(\varphi, \psi) = L_1\varphi \cup R_1\psi$.

Each number s has a unique presentation as $s = t01^n$, 1^n standing for $1 \ldots 1$, n times

Proposition 3.2. Let $\mathscr{S}$ be the IOS of example 3.1. Then

$$\langle\varphi\rangle(s01^n) = \varphi(s)01^n,$$

$[\varphi](s) = t$ iff $\exists n r_0 \ldots r_n(r_0 = s \& \forall i < n(r_i \text{ is odd } \& r_{i+1} = \varphi(R_1(r_i))) \& r_n = t0)$.

Proof. Multiplying the equality $\langle\varphi\rangle = (\varphi L, \langle\varphi\rangle R)$ on the left by $\bar{n}(= LR^n)$ one gets $\bar{n}\langle\varphi\rangle = \varphi\bar{n}$ for all n. Therefore,

$$\langle\varphi\rangle(s01^n) = \bar{n}\langle\varphi\rangle(s) = \varphi\bar{n}(s) = \varphi(s)01^n.$$

Take $\theta_0 = O$ and $\theta_{n+1} = (I, \varphi\theta_n)$ as in the proof of 3.1. Obviously, $\theta_0 = \bigcup_{i<0}(R_1\varphi)^i L_1$. If $\theta_n = \bigcup_{i<n}(R_1\varphi)^i L_1$, then

$$\theta_{n+1} = L_1 \cup R_1\varphi\theta_n = \bigcup_{i<n+1}(R_1\varphi)^i L_1,$$

hence $[\varphi] = \sup_n\theta_n = \bigcup_n(R_1\varphi)^n L_1$. If there exists an n such that $(R_1\varphi)^n L_1(s)\downarrow$, then it is unique (being the least m for which $(R_1\varphi)^m(s)$ is even), hence $[\varphi](s) = (R_1\varphi)^n L_1(s)$. Otherwise $[\varphi](s)\uparrow$. This yields the desired characterization of $[\ \ ]$, which completes the proof.

It follows from 3.2 that translation is a *primitive recursive operation*, which means that $\langle\varphi\rangle$ can be constructed by clauses 1–3 stated in the previous chapter with φ added to the initial functions in clause 1, employing one and the same construction for all φ. In other words, $\langle\varphi\rangle$ is uniformly primitive recursive in φ.

As for iteration, 3.2 shows that it is a *μ-recursive operation*. That is, for all φ the function $[\varphi]$ can be constructed by clauses 1–4 with φ allowed in clause 1, using one and the same construction for all φ; in other words, $[\varphi]$ is uniformly μ-recursive in φ. (Why not say 'partial recursive in φ'? Because it differs from 'μ-recursive in φ'; see the comments to exercise 8.2***.)

On the other hand, it will be shown in chapter 22 that all μ-recursive operations over $\mathscr{F}$ can be expressed by means of $\circ, \Pi\langle\ \ \rangle, [\ \ ]$. In particular, the unary partial recursive functions are exactly the members of $\mathscr{F}$ IOS-recursive in Z, where $Z(0) = 0$ and $Z(s) = s1$ otherwise.

The following example is based on the class $\mathscr{F}_1$ of chapter 2.

Proposition 3.3 (Example 3.2). Let $\mathscr{F}_0 = \{\varphi/\varphi:\omega \dashrightarrow \omega\}$. Take

$$\begin{aligned}
&\mathscr{F} = \{\phi/\phi:\omega \times \mathscr{F}_0 \dashrightarrow \omega \,\&\, \phi \text{ is monotonic}\},\\
&\phi \le \Psi \text{ iff } \Psi \text{ is an extension of } \phi,\\
&\phi\Psi = \lambda s\varphi.\phi(s, \lambda t.\Psi(t, \varphi)),\\
&(\phi, \Psi)(s0, \varphi) = \phi(s, \varphi),\\
&(\phi, \Psi)(s1, \varphi) = \Psi(s, \varphi),\\
&I = \lambda s\varphi.\varphi(s),\ L = \lambda s\varphi.\varphi(s0),\ R = \lambda s\varphi.\varphi(s1)
\end{aligned}$$

Then $\mathscr{S} = (\mathscr{F}, I, \Pi, L, R)$ is an IOS.

Proof. Verifying the axioms of OS is left to the reader as an easy exercise. We shall establish the validity of μA_0 by making use of the fact that whenever $\{\phi_\eta\}_{\eta<\xi}$ is an increasing transfinite sequence in $\mathscr{F}$, then $\phi = \bigcup_{\eta<\xi}\phi_\eta \in \mathscr{F}$, $\phi = \sup_{\eta<\xi}\phi_\eta$, $\phi\Psi = \sup_{\eta<\xi}(\phi_\eta\Psi)$ and $(\Psi, \phi) = \sup_{\eta<\xi}(\Psi, \phi_\eta)$ for all Ψ.

Translation is treated just as in 3.1, defining $\langle\phi\rangle$ as $\sup_n\Theta_n$, where $\Theta_0 = O = \lambda s\varphi.\uparrow$ and $\Theta_{n+1} = (\phi L, \Theta_n R)$.

In the case of iteration, however, one has to consider transfinite sequences, for the equality $\phi\sup_n\Theta_n = \sup_n\phi\Theta_n$ may fail for discontinuous ϕ.

Suppose that elements Θ_η have been constructed such that $\Theta_\eta = \sup_{\zeta<\eta}(I, \phi\Theta_\zeta)$ for all $\eta < \xi$. Then the sequence $\{\Theta_\eta\}_{\eta<\xi}$ is obviously increasing, hence so is $\{(I, \phi\Theta_\eta)\}_{\eta<\xi}$ and the element $\Theta_\xi = \sup_{\eta<\xi}(I, \phi\Theta_\eta)$ exists. In this way one gets an increasing transfinite sequence $\{\Theta_\xi\}$. Because of cardinality reasons it cannot consist of distinct members, hence there is a ζ such that

$\Theta_{\zeta+1} = \Theta_\zeta$; any cardinal $\zeta > \mathrm{Card}(\mathscr{F})$ will do. Now put $[\phi] = \Theta_\zeta$, then

$$(I, \phi[\phi]) = (I, \phi\Theta_\zeta) = \Theta_{\zeta+1} = [\phi].$$

It remains to show that $(\Psi, \phi\Sigma) \leq \Sigma$, implies $[\phi]\Psi \leq \Sigma$. Supposing $\Theta_\eta\Psi \leq \Sigma$ for all $\eta < \xi$, one gets

$$\Theta_\xi\Psi = \left(\sup_{\eta<\xi}(I, \phi\Theta_\eta)\right)\Psi = \sup_{\eta<\xi}((I, \phi\Theta_\eta)\Psi) = \sup_{\eta<\xi}(\Psi, \phi\Theta_\eta\Psi)$$
$$\leq (\Psi, \phi\Sigma) \leq \Sigma.$$

Therefore, $\Theta_\xi\Psi \leq \Sigma$ for all ξ, hence $[\phi]\Psi \leq \Sigma$. The proof is complete.

The following statement characterizes the operations $\langle\ \ \rangle, [\ \ ]$ of example 3.2 explicitly. Being by definition a particular instance of the least fixed point operator over $\mathscr{F}$, $[\phi] = \mu\Theta.(I, \phi\Theta)$, iteration turns out to be at the same time the general least fixed point operator over $\mathscr{F}_0$.

Proposition 3.4. Let $\mathscr{S}$ be the space of example 3.2. Then

$$\langle\phi\rangle(s01^n, \varphi) = \phi(s, \lambda t.\varphi(t01^n)),$$

while for all φ the function $\sigma = \lambda s.[\phi](s, \varphi)$ is the least solution of the equality

(1)
$$\theta(s\varepsilon) = \begin{cases} \varphi(s), & \text{if } \varepsilon = 0, \\ \phi(s, \theta), & \text{if } \varepsilon = 1, \end{cases}$$

or $\sigma = \mu\theta.(\varphi, \lambda s.\phi(s, \theta))$ in terms of example 3.1.

Proof. The equality $\bar{n}\langle\phi\rangle = \phi\bar{n}$ implies

$$\langle\phi\rangle(s01^n, \varphi) = \bar{n}\langle\phi\rangle(s, \varphi) = \phi\bar{n}(s, \varphi) = \phi(s, \lambda t.\bar{n}(t, \varphi)) = \phi(s, \lambda t.\varphi(t01^n)).$$

As for iteration, notice first that for all $\psi \in \mathscr{F}_0$ the functional $\check{\psi} = \lambda s\varphi.\psi(s)$ is in $\mathscr{F}$ and $\check{\sigma} = [\phi]\check{\varphi}$.

Multiplying the equality $[\phi] = (I, \phi[\phi])$ by $\check{\varphi}$, one gets $\check{\sigma} = (\check{\varphi}, \phi\check{\sigma})$, hence σ satisfies the equality (1) which can also be written as $\check{\theta} = (\check{\varphi}, \phi\check{\theta})$. If τ is another solution to (1), then $\check{\tau} = (\check{\varphi}, \phi\check{\tau})$ implies $[\phi]\check{\varphi} \leq \check{\tau}$ by (££). Therefore, $\check{\sigma} \leq \check{\tau}$, hence $\sigma \leq \tau$, which completes the proof.

Recursion in specific functionals. It follows from 12.38*, 22.4 that $\Gamma: \mathscr{F}_0 \to \mathscr{F}_0$ is a μ-recursive operation iff $\lambda s\theta.\Gamma(\theta)(s)$ is an element of $\mathscr{F}$ recursive in $\check{Z}, Id, Ml$, where $Id = \lambda s\theta.s$, $Ml = \lambda s\theta.\theta(\theta(s0)1)$.

A function $\varphi \in \mathscr{F}_0$ can be shown to have a Π^1_1 graph iff the element $\check{\varphi}$ of $\mathscr{F}$ is recursive in $\check{Z}, Id, Ml, E^\#$ (Kleene [1959], Kechris and Moschovakis [1977]), where $E^\#(s, \theta) = 0$, if $\exists t(\theta(s01^t) = 0)$, $E^\#(s, \theta) = 1$, if $\forall t(\theta(s01^t) > 0)$, and $E^\#(s, \theta)\uparrow$ otherwise. If φ is total, then one can replace 'Π^1_1' by 'Δ^1_1, i.e. hyperarithmetical'.

PART B

Pure recursion theory on operative spaces

CHAPTER 4

Operative spaces

The algebraic system of operative space is introduced in this chapter and its operations of multiplication and pairing are studied.

Suppose we are given a partially ordered semigroup $\mathscr{F}$ with a unit I, an operation $\Pi:\mathscr{F}^2 \to \mathscr{F}$ called *pairing* and two distinct elements $L, R \in \mathscr{F}$. Some notation: small greek letters range over $\mathscr{F}$, while certain capital latin letters stand for fixed elements of $\mathscr{F}$. For the sake of brevity we write (φ, ψ) for $\Pi(\varphi, \psi)$ and $(\varphi_1, \ldots, \varphi_n)$ for $(\varphi_1, (\varphi_2, \ldots, \varphi_n))$, provided $n > 2$. Occasionally $\circ$ denotes the semigroup *multiplication* of $\mathscr{F}$. Natural numbers are represented in $\mathscr{F}$ by elements of the form $\bar{n} = LR^n$.

The 5-tuple $\mathscr{S} = (\mathscr{F}, I, \Pi, L, R)$ is said to be an *operative space* (OS for short) iff the following three axioms are also satisfied.

$$\mathrm{A}_1.\quad \varphi \le \varphi_1, \psi \le \psi_1 \Rightarrow (\varphi, \psi) \le (\varphi_1, \psi_1).$$
$$\mathrm{A}_2.\quad (\varphi, \psi)\chi = (\varphi\chi, \psi\chi).$$
$$\mathrm{A}_3.\quad L(\varphi, \psi) = \varphi, R(\varphi, \psi) = \psi.$$

A *subspace* of $\mathscr{S}$ is a 5-tuple $\mathscr{S}_1 = (\mathscr{F}_1, I, \Pi \restriction \mathscr{F}_1^2, L, R)$ such that $\mathscr{F}_1 \subseteq \mathscr{F}$, $I, L, R \in \mathscr{F}_1$ and $\mathscr{F}_1$ is closed under the operations $\circ, \Pi$. Of course, any subspace of $\mathscr{S}$ is itself an OS.

We give several examples of OS to illustrate the above definitions.

Example 4.1. The OS $\mathscr{S} = (\mathscr{F}, I, \Pi, L, R)$ of example 3.1.

Example 4.2. The subspace $\mathscr{S}_1 = (\mathscr{F}_1, I, \Pi \restriction \mathscr{F}_1^2, L, R)$ of the preceding OS, based on $\mathscr{F}_1 = \{\varphi / \varphi:\omega \to \omega\}$. In other words, $\mathscr{F}_1$ consists of those functions in $\mathscr{F}$ which are total.

Example 4.3. Example 4.1 with a *pairing scheme* Π, L, R modified as follows: Take $(\varphi, \psi)(3s) = \varphi(s)$, $(\varphi, \psi)(3s + 1) = \psi(s)$, $(\varphi, \psi)(3s + 2)\uparrow$, $L = \lambda s.3s$ and $R = \lambda s.3s + 1$.

Example 4.4. Let M be an arbitrary infinite set, let $J:M^2 \to M$ be injective (such a *pairing function* J exists since M is infinite), and let $L, R:M \to M$ satisfy $L(J(s,t)) = s$, $R(J(s,t)) = t$ for all $s, t \in M$. Then take $\mathscr{F} = \{\varphi / \varphi:M \to M\}$, $I = \lambda s.s$, $\varphi \le \psi$ iff $\varphi = \psi$, $\varphi\psi = \lambda s.\varphi(\psi(s))$ and $(\varphi, \psi) = \lambda s.J(\varphi(s), \psi(s))$.

Example 4.5. Example 4.4 with $M=\omega$; in particular, one may use the pairing functions $J=\lambda st.2^s(2t+1)-1$, $\lambda st.(2s+1)2^t$, $\lambda st.\frac{1}{2}(s+t)(s+t+1)+s$ (Cantor's function) or $\lambda st.2^s3^t$.

Example 4.6. The OS of example 3.2.

Assume from now on that on OS $\mathscr{S}$ is given. Let $\mathscr{D}=cl(I, L, R/\circ)$ and α, β range over $\mathscr{D}$.

Proposition 4.1. $\mathscr{D}=cl(I/\lambda\theta.\theta L, \lambda\theta.\theta R)$.

Proof. Write $\mathscr{D}_1$ for $cl(I/\lambda\theta.\theta L, \lambda\theta.\theta R)$. The inclusion $\mathscr{D}_1\subseteq\mathscr{D}$ is obvious. In order to get $\mathscr{D}_1=\mathscr{D}$ it suffices to show that whenever $\alpha,\beta\in\mathscr{D}_1$, then $\alpha\beta\in\mathscr{D}_1$. This can be done by induction on the construction of β. If $\beta=I$ and $\alpha\in\mathscr{D}_1$, then $\alpha\beta=\alpha\in\mathscr{D}_1$. Suppose that $\beta=\beta_1 L$ and $\alpha\beta_1\in\mathscr{D}_1$ for all $\alpha\in\mathscr{D}_1$. Then $\alpha\beta_1 L\in\mathscr{D}_1$, i.e. $\alpha\beta\in\mathscr{D}_1$ for all $\alpha\in\mathscr{D}_1$. The case $\beta=\beta_1 R$ is treated similarly.

Proposition 4.2. For every α there is an n such that $\alpha(I,I)^n=I$.

Proof. If $\alpha=I$, then $\alpha(I,I)^0=\alpha I=I$.

Let $\alpha=\alpha_1 L$ or $\alpha=\alpha_1 R$ and suppose that the assertion holds for α_1. Then there is a n such that $\alpha_1(I,I)^n=I$, hence $\alpha(I,I)^{n+1}=\alpha_1(I,I)^n=I$, which completes the proof by 4.1.

Proposition 4.3. $\alpha L\not\leq\beta R$, $\alpha R\not\leq\beta L$ (for all α,β: we often omit these words).

Proof. Suppose $\alpha L\leq\beta R$. Proposition 4.2 implies that there are φ,ψ such that $\alpha\varphi=\beta\psi=I$. Multiplying the inequality $\alpha L\leq\beta R$ by $(\varphi L,\psi R)$, we get $L\leq R$. Multiplying the last inequality by (R, L), we get $R\leq L$, hence $L=R$, which is not the case.

Supposing $\alpha R\leq\beta L$ and multiplying by (R, L), we get $\alpha L\leq\beta R$ contrary to the assertion just proved. This completes the proof.

Consider the alphabet $\mathbb{L}\mathbb{R}$. To each word a in it we assign an element α. Namely, say that the empty word Λ *represents* I and whenever a represents α, then $a\mathbb{L}$, $a\mathbb{R}$ represent respectively αL, αR.

Proposition 4.4. Let a,b represent respectively α,β and $\alpha\leq\beta$. Then $a=b$ (a and b are graphically identical), which in particular implies $\alpha=\beta$.

Proof. By induction on the construction of a.

Let $a=\Lambda$. Suppose that $b=b_1\mathbb{L}$. Then $\beta=\beta_1 L$, where b_1 represents β_1. Multiplying the inequality $I\leq\beta_1 L$ on the left by R, we get $R\leq R\beta_1 L$ contrary to 4.3. The possibility $b=b_1\mathbb{R}$ is also ruled out. Therefore, $b=\Lambda=a$.

Let $a=a_1\mathbb{L}$ and a_1 meet the induction hypothesis. Supposing $b=\Lambda$, we get a contradiction as above. Suppose that $b=b_1\mathbb{R}$. Then $\alpha=\alpha_1 L$ and $\beta=\beta_1 R$, where a_1,b_1 represent respectively α_1,β_1. We get $\alpha_1 L\leq\beta_1 R$ contrary to 4.3. Therefore, $b=b_1\mathbb{L}$, hence $\alpha=\alpha_1 L$, $\beta=\beta_1 L$ with a_1,b_1 representing respectively α_1,β_1. The inequality $\alpha_1 L\leq\beta_1 L$ multiplied by (I,I) gives $\alpha_1\leq\beta_1$, which implies $a_1=b_1$ by the induction hypothesis. Therefore, $a=b$. The case of $a=a_1\mathbb{R}$ is treated similarly.

Proposition 4.5. If $\exists\chi(\chi\varphi\in\mathscr{D})$, then $\varphi L\neq\psi R$ and $\varphi R\neq\psi L$ for all ψ.

Proof. Let $\chi\varphi=\alpha$. Then $\varphi L=\psi R$ implies $\alpha L=\chi\psi R$. Multiplying by (R, R), we get $\alpha R=\chi\psi R$, hence $\alpha L=\alpha R$ contrary to 4.3.

The equality $\varphi R=\psi L$ multiplied by (R, L) gives $\varphi L=\psi R$, which is not the case. Thus the proof is complete.

It follows by 4.5 that $\alpha L\neq\psi R$ and $\alpha R\neq\psi L$ for all α, ψ.

Proposition 4.6. $\varphi L\neq I$, $\varphi R\neq I$.

Proof. The equality $\varphi L=I$ multiplied on the left by R gives $R=R\varphi L$ contrary to 4.5. Similarly $\varphi R\neq I$.

Proposition 4.7. Whenever $\varphi\alpha\in\mathscr{D}$, then $\varphi\in\mathscr{D}$.

Proof. The case $\alpha=I$ is immediate. Let $\alpha=\alpha_1 L$ and whenever $\varphi\alpha_1\in\mathscr{D}$, then $\varphi\in\mathscr{D}$. Let $\varphi\alpha=\beta$. Then $\varphi\alpha_1 L=\beta$, hence $\beta=\beta_1 L$ for a certain β_1 by 4.5, 4.6. Multiplying by (I, I), we get $\varphi\alpha_1=\beta_1$, hence $\varphi\in\mathscr{D}$. The case of $\alpha=\alpha_1 R$ is treated similarly.

Proposition 4.8. $(I, I)\notin\mathscr{D}$. Therefore, $\mathscr{D}$ is not closed under Π.

Proof. Suppose that $(I, I)=\alpha$. Then $L\alpha=L(I, I)=I$ contrary to 4.4.

Proposition 4.9. If $(\varphi, \psi)\in\mathscr{D}$, then $\varphi, \psi\in\mathscr{D}$.

Proof. If $(\varphi, \psi)=\alpha$, then $\varphi=L\alpha$ and $\psi=R\alpha$, hence $\varphi, \psi\in\mathscr{D}$.

Moreover, it follows that

$$\alpha=(\varphi, \psi)=(L\alpha, R\alpha)=(L, R)\alpha,$$

hence $(L, R)=I$ by 4.2. It is worth mentioning however that the last equality fails in some OS, e.g. example 4.3. Therefore, $(\varphi, \psi)\notin\mathscr{D}$ for all φ, ψ in such cases.

Proposition 4.10. If $\bar{m}\leq\bar{n}$, then $m=n$.

This follows from 4.4. Consequently, our way of representing natural numbers in $\mathscr{F}$ is correct.

Proposition 4.11. Let $n\geq 1$. Then $\bar{k}(\varphi_0, \ldots, \varphi_n)=\varphi_k$ for all $k<n$, while $R^n(\varphi_0, \ldots, \varphi_n)=\varphi_n$.

This is proved by an easy induction on n.

Proposition 4.12. Let $\varphi\in cl(I, L, R/\circ, \Pi)$. Then there is a natural number m and elements $\varphi_0, \ldots, \varphi_m$ such that $\varphi_m=\varphi$, for all $i\leq m$ either $\varphi_i\in\mathscr{D}$ or $\varphi_i=(\varphi_k, \varphi_l)$ with $k, l<i$, and for all α either $\alpha\varphi\in\mathscr{D}$ or $\alpha\varphi=\varphi_i$ for a certain $i\leq m$.

Proof. If $\varphi\in\{I, L, R\}$, then take $m=0$, $\varphi_0=\varphi$.

Let $\varphi_0, \ldots, \varphi_m$ and $\psi_0, \ldots, \psi_n$ correspond to φ, ψ respectively.

Let $\chi=(\varphi, \psi)$. Then take $k=m+n+2$, $\chi_i=\varphi_i$ for $i\leq m$, $\chi_{m+i+1}=\psi_i$ for $i\leq n$ and $\chi_{m+n+2}=\chi$. Let $\alpha\in\mathscr{D}$. If $\alpha=I$, then $\alpha\chi=\chi=\chi_{m+n+2}$. If $\alpha=\alpha_1 L$, then $\alpha\chi=\alpha_1\varphi$ and the induction assumption for φ applies. If $\alpha=\alpha_1 R$, then $\alpha\chi=\alpha_1\psi$ and we use the assumption for ψ.

Let $\chi = \varphi\psi$. We prove by induction on $i, i \leq m$, that $\varphi_i\psi$ has the required properties. If $\varphi_i \in \mathscr{D}$, then the assumption for ψ yields that either $\varphi_i\psi \in \mathscr{D}$ or $\varphi_i\psi = \psi_j$ with a certain $j \leq n$, whence $\varphi_i\psi$ has the required properties. If $\varphi_i = (\varphi_k, \varphi_l)$ for some $k, l < i$, then $\varphi_i\psi = (\varphi_k\psi, \varphi_l\psi)$. The assertion holds for both $\varphi_k\psi$, $\varphi_l\psi$ and we proceed as in the case of $\chi = (\varphi, \psi)$. This completes the proof.

Proposition 4.12 reflects the intuitively clear fact that $cl(I, L, R/\circ, \Pi) = cl(\mathscr{D}/\Pi)$. If $\{I, L, R\}$ is replaced by an arbitrary subset $\mathscr{B}$ of $\mathscr{F}$, then something weaker but still useful can be proved.

Proposition 4.13.

$$cl(\mathscr{B}/\circ, \Pi) = cl(\mathscr{B}/\Pi, \lambda\theta.\psi\theta \text{ for all } \psi \in \mathscr{B}).$$

Proof. Let $\mathscr{B}^* = cl(\mathscr{B}/\Pi, \lambda\theta.\psi\theta$ for all $\psi \in \mathscr{B})$. It suffices to show that whenever $\varphi \in cl(\mathscr{B}/\circ, \Pi)$, then $\varphi \in \mathscr{B}^*$ and $\varphi\psi \in \mathscr{B}^*$ for all $\psi \in \mathscr{B}^*$.

Certainly, all the members of $\mathscr{B}$ have this property.

Suppose that φ_1, φ_2 have the property in question. Then $\varphi_1, \varphi_2 \in \mathscr{B}^*$; hence $\varphi_1\varphi_2 \in \mathscr{B}^*$ and $(\varphi_1, \varphi_2) \in \mathscr{B}^*$. Let $\psi \in \mathscr{B}^*$. Then $\varphi_2\psi \in \mathscr{B}^*$, hence $\varphi_1\varphi_2\psi \in \mathscr{B}^*$. It also follows that $\varphi_1\psi \in \mathscr{B}^*$, hence $(\varphi_1, \varphi_2)\psi = (\varphi_1\psi, \varphi_2\psi) \in \mathscr{B}^*$. Therefore, both $\varphi_1\varphi_2$ and (φ_1, φ_2) have the required property, which completes the proof.

EXERCISES TO CHAPTER 4

The first two exercises introduce new examples of OS which correspond to examples 1, 2 in Skordev [1980], chapter 2.

Exercise 4.1 (Example 4.7). Let M be an infinite set, $L, R: M \to M$ be injective and $L(M) \cap R(M) = \varnothing$. (Such L, R exist since M is infinite.) Take $\mathscr{F} = \{\varphi/\varphi: M \longrightarrow M\}$, $I = \lambda s.s$, $\varphi \leq \psi$ iff $\varphi \subseteq \psi$, $\varphi\psi = \lambda s.\psi(\varphi(s))$, $(\varphi, \psi)(L(s)) = \varphi(s)$, $(\varphi, \psi)(R(s)) = \psi(s)$ and $(\varphi, \psi)(s)\uparrow$ otherwise. Prove that $\mathscr{S} = (\mathscr{F}, I, \Pi, L, R)$ is an OS.

Hint. Follow the proof of 3.1.

Example 3.1 is a particular instance of example 4.7 with $M = \omega$.

Exercise 4.2 (Example 4.8). Let M, L, R, I be the same as in the previous exercise. Take $\mathscr{F} = \{\varphi/\varphi: M \to 2^M\}$, $\varphi \leq \psi$ iff $\forall s(\varphi(s) \subseteq \psi(s))$ and introduce $\circ, \Pi$ as above. Show that $\mathscr{S} = (\mathscr{F}, I, \Pi, L, R)$ is an OS and the OS of example 4.7 is a subspace of it.

Here $\mathscr{F}$ consists of unary partial multiple-valued functions which can also be regarded as binary relations, i.e. $\mathscr{F} = \{\varphi/\varphi \subseteq M^2\}$.

The following exercise establishes the independence of the axioms A_1, A_2, A_3.

Exercise 4.3. Construct a 5-tuple $\mathscr{S} = (\mathscr{F}, I, \Pi, L, R)$ to meet all the OS-axioms but A_1 (respectively A_2, A_3).

Hint. Let J, L, R be as in example 4.5 and $J(0,0) = 0$. (The first function J suggested there would do.) Take $\mathscr{F} = \{\varphi/\varphi: \omega \to \omega \,\&\, \varphi(0) = 0\}$, $I = \lambda s.s$,

$\varphi \le \psi$ iff $\forall s(\varphi(s)=0 \vee \varphi(s)=\psi(s))$, $\varphi\psi=\lambda s.\varphi(\psi(s))$ and $(\varphi,\psi)=\lambda s.J(\varphi(s), \psi(s))$, then take $\varphi=\lambda s.0$ and show that $\varphi \le I$ but $(\varphi, I) \not\le (I, I)$. In the case of A_2 take an OS and replace Π, L, R by $\Pi_1=\lambda\varphi\psi.(\varphi, I, \psi)$, $L_1=L$ and $R_1=R^2$. In the case of A_3 take $\Pi_1=\lambda\varphi\psi.\varphi$.

Exercise 4.4. Show that the axiom A_1 can be replaced by its particular instance $\varphi \le \varphi_1 \Rightarrow (I,\varphi) \le (I, \varphi_1)$.

Hint. Make use of the equality

$$(\varphi,\psi)=(RL,R)(I,\psi L)(I,\varphi).$$

Exercise 4.5. Let $\mathscr{S}$ be an OS and $\mathscr{S}_1$ be obtained from it by introducing a new partial order, $\varphi \le_1 \psi$ iff $\varphi=\psi$. Show that $\mathscr{S}_1$ is also an OS.

It may well happen that a set $\mathscr{F}$ admits different operations $\circ, \Pi$ satisfying the OS-axioms, cf. examples 4.2, 4.5. If $\circ$ is fixed however, then the following exercise shows that Π is in a sense unique.

Exercise 4.6. Let $\mathscr{F}$ be a semigroup and both Π, L, R and Π_1, L_1, R_1 meet the axioms A_2, A_3. Prove that there are elements ρ, ρ_1 such that $(\varphi,\psi)_1=\rho(\varphi,\psi)$ and $(\varphi,\psi)=\rho_1(\varphi,\psi)_1$ for all φ,ψ.

Hint. Take $\rho=(L,R)_1$, $\rho_1=(L_1,R_1)$.

It follows by exercise 4.6 that all operations Π to meet A_2, A_3 are simultaneously monotonic or nonmonotonic, hence the semigroup suggested in the hint to exercise 4.3 can not be augmented with a pairing operation to become an OS.

Assume from now on that an OS $\mathscr{S}=(\mathscr{F}, I, \Pi, L, R)$ is given.

Exercise 4.7. Show that the operations $\circ, \Pi$ are non-commutative, Π is non-associative and there is an element φ such that $\psi\varphi=I$ for no ψ.

Hint. $LR \ne RL$, $(I,R,I) \ne ((I,R),I)$ and $\psi L \ne I$ for all ψ.

Exercise 4.8. Prove the following equalities.

a. $RLR(R^2,(LR,RL)^4L(I,I)^5,L)((I,R),L)=L.$
b. $R(\varphi R(\psi R,L)^3,L)^3(\varphi,\psi)=(\varphi\psi)^2\varphi.$
c. $L(\varphi LR,(\psi R,\psi L)(L,R^2))^{3n}=(\varphi\psi^2)^nL.$
d. $RL((L\chi,R)(\psi R,\varphi(R,L)),\varphi L)(\psi,R)(\chi,R)=\varphi R(L^2,R^2,\psi(\chi L,R^2))(I,I).$

Exercise 4.9. Let $K_0=(R,L), K_1=(I,I)$, $\langle\ \rangle=\lambda\theta.(L,\theta R)$ and $\mathscr{B}_0=\{R, K_1, \langle K_0\rangle, \langle K_1\rangle, \langle\langle L\rangle\rangle, \langle\langle R\rangle\rangle\}$. Prove that $cl(\{L,R\}\cup\mathscr{B}/\circ,\Pi)=cl(\mathscr{B}_0\cup\langle\mathscr{B}\rangle/\circ)$. (Of course, $\langle\mathscr{B}\rangle$ stands for $\{\langle\psi\rangle/\psi\in\mathscr{B}\}$.)

Hint: Use the equalities $\varphi=R\langle\varphi\rangle K_1$, $\langle\varphi\psi\rangle=\langle\varphi\rangle\langle\psi\rangle$, $(\varphi,\psi)=\langle\psi\rangle K_0\langle\varphi\rangle K_1$ and $\langle\langle\varphi\rangle\rangle=K_3\langle\varphi\rangle K_4$, where $K_3=\langle\langle R\rangle K_0\langle RL\rangle\circ K_1\rangle K_0\langle L^2\rangle K_1$, $K_4=\langle R^2\rangle K_0\langle\langle L\rangle\rangle K_1$.

It is possible to consider an algebraic system alternative to OS, replacing Π by $\langle\ \rangle$ and introducing suitable axioms.

CHAPTER 5

Iterative operative spaces

The aim of this chapter is to formally introduce the algebraic system of iterative operative space as the natural framework for the development of a general recursion theory.

Let $\mathscr{S} = (\mathscr{F}, I, \Pi, L, R)$ be an OS. Consider the following axiom.

Axiom μA_0. Two additional unary operations $\langle\ \rangle, [\ ]$ called *translation* and *iteration* are given such that

$$(£)\quad (\varphi L, \langle\varphi\rangle R) \leq \langle\varphi\rangle,$$
$$R\psi \leq \psi\psi_1 \& (\varphi L\psi, \tau\psi_1) \leq \tau \Rightarrow \langle\varphi\rangle\psi \leq \tau.$$
$$(££)\quad (I, \varphi[\varphi]) \leq [\varphi],$$
$$(\psi, \varphi\tau) \leq \tau \Rightarrow [\varphi]\psi \leq \tau.$$

Taking $\psi = I, \psi_1 = R$, one gets immediately that $\langle\varphi\rangle = \mu\theta.(\varphi L, \theta R)$ and $[\varphi] = \mu\theta.(I, \varphi\theta)$.

If the axiom μA_0 is satisfied, then $\mathscr{S}$ is said to be an *iterative operative space* (IOS). The operations $\langle\ \rangle, [\ ]$ are not included in the signature of IOS because their semantics is unambiguously implied by that of $\leq, \circ, \Pi, L, R$.

For instance, the OS of examples 4.1, 4.6 are iterative respectively by 3.1, 3.3. The OS of example 4.3 is also iterative. It will be shown in the exercises that the OS of examples 4.7, 4.8 are iterative, while those of examples 4.2, 4.4, 4.5 are not.

Instead of the axiom μA_0, we shall use sometimes the stronger axioms $\mu A_1, \mu A_2$ or μA_3. Some auxiliary notions are needed to formulate them. To begin with, we introduce the notion of *inductive mapping* inductively as follows.

1. The mappings $\Gamma = \lambda\theta_1\ldots\theta_n.\theta_i$, $1 \leq i \leq n$, and $\Gamma = \lambda\theta_1\ldots\theta_n.\psi$, $\psi \in \{I, L, R\}$, are inductive.

2. If $\Gamma_1, \Gamma_2 : \mathscr{F}^n \to \mathscr{F}$ are inductive, then so are $\Gamma = \lambda\theta_1\ldots\theta_n.\Gamma_1(\theta_1,\ldots,\theta_n) \circ \Gamma_2(\theta_1,\ldots,\theta_n)$ and $\Gamma = \lambda\theta_1\ldots\theta_n.(\Gamma_1(\theta_1,\ldots,\theta_n), \Gamma_2(\theta_1,\ldots,\theta_n))$.

3. If $\Gamma_1 : \mathscr{F}_1^{n+1} \to \mathscr{F}$ is inductive and for all $\theta_1,\ldots,\theta_n$ the element $\mu\theta.\Gamma_1(\theta_1,\ldots,\theta_n,\theta)$ exists, then $\Gamma = \lambda\theta_1\ldots\theta_n.\mu\theta.\Gamma_1(\theta_1,\ldots,\theta_n,\theta)$ is inductive.

A *regular segment* is a subset of $\mathscr{F}$ of the form $\{\theta / \alpha\theta\chi \leq \tau \text{ for all } \langle\alpha,\chi,\tau\rangle \in \mathscr{A}\}$, where $\mathscr{A} \subseteq \mathscr{D} \times \mathscr{F}^2$. A *normal segment* is a set of the form $\{\theta/\theta\chi \leq \tau \text{ for all } \langle\chi,\tau\rangle \in \mathscr{A}\}$, where $\mathscr{A} \subseteq \mathscr{F}^2$. In order to state the axiom μA_1 assume that the element $\langle\varphi\rangle = \mu\theta.(\varphi L, \theta R)$ exists for all φ and call *simple segments* the sets of the form $\{\theta/\theta\chi \leq \tau\}$ or $\{\theta / \langle\theta\rangle \leq \langle I\rangle\tau\}$.

Axiom μA_1 *(*$\mu A_2, \mu A_3$*).* For any $n+1$-ary inductive mapping Γ and any

$\theta_1,\ldots,\theta_n$ the inequality $\Gamma(\theta_1,\ldots,\theta_n,\theta)\leq\theta$ has a solution which is a member of all simple (respectively normal, regular) segments closed under the mapping $\lambda\theta.\Gamma(\theta_1,\ldots,\theta_n,\theta)$.

All normal segments are regular since $I\in\mathscr{D}$; thus axiom μA_3 implies axiom μA_2. Further connections between the four μ-axioms are established below. In each particular instance the μ-axiom used will be indicated by the corresponding number of asterisks, e.g. the proof of 5.7** makes use of μA_2.

Proposition 5.1. All inductive mappings are monotonic.

Proof. The mappings $\Gamma=\lambda\theta_1\ldots\theta_n.\theta_i$ and $\Gamma=\lambda\theta_1\ldots\theta_n.\psi$ are monotonic. If $\Gamma_1,\Gamma_2:\mathscr{F}^n\to\mathscr{F}$ are monotonic, then so are

$$\Gamma=\lambda\theta_1\ldots\theta_n.\Gamma_1(\theta_1,\ldots,\theta_n)\Gamma_2(\theta_1,\ldots,\theta_n)$$

and

$$\Gamma=\lambda\theta_1\ldots\theta_n.(\Gamma_1(\theta_1,\ldots,\theta_n),\Gamma_2(\theta_1,\ldots,\theta_n))$$

since the monotonicity of $\circ,\Pi$.

Let $\Gamma_1:\mathscr{F}^{n+1}\to\mathscr{F}$ be monotonic and

$$\Gamma=\lambda\theta_1\ldots\theta_n.\mu\theta.\Gamma_1(\theta_1,\ldots,\theta_n,\theta).$$

If $\theta_1\leq\tau_1,\ldots,\theta_n\leq\tau_n$, then

$$\Gamma_1(\theta_1,\ldots,\theta_n,\Gamma(\tau_1,\ldots,\tau_n))\leq\Gamma_1(\tau_1,\ldots,\tau_n,\Gamma(\tau_1,\ldots,\tau_n))\leq\Gamma(\tau_1,\ldots,\tau_n),$$

hence $\Gamma(\theta_1,\ldots,\theta_n)\leq\Gamma(\tau_1,\ldots,\tau_n)$ since $\Gamma(\theta_1,\ldots,\theta_n)=\mu\theta.\Gamma_1(\theta_1,\ldots,\theta_n,\theta)$. Therefore, Γ is monotonic. The proof is complete.

Proposition 5.2* (5.2).** Let θ_0 be the element assumed in the axiom μA_1 (respectively $\mu A_2,\mu A_3$). Then $\theta_0=\mu\theta.\Gamma(\theta_1,\ldots,\theta_n,\theta)$.

Proof. Since $\Gamma(\theta_1,\ldots,\theta_n,\theta_0)\leq\theta_0$, it suffices to show that whenever $\Gamma(\theta_1,\ldots,\theta_n,\tau)\leq\tau$, then $\theta_0\leq\tau$.

Suppose that $\Gamma(\theta_1,\ldots,\theta_n,\tau)\leq\tau$ and consider the simple (respectively normal) segment $\mathscr{E}=\{\theta/\theta\leq\tau\}$. Whenever $\theta\in\mathscr{E}$, then

$$\Gamma(\theta_1,\ldots,\theta_n,\theta)\leq\Gamma(\theta_1,\ldots,\theta_n,\tau)\leq\tau$$

by the monotonicity of Γ, hence $\mathscr{E}$ is closed under the mapping $\lambda\theta.\Gamma(\theta_1,\ldots,\theta_n,\theta)$, which implies $\theta_0\in\mathscr{E}$ by μA_1 (by μA_2). Thereby the proof is completed.

Notice that the element $\mu\theta.\Gamma(\theta_1,\ldots,\theta_n,\theta)$ above is also a least fixed point of $\lambda\theta.\Gamma(\theta_1,\ldots,\theta_n,\theta)$ since the monotonicity of this mapping.

It follows from this last statement that each of the axioms $\mu A_1,\mu A_2,\mu A_3$ ensures the existence of the elements $\langle\varphi\rangle=\mu\theta.(\varphi L,\theta R)$ and $[\varphi]=\mu\theta.(I,\varphi\theta)$ for all φ.

The following two statements show that the axiom μA_1 implies both (£) and (££), i.e. μA_1 implies μA_0.

Proposition 5.3*. If $R\psi\leq\psi\psi_1$ and $(\varphi L\psi,\tau\psi_1)\leq\tau$, then $\langle\varphi\rangle\psi\leq\tau$.

Proof. The set $\mathscr{E}=\{\theta/\theta\psi\leq\tau\}$ is a simple segment. If $\theta\in\mathscr{E}$, then

$$(\varphi L,\theta R)\psi=(\varphi L\psi,\theta R\psi)\leq(\varphi L\psi,\theta\psi\psi_1)\leq(\varphi L\psi,\tau\psi_1)\leq\tau;$$

hence $\langle\varphi\rangle\in\mathscr{E}$ by μA_1. Thus the proof is complete.

Proposition 5.4*. If $(\psi, \varphi\tau) \leq \tau$, then $[\varphi]\psi \leq \tau$.

Proof. The set $\mathscr{E} = \{\theta/\theta\psi \leq \tau\}$ is a simple segment and if $\theta \in \mathscr{E}$, then

$$(I, \varphi\theta)\psi = (\psi, \varphi\theta\psi) \leq (\psi, \varphi\tau) \leq \tau.$$

Therefore, $[\varphi] \in \mathscr{E}$ by μA_1, which completes the proof.

Our next aim is to show that μA_2 implies μA_1.

Proposition 5.5. $L\langle\varphi\rangle = \varphi L$ and $R\langle\varphi\rangle = \langle\varphi\rangle R$.

This follows from the equality $\langle\varphi\rangle = (\varphi L, \langle\varphi\rangle R)$.

Proposition 5.6. $\bar{n}\langle\varphi\rangle = \varphi\bar{n}$ for all n.

Proof.

$$\bar{n}\langle\varphi\rangle = LR^n\langle\varphi\rangle = L\langle\varphi\rangle R^n = \varphi LR^n = \varphi\bar{n}.$$

Proposition 5.7.** If $\forall n(\varphi\bar{n}\psi \leq \rho\bar{n}\sigma)$, then $\langle\varphi\rangle\psi \leq \langle\rho\rangle\sigma$.

Proof. Consider the normal segment

$$\mathscr{E} = \{\theta/\forall n(\theta R^n\psi \leq \langle\rho\rangle R^n\sigma)\}.$$

If $\theta \in \mathscr{E}$, then

$$\begin{aligned}(\varphi L, \theta R)R^n\psi &= (\varphi\bar{n}\psi, \theta R^{n+1}\psi) \leq (\rho\bar{n}\sigma, \langle\rho\rangle R^{n+1}\sigma)\\ &= (\rho L, \langle\rho\rangle R)R^n\sigma = \langle\rho\rangle R^n\sigma,\end{aligned}$$

hence $\langle\varphi\rangle \in \mathscr{E}$ by μA_2. Taking $n = 0$, one gets $\langle\varphi\rangle\psi \leq \langle\rho\rangle\sigma$.

Proposition 5.8.** $\langle\varphi\rangle\psi \leq \langle\rho\rangle\sigma$ iff $\forall n(\varphi\bar{n}\psi \leq \rho\bar{n}\sigma))$.

This follows from 5.6, 5.7**.

Proposition 5.9.** $\langle\varphi\rangle\psi = \langle\rho\rangle\sigma$ iff $\forall n(\varphi\bar{n}\psi = \rho\bar{n}\sigma)$.

This follows from 5.8**.

Proposition 5.10.** All simple segments are normal.

Proof. If $\mathscr{E} = \{\theta/\theta\chi \leq \tau\}$, then $\mathscr{E}$ is a normal segment. If $\mathscr{E} = \{\theta/\langle\theta\rangle \leq \langle I\rangle\tau\}$, then $\mathscr{E} = \{\theta/\forall n(\theta\bar{n} \leq \bar{n}\tau)\}$ by 5.8**, hence $\mathscr{E}$ is again a normal segment. The proof is complete.

As an immediate corollary to 5.10** one obtains that μA_2 implies μA_1, thus

$$\mu A_3 \Rightarrow \mu A_2 \Rightarrow \mu A_1 \Rightarrow \mu A_0.$$

The sufficient condition $(*)_0$ given in chapter 18 verifies μA_2 but not μA_3, hence $\mu A_2 \not\Rightarrow \mu A_3$. I conjecture that $\mu A_1 \not\Rightarrow \mu A_2$ and $\mu A_0 \not\Rightarrow \mu A_1$.

The proof of the Normal Form Theorem and the First Recursion Theorem given in chapter 9 imply that it is sufficient to allow in $\mu A_1, \mu A_2, \mu A_3$ the mappings $\lambda\theta_1\theta.(\theta_1 L, \theta R)$, $\lambda\theta_1\theta.(I, \theta_1\theta)$ and $\lambda\theta_1\theta.\bar{1}[\theta_1(I, \langle\theta\rangle)]$ only. Therefore, axiom scheme μA_1 is equivalent to a single first order axiom. The axioms $\mu A_2, \mu A_3$ are second order ones, though the normal and regular segments we consider will often be introduced by means of countable and effectively given sets $\mathscr{A}$.

It almost always suffices to allow in $\mu A_2, \mu A_3$ just the mappings $\lambda\theta_1\theta.(\theta_1 L, \theta R)$ and $\lambda\theta_1\theta.(I, \theta_1\theta)$; axiom μA_3 restricted this way is exactly the basic μ-axiom in Ivanov [1980]. The first version of a μ-axiom for OS was μA_3, while $\mu A_0, \mu A_1$ were invented later. (The point (££) was borrowed from Skordev [1982].)

The requirements of (£) can be restricted to the element $\langle I \rangle$, provided the equality $(\varphi L, \theta R) = \theta$ has a solution for all φ.

Proposition 5.11. Let $\langle I \rangle$ be an element satisfying (£), i.e. $(L, \langle I \rangle R) \le \langle I \rangle$ and whenever $R\psi \le \psi\psi_1$, $(L\psi, \tau\psi_1) \le \tau$, then $\langle I \rangle\psi \le \tau$. Let $\langle\ \ \rangle_1$ be a unary operation over $\mathscr{F}$ such that $(\varphi L, \langle \varphi \rangle_1 R) = \langle \varphi \rangle_1$ for all φ. Then the operation $\langle\ \ \rangle = \lambda\varphi.\langle I \rangle\langle \varphi \rangle_1$ satisfies (£).

Proof. It follows that

$$(\varphi L, \langle \varphi \rangle R) = (\varphi L, \langle I \rangle\langle \varphi \rangle_1 R) = (L, \langle I \rangle R)(\varphi L, \langle \varphi \rangle_1 R) = \langle I \rangle\langle \varphi \rangle_1 = \langle \varphi \rangle.$$

If $R\psi \le \psi\psi_1$ and $(\varphi L\psi, \tau\psi_1) \le \tau$, then $R\langle \varphi \rangle_1\psi \le \langle \varphi \rangle_1\psi\psi_1$ and $(L\langle \varphi \rangle_1\psi, \tau\psi_1) \le \tau$; hence $\langle I \rangle\langle \varphi \rangle_1\psi \le \tau$, i.e. $\langle \varphi \rangle\psi \le \tau$. This completes the proof.

Proposition 5.12. $L[\varphi] = I$, $R[\varphi] = \varphi[\varphi]$.

This follows from the equality $[\varphi] = (I, \varphi[\varphi])$.

Proposition 5.13. Suppose that axiom μA_3 is satisfied at least by the mapping $\lambda\theta_1\theta.(I, \theta_1\theta)$ and let $\langle\ \ \rangle_1$ be a unary operation such that $\bar{n}\langle \varphi \rangle_1 = \varphi\bar{n}$ for all n. (In particular, the last equality will hold whenever $(\varphi L, \langle \varphi \rangle_1 R) = \langle \varphi \rangle_1$.) Then axiom μA_0 and the assertion of 5.7** hold.

Proof. Take $\sigma = \langle(\varphi L^2, LR^2)\rangle_1[\langle(RL, R^2)\rangle_1]$ and $\langle \varphi \rangle = \bar{1}[\sigma]$ by definition. Notice that $\bar{n}\sigma = (\varphi\bar{n}L, \overline{n+2})$ for all n.

Consider the regular segment

$$\mathscr{E} = \{\theta / L\theta \le I \& \forall n(\overline{n+1}\theta R \le \overline{n+2}[\sigma])\}.$$

If $\theta \in \mathscr{E}$, then $L(I, \sigma\theta) = I$ and

$$\begin{aligned}\overline{n+1}(I, \sigma\theta)R &= \bar{n}\sigma\theta R = (\varphi\bar{n}L, \overline{n+2})\theta R = (\varphi\bar{n}L\theta R, \overline{n+2}\theta R)\\ &\le (\varphi\overline{n+1}, \overline{n+3}[\sigma]) = (\varphi\overline{n+1}L, \overline{n+3})[\sigma]\\ &= \overline{n+1}\sigma[\sigma] = \overline{n+2}[\sigma];\end{aligned}$$

hence $[\sigma] \in \mathscr{E}$. We get in particular $\bar{1}[\sigma]R \le \bar{2}[\sigma]$; hence

$$(\varphi L, \langle \varphi \rangle R) = (\varphi L, \bar{1}[\sigma]R) \le (\varphi L, \bar{2}[\sigma]) = (\varphi L^2, \bar{2})[\sigma] = \bar{0}\sigma[\sigma] = \bar{1}[\sigma] = \langle \varphi \rangle.$$

Let $R\psi \le \psi\psi_1$ and $(\varphi L\psi, \tau\psi_1) \le \tau$. Consider the regular segment

$$\mathscr{E} = \{\theta / L\theta \le I \& \forall n(\overline{n+1}\theta\psi \le \tau\psi_1^n)\}.$$

If $\theta \in \mathscr{E}$, then

$$\begin{aligned}\overline{n+1}(I, \sigma\theta)\psi &= \bar{n}\sigma\theta\psi = (\varphi\bar{n}L, \overline{n+2})\theta\psi = (\varphi\bar{n}L\theta\psi, \overline{n+2}\theta\psi)\\ &\le (\varphi\bar{n}\psi, \tau\psi_1^{n+1}) \le (\varphi L\psi\psi_1^n, \tau\psi_1^{n+1}) = (\varphi L\psi, \tau\psi_1)\psi_1^n \le \tau\psi_1^n,\end{aligned}$$

hence $[\sigma]\in\mathscr{E}$. Taking $n=0$, we get $1[\sigma]\psi\leq\tau$, i.e. $\langle\varphi\rangle\psi\leq\tau$. Therefore, (£) is valid, while (££) follows by the proof of 5.4*.

Suppose that $\varphi\bar{n}\psi\leq\rho\bar{n}\tau$ for all n. In order to show that $\langle\varphi\rangle\psi\leq\langle\rho\rangle\tau$ take the regular segment

$$\mathscr{E}=\{\theta/L\theta\leq I\&\forall n(\overline{n+1}\theta\psi\leq\langle\rho\rangle R^n\tau)\}.$$

If $\theta\in\mathscr{E}$, then

$$\begin{aligned}\overline{n+1}(I,\sigma\theta)\psi&=(\varphi\bar{n}L,\overline{n+2})\theta\psi\leq(\varphi\bar{n}\psi,\langle\rho\rangle R^{n+1}\tau)\leq(\rho\bar{n}\tau,\langle\rho\rangle R^{n+1}\tau)\\&=(\rho L,\langle\rho\rangle R)R^n\tau=\langle\rho\rangle R^n\tau,\end{aligned}$$

hence $[\sigma]\in\mathscr{E}$. Taking $n=0$, we get the desired inequality, which completes the proof.

It may happen that an OS is iterative but a subspace of it is not, e.g. the OS of example 4.1 and its subspace of example 4.2. However, the following statement shows that the axiom μA_0 is in a sense hereditary.

Proposition 5.14. Let $\mathscr{S}$ be an OS, let $\mathscr{S}_1$ be a subspace of it. Suppose that $\mathscr{S}$ satisfies μA_0 and the semigroup of $\mathscr{S}$ is closed under $\langle\ \rangle$ and $[\ ]$. Then $\mathscr{S}_1$ satisfies μA_0 and is a subspace of $\mathscr{S}$ as an IOS.

The proof is trivial. Both (£) and (££) are hereditary.

EXERCISES TO CHAPTER 5

Exercise 5.1. Prove that the OS of example 4.7 is iterative. Writing L_1, R_1 respectively for L^{-1}, R^{-1}, show that

$$\langle\varphi\rangle=\bigcup_n R_1^n L_1\varphi L R^n \text{ and } [\varphi]=\bigcup_n (R_1\varphi)^n L_1.$$

Hint. Follow the proofs of 3.1, 3.2.

Exercise 5.2. Prove that the OS $\mathscr{S}$ of example 4.8 is iterative and $\langle\ \rangle, [\ ]$ can be characterized as in the previous exercise. Show that the IOS of example 4.7 is a subspace of $\mathscr{S}$.

Hint. Follow the proofs of 3.1, 3.2 again.

Exercise 5.3. Let $\mathscr{S}$ be the IOS of example 4.7 or example 4.8. Prove that whenever $\Gamma:\mathscr{F}\to\mathscr{F}$ is monotonic, then it has a least fixed point which is a member of all regular segments $\mathscr{E}$ closed under Γ. Therefore, $\mathscr{S}$ is μA_3-iterative.

Hint. Take $\theta_\xi=\sup\{\Gamma(\theta_\eta)/\eta<\xi\}$ for all ξ. The increasing transfinite sequence $\{\theta_\xi\}$ can not consist of distinct members, hence there is a ζ such that $\theta_\zeta=\theta_{\zeta+1}$. Therefore, θ_ζ is a fixed point of Γ. Using the fact that $\mathscr{E}$ is closed under least upper bounds, show by transfinite induction that $\theta_\xi\in\mathscr{E}$ for all ξ, hence $\theta_\zeta\in\mathscr{E}$. (It suffices to consider the sequence $\{\theta_\xi\}_{\xi\leq\omega}$, provided Γ is continuous with respect to least upper bounds of increasing countable sequences. And the inductive mappings are continuous in this particular instance.)

Exercise 5.4. Prove that the IOS of example 3.2 has the property considered in the previous exercise.

Exercise 5.5. Prove that if (££) holds, then the axiom $\psi \leq \psi_1 \Rightarrow \varphi\psi \leq \varphi\psi_1$ can be replaced by the weaker one $\psi \leq \psi_1 \Rightarrow L\psi \leq L\psi_1$, $R\psi \leq R\psi_1$.

Hint. Using (££), show that $[\varphi L] \leq (I, \varphi)$. The inequality $(I, \varphi L[\varphi L]) \leq [\varphi L]$ implies both $I \leq L[\varphi L]$ and $\varphi L[\varphi L] \leq R[\varphi L]$, hence $I = L[\varphi L]$ and $\varphi = \varphi L[\varphi L] \leq R[\varphi L]$. Therefore $\varphi = R[\varphi L]$. Show that $\psi \leq \psi_1$ implies $[\varphi L]\psi \leq [\varphi L]\psi_1$ by (££); multiplying on the left by R, obtain $\varphi\psi \leq \varphi\psi_1$.

Exercise 5.6. Let $\mathscr{S}$ be an OS with $\varphi \leq \psi$ iff $\varphi = \psi$. Show that $\mathscr{S}$ does not satisfy (££).

Hint. Suppose that the equality $(I, R\theta) = \theta$ has a least solution denoted by $[R]$. However, 'least' means 'unique' in this space, hence $(I, R(I, \varphi)) = (I, \varphi)$ implies $(I, \varphi) = [R]$ for all φ, which is not the case.

It follows from this last exercise that the OS of examples 4.2, 4.4, 4.5 are not iterative.

CHAPTER 6

Translation and iteration lemmas

In this chapter we study the operations of translation and iteration. A new operation called primitive recursion is also introduced. The independence of the IOS-operations is examined in the exercises.

Assume that an IOS $\mathscr{S} = (\mathscr{F}, I, \Pi, L, R)$ is given. Recall that $\langle \quad \rangle = \lambda\varphi.\mu\theta.(\varphi L, \theta R)$ and $[\quad] = \lambda\varphi.\mu\theta.(I, \varphi\theta)$, in particular, $(\varphi L, \langle\varphi\rangle R) = \langle\varphi\rangle$ and $(I, \varphi[\varphi]) = [\varphi]$ for all φ.

Proposition 6.1. Iteration is monotonic.
Follows by 5.1.

Proposition 6.2. $[L] = (I, I)$.
Proof. Using 5.12, we get

$$[L] = (I, L[L]) = (I, I).$$

Proposition 6.3. $\bar{n}[I] = I$ for all n.
Proof. We have $\bar{0}[I] = L[I] = I$ and

$$\overline{n+1}\,[I] = \bar{n}R[I] = \bar{n}I[I] = \bar{n}[I],$$

which completes the proof.

The partially ordered set $\mathscr{F}$ has a least member, namely $O = R[R]$.

Proposition 6.4. $O \leq \varphi$ for all φ.
Proof. The equality $(I, R(I, \varphi)) = (I, \varphi)$ implies $[R] \leq (I, \varphi)$ by (££). Therefore, $O = R[R] \leq \varphi$.

Proposition 6.5. $LO = RO = O$.
Proof. Using 6.4, we get

$$LO \leq L(O, O) = O,$$

hence $LO = O$. Similarly, $RO = O$. This direct proof is due to N. Georgieva.

As an immediate corollary to 6.5 $\alpha O = O$ for all $\alpha \in \mathscr{D}$. However, the statement $\forall\varphi(\varphi O = O)$ is not a theorem; it is valid in examples 4.1, 4.3, 4.7, 4.8 but fails in example 4.6.

Proposition 6.6. $O\varphi = O$ for all φ.

Proof. It suffices to show that $O\varphi \leq O$. The equality $(\varphi, R(\varphi, O)) = (\varphi, O)$ implies $[R]\varphi \leq (\varphi, O)$ by (££), hence $O\varphi \leq O$.

Proposition 6.7. $[\varphi] = (I, R[\varphi])$.

This follows from 5.12.

Proposition 6.8. $\varphi = R[\varphi L]$.

Proof.

$$R[\varphi L] = \varphi L[\varphi L] = \varphi I = \varphi.$$

Proposition 6.9. $(\varphi, \psi) = A[\varphi L^2][\psi L]$, where $A = (R, RL)$.

Proof.

$$A[\varphi L^2][\psi L] = (\varphi L^2[\varphi L^2], R)[\psi L] = (\varphi L[\psi L], R[\psi L]) = (\varphi, \psi).$$

The above equality is due to N. Georgieva and corresponds to a similar one of Skordev [1982].

Proposition 6.10. $\chi[\varphi\chi]\psi = \mu\theta.\chi(\psi, \varphi\theta)$. In particular, $[\varphi]\psi = \mu\theta.(\psi, \varphi\theta)$.

Proof. We have

$$\chi(\psi, \varphi\chi[\varphi\chi]\psi) = \chi(I, \varphi\chi[\varphi\chi])\psi = \chi[\varphi\chi]\psi.$$

Whenever $\chi(\psi, \varphi\tau) \leq \tau$, then $(\psi, \varphi\chi(\psi, \varphi\tau)) \leq (\psi, \varphi\tau)$; hence $[\varphi\chi]\psi \leq (\psi, \varphi\tau)$ by (££). Therefore, $\chi[\varphi\chi]\psi \leq \chi(\psi, \varphi\tau) \leq \tau$, which completes the proof.

Proposition 6.11. $R[\varphi]\psi = \mu\theta.\varphi(\psi, \theta)$.

This follows from 5.12, 6.10.

The operation $[\ \ ]_1 = \lambda\varphi.\mu\theta.\varphi(I, \theta)$ is a natural alternative to iteration, replacing (££) by $(££)_1$:

$$\varphi(I, [\varphi]_1) \leq [\varphi]_1,$$
$$\varphi(\psi, \tau) \leq \tau \Rightarrow [\varphi]_1\psi \leq \tau.$$

Proposition 6.11 gives $[\varphi]_1 = R[\varphi]$; hence $[\varphi] = (I, [\varphi]_1)$.

Proposition 6.12. $R[\varphi(\psi L, \chi R)]\rho = \mu\theta.\varphi(\psi\rho, \chi\theta)$.

This follows from 6.11 and the equality

$$\varphi(\psi\rho, \chi\theta) = \varphi(\psi L, \chi R)(\rho, \theta).$$

Proposition 6.13. $[\varphi]\psi = R[\sigma]$, where $\sigma = (\psi L, \varphi R)$.

Proof.

$$[\varphi]\psi = \mu\theta.(\psi, \varphi\theta) = R[(\psi L, \varphi R)]$$

by 6.10, 6.12.

Proposition 6.14. $\varphi[\psi] = \bar{1}[\sigma]$, where $\sigma = (\varphi R^2, \bar{0}, \psi R^2)$.

Proof. We have

$$(I, \psi R^2[\sigma]) = (\bar{0}, \psi R^2)[\sigma] = R\sigma[\sigma] = R^2[\sigma];$$

hence $[\psi] \leq R^2[\sigma]$. Multiplying on the left by φ, we get

$$\varphi[\psi] \leq \varphi R^2[\sigma] = L\sigma[\sigma] = \bar{1}[\sigma].$$

Let $\tau = (I, \varphi[\psi], [\psi])$. Then

$$(I, \sigma\tau) = (I, \varphi[\psi], I, \psi[\psi]) = \tau,$$

hence $[\sigma] \leq \tau$. Therefore, $\bar{1}[\sigma] \leq \bar{1}\tau = \varphi[\psi]$. The proof is complete.

Proposition 6.15. $[\varphi]\psi[\chi] = \bar{1}[\sigma]$, where $\sigma = ((\psi R^2, \varphi\bar{1}), \bar{0}, \chi R^2)$.

Proof. To begin with, $(I, \chi R^2[\sigma]) = (\bar{0}, \chi R^2)[\sigma] = R\sigma[\sigma] = R^2[\sigma]$ implies $[\chi] \leq R^2[\sigma]$. Using this inequality, we get

$$(\psi[\chi], \varphi\bar{1}[\sigma]) \leq (\psi R^2[\sigma], \varphi\bar{1}[\sigma]) = (\psi R^2, \varphi\bar{1})[\sigma] = L\sigma[\sigma] = \bar{1}[\sigma],$$

hence $[\varphi]\psi[\chi] \leq \bar{1}[\sigma]$ by (££).

On the other hand, writing τ for $(I, [\varphi]\psi[\chi], [\chi])$ we get

$$(I, \sigma\tau) = (I, (\psi[\chi], \varphi[\varphi]\psi[\chi]), I, \chi[\chi]) = (I, (I, \varphi[\varphi])\psi[\chi], [\chi]) = \tau;$$

hence $[\sigma] \leq \tau$. Therefore, $\bar{1}[\sigma] \leq [\varphi]\psi[\chi]$, which completes the proof.

Proposition 6.16. $[\varphi[\psi]] = \bar{1}[\sigma]$, where $\sigma = ((\bar{0}, \varphi R^2), \bar{1}, \psi R^2)$.

Proof. We have $(\bar{1}[\sigma], \psi R^2[\sigma]) = (\bar{1}, \psi R^2)[\sigma] = R\sigma[\sigma] = R^2[\sigma]$; hence $[\psi]\bar{1}[\sigma] \leq R^2[\sigma]$. Using this inequality, we get

$$(I, \varphi[\psi]\bar{1}[\sigma]) \leq (I, \varphi R^2[\sigma]) = (\bar{0}, \varphi R^2)[\sigma] = L\sigma[\sigma] = \bar{1}[\sigma],$$

hence $[\varphi[\psi]] \leq \bar{1}[\sigma]$.

Let $\tau = (I, [\varphi[\psi]], [\psi][\varphi[\psi]])$. Then

$$\begin{aligned}(I, \sigma\tau) &= (I, (I, \varphi[\psi][\varphi[\psi]]), [\varphi[\psi]], \psi[\psi][\varphi[\psi]]) \\ &= (I, [\varphi[\psi]], (I, \psi[\psi])[\varphi[\psi]]) = \tau,\end{aligned}$$

hence $[\sigma] \leq \tau$. Therefore, $\bar{1}[\sigma] \leq [\varphi[\psi]]$. The proof is complete.

Proposition 6.17. Translation is monotonic.

This follows from 5.1.

Proposition 6.18. Translation is injective.

Proof. $\varphi = L\langle\varphi\rangle(I, I)$.

By contrast, iteration is not injective. For instance, $[R] = [R^2]$, while $R \neq R^2$ by 4.4. Actually, $[R] = (I, O) \leq [R^2]$. On the other hand,

$$(I, R^2[R]) = (I, RO) = (I, O) = [R]$$

implies $[R^2] \leq [R]$.

Proposition 6.19. If $\varphi L\psi \leq \rho L\tau$, $R\psi \leq \psi\psi_1$ and $\tau\psi_1 \leq R\tau$ for a certain ψ_1, then $\langle\varphi\rangle\psi \leq \langle\rho\rangle\tau$.

Proof. We have $R\psi \leq \psi\psi_1$ and

$$(\varphi L\psi, \langle\rho\rangle\tau\psi_1) \leq (\rho L\tau, \langle\rho\rangle R\tau) = (\rho L, \langle\rho\rangle R)\tau = \langle\rho\rangle\tau,$$

hence $\langle\varphi\rangle\psi \leq \langle\rho\rangle\tau$ by (£).

Proposition 6.20. If $\varphi L\psi = \rho L\tau$, $R\psi = \psi\psi_1$ and $\tau\psi_1 = R\tau$ for some ψ_1, then $\langle\varphi\rangle\psi = \langle\rho\rangle\tau$.

This follows from 6.19.

Proposition 6.21. $\langle\varphi\rangle\langle\psi\rangle = \langle\varphi\psi\rangle$.

Proof. Take $\psi_1 = R$, $\rho = \varphi\psi$ and $\tau = I$. Then the equalities $\varphi L\langle\psi\rangle = \varphi\psi L$, $R\langle\psi\rangle = \langle\psi\rangle R$ imply $\langle\varphi\rangle\langle\psi\rangle = \langle\varphi\psi\rangle$ by 6.20.

Proposition 6.22. If $(\varphi L, \tau R) = \tau$, then $\langle\varphi\rangle = \langle I\rangle\tau$.

This follows from 6.20.

We get in particular that $\langle\varphi\rangle$ is a unique solution to the equation $(\varphi L, \theta R) = \theta$, provided $\langle I\rangle = I$. (The last equality holds in examples 4.1, 4.6 but fails in example 4.3.) By contrast, there is no space in which the equation $(I, \varphi\theta) = \theta$ has a unique solution for each φ. Indeed, both $[L]$ and $[R]$ satisfy the equality $(I, R\theta) = \theta$ by 6.7 but are not equal.

Proposition 6.23. $R[\langle\varphi\rangle] = [\langle\varphi\rangle]\varphi$.

Proof. Using 6.13, we get

$$[\langle\varphi\rangle]\varphi = R[(\varphi L, \langle\varphi\rangle R)] = R[\langle\varphi\rangle].$$

Proposition 6.24. $\bar{n}[\langle\varphi\rangle] = \varphi^n$ for all n.

Proof.

$$\bar{n}[\langle\varphi\rangle] = LR^n[\langle\varphi\rangle] = L[\langle\varphi\rangle]\varphi^n = \varphi^n.$$

The binary operation $\Delta = \lambda\varphi\psi.\langle\varphi\rangle[\langle\psi\rangle]$ is called *primitive recursion.*

Proposition 6.25. $\Delta(\varphi, \psi) = \langle\varphi\rangle\Delta(I, \psi)$.

Proof.

$$\Delta(\varphi, \psi) = \langle\varphi\rangle[\langle\psi\rangle] = \langle\varphi\rangle\langle I\rangle[\langle\psi\rangle] = \langle\varphi\rangle\Delta(I, \psi).$$

Proposition 6.26. $\langle I\rangle\Delta(\varphi, \psi) = \Delta(\varphi, \psi)$.

This follows from the equality $\langle I\rangle\langle\varphi\rangle = \langle\varphi\rangle$.

Proposition 6.27. $L\Delta(\varphi, \psi) = \varphi$, $R\Delta(\varphi, \psi) = \Delta(\varphi, \psi)\psi$.

Proof.

$$L\Delta(\varphi, \psi) = \varphi L[\langle\psi\rangle] = \varphi, \quad R\Delta(\varphi, \psi) = \langle\varphi\rangle R[\langle\psi\rangle]$$
$$= \langle\varphi\rangle[\langle\psi\rangle]\psi = \Delta(\varphi, \psi)\psi.$$

Proposition 6.28. $\bar{n}\Delta(\varphi, \psi) = \varphi\psi^n$ for all n.

This follows from 6.27.

Proposition 6.29. $(\varphi, \psi) = B\Delta(L, B\Delta(\varphi, R))B\Delta(I, \psi)$, where $B = (LR, L)$.
Proof.

$$B\Delta(L, B\Delta(\varphi, R))B\Delta(I, \psi) = B\Delta(L, (\varphi R, \varphi))(\psi, I) = (\varphi R, L)(\psi, I) = (\varphi, \psi).$$

Proposition 6.30. If $\psi\chi \leq \chi\psi_1$ and $(\varphi\chi, \tau\psi_1) \leq \tau$ for some ψ_1, then $\Delta(\varphi, \psi)\chi \leq \tau$.
Proof. The inequalities

$$R[\langle\psi\rangle]\chi = [\langle\psi\rangle]\psi\chi \leq [\langle\psi\rangle]\chi\psi_1, \quad (\varphi L[\langle\psi\rangle]\chi, \tau\psi_1) = (\varphi\chi, \tau\psi_1) \leq \tau$$

imply $\langle\varphi\rangle[\langle\psi\rangle]\chi \leq \tau$ by (£).

Proposition 6.31. If $\varphi\chi = L\tau$, $\psi\chi = \chi\psi_1$ and $\tau\psi_1 = R\tau$ for some ψ_1, then $\Delta(\varphi, \psi)\chi = \langle I\rangle\tau$.
Proof. The equalities

$$\varphi L[\langle\psi\rangle]\chi = \varphi\chi = L\tau, \quad R[\langle\psi\rangle]\chi = [\langle\psi\rangle]\psi\chi = [\langle\psi\rangle]\chi\psi_1$$

and $\tau\psi_1 = R\tau$ imply $\langle\varphi\rangle[\langle\psi\rangle]\chi = \langle I\rangle\tau$ by 6.20.

One gets $\Delta(\varphi, \psi)\chi = \tau$, provided τ is of the form $\langle\tau_1\rangle\tau_2$. In particular, $\Delta(\varphi, \psi)\chi = \Delta(\varphi\chi, \psi_1)$ whenever $\psi\chi = \chi\psi_1$.

The following result shows that translation is a particular instance of primitive recursion.

Proposition 6.32. $\langle\varphi\rangle = \Delta(\varphi L, R)$.
This follows from 6.31. Take $\chi = I$, $\psi_1 = R$, $\tau = \langle\varphi\rangle$.

Proposition 6.33. $\Delta(\varphi, \psi) = \mu\theta.(\varphi, \theta\psi)$.
Proof. Using 6.25, 6.27, we get

$$\begin{aligned}(\varphi, \Delta(\varphi, \psi)\psi) &= (\varphi, \langle\varphi\rangle\Delta(I, \psi)\psi) = (\varphi L, \langle\varphi\rangle R)\Delta(I, \psi) \\ &= \langle\varphi\rangle\Delta(I, \psi) = \Delta(\varphi, \psi).\end{aligned}$$

If $(\varphi, \tau\psi) \leq \tau$, then $\Delta(\varphi, \psi) \leq \tau$ by 6.30.

Proposition 6.34. If $(\varphi, \tau\psi) = \tau$, then $\Delta(\varphi, \psi) = \langle I\rangle\tau$.
This follows from 6.31.

In particular, $\Delta(\varphi, \psi)$ is a unique solution to the equality $(\varphi, \theta\psi) = \theta$, provided $\langle I\rangle = I$.

Proposition 6.35. Let $C = \Delta((L^2, LR), \ (RL, R^2))$. Then $\bar{n}C = (\bar{n}L, \bar{n}R) = (\bar{n}L, \overline{n+1})$ for all n.
Proof.

$$\bar{n}C = (L^2, LR)(RL, R^2)^n = (L^2, LR)(R^nL, R^{n+1}) = (LR^nL, LR^{n+1}).$$

It is worth mentioning that $\langle I\rangle C = C$ by 6.26. Notice also that $C = \Delta(B^2, A^2)$.

Proposition 6.36. $\langle(\varphi,\psi)\rangle = C(\langle\varphi\rangle,\langle\psi\rangle)$.

Proof. The equalities

$$(L^2, LR)(\langle\varphi\rangle,\langle\psi\rangle) = (\varphi,\psi)L = L\langle(\varphi,\psi)\rangle,$$
$$(RL, R^2)(\langle\varphi\rangle,\langle\psi\rangle) = (\langle\varphi\rangle,\langle\psi\rangle)R,$$
$$\langle(\varphi,\psi)\rangle R = R\langle(\varphi,\psi)\rangle$$

imply $C(\langle\varphi\rangle,\langle\psi\rangle) = \langle(\varphi,\psi)\rangle$ by 6.31.

Proposition 6.37. $\langle[\varphi]\rangle = C[\langle\varphi\rangle C]$.

Proof. We have

$$(L, \varphi(L^2, LR)[\langle\varphi\rangle C]) = (L, \varphi LC[\langle\varphi\rangle C]) = (L, L\langle\varphi\rangle C[\langle\varphi\rangle C])$$
$$= (L^2, LR)[\langle\varphi\rangle C],$$

hence $[\varphi]L \leq (L^2, LR)[\langle\varphi\rangle C]$ by (££). Similarly,

$$(R, \langle\varphi\rangle C(RL, R^2)[\langle\varphi\rangle C]) = (R, R\langle\varphi\rangle C[\langle\varphi\rangle C]) = (RL, R^2)[\langle\varphi\rangle C]$$

gives $[\langle\varphi\rangle C]R \leq (RL, R^2)[\langle\varphi\rangle C]$.

Taking $\tau = (I, \varphi[\varphi]L, \langle\varphi\rangle C[\langle\varphi\rangle C]R)$, we get

$$(I, \langle\varphi\rangle C\tau) = (I, (\varphi L, \langle\varphi\rangle R)C\tau) = (I, \varphi(L^2, LR)\tau, \langle\varphi\rangle C(RL, R^2)\tau)$$
$$= (I, \varphi(L, \varphi[\varphi]L), \langle\varphi\rangle C(R, \langle\varphi\rangle C[\langle\varphi\rangle C]R)) = \tau,$$

hence $[\langle\varphi\rangle C] \leq \tau$. Therefore, $(L^2, LR)[\langle\varphi\rangle C] \leq [\varphi]L$ and $(RL, R^2)\circ [\langle\varphi\rangle C] \leq [\langle\varphi\rangle C]R$.

We conclude that $(L^2, LR)[\langle\varphi\rangle C] = [\varphi]L$ and $(RL, R^2)[\langle\varphi\rangle C] = [\langle\varphi\rangle C]R$, which implies $C[\langle\varphi\rangle C] = \langle[\varphi]\rangle$ by 6.31.

Proposition 6.38 (First Recursion Lemma). Let $\sigma = \Delta((L, \varphi\bar{2}), (\psi L, R^2))$. Then $\bar{1}[\sigma] = \mu\theta.(I, \varphi\theta\psi)$.

Proof. We have

$$(\psi, \sigma(\psi, R^2[\sigma])) = (\psi, \sigma(\psi L, R^2)[\sigma]) = (\psi, R\sigma[\sigma]) = (\psi, R^2[\sigma]),$$

hence $[\sigma]\psi \leq (\psi, R^2[\sigma])$. Therefore,

$$(I, \varphi\bar{1}[\sigma]\psi) \leq (I, \varphi\bar{1}(\psi, R^2[\sigma])) = (I, \varphi\bar{2}[\sigma]) = (L, \varphi\bar{2})[\sigma] = \bar{0}\sigma[\sigma] = \bar{1}[\sigma].$$

Let $(I, \varphi\tau\psi) \leq \tau$. Then the inequalities

$$(\psi L, R^2)(I, \Delta(\tau,\psi)) = (\psi, \Delta(\tau,\psi)\psi) = (I, \Delta(\tau,\psi))\psi,$$
$$((L, \varphi\bar{2})(I, \Delta(\tau,\psi)), \Delta(\tau,\psi)\psi) = ((I, \varphi\tau\psi), \Delta(\tau,\psi)\psi) \leq (\tau, \Delta(\tau,\psi)\psi)$$
$$= \Delta(\tau,\psi)$$

imply $\sigma(I, \Delta(\tau,\psi)) \leq \Delta(\tau,\psi)$ by 6.30, hence $R[\sigma] \leq \Delta(\tau,\psi)$ by 6.11. Therefore, $\bar{1}[\sigma] \leq \bar{0}\Delta(\tau,\psi) = \tau$, which completes the proof.

Proposition 6.39 (Second Recursion Lemma). Let $\varphi_1 = \varphi(\psi L, \chi R)$ and $\theta_1 = \mu\theta.(I, \varphi_1\theta\rho)$; the element θ_1 exists by 6.38. Then $\varphi_1\theta_1 = \mu\theta.\varphi(\psi, \chi\theta\rho)$.

Proof. We have

$$\varphi(\psi, \chi\varphi_1\theta_1\rho) = \varphi_1(I, \varphi_1\theta_1\rho) = \varphi_1\theta_1.$$

If $\varphi(\psi, \chi\tau\rho) \leq \tau$, then

$$(I, \varphi_1(I, \tau\rho)\rho) = (I, \varphi(\psi, \chi\tau\rho)\rho) \leq (I, \tau\rho),$$

hence $\theta_1 \leq (I, \tau\rho)$. Therefore,

$$\varphi_1\theta_1 \leq \varphi_1(I, \tau\rho) = \varphi(\psi, \chi\tau\rho) \leq \tau,$$

which completes the proof.

Proposition 6.40. Let $\rho = \Delta(L, R^2)$, $P = \Delta(\rho R, \rho)$, $\sigma = \Delta(L^2, A)$ and $Q = \mu\theta.\sigma(\theta R, L)$. (The element Q exists by 6.39.) Then $\langle\!\langle \varphi \rangle\!\rangle = P\langle \varphi \rangle Q$ for all φ.

Proof. It follows that

$$L\sigma = L^2, R^2\sigma = \sigma A^2, LA^2 = RL,$$

hence $\rho\sigma = \langle I \rangle L$ by 6.31. Therefore,

$$\rho R\sigma = \rho\sigma A = \langle I \rangle LA = \langle I \rangle R.$$

The equality $Q = \sigma(QR, L)$ multiplied on the left by ρR and ρ gives respectively $\rho RQ = \langle I \rangle L$, $\rho Q = QR$.

The equalities

$$L\langle \varphi \rangle = \varphi L = L\langle \varphi \rangle \rho, R^2\langle \varphi \rangle = \langle \varphi \rangle R^2, \langle \varphi \rangle \rho R^2 = \langle \varphi \rangle R\rho = R\langle \varphi \rangle \rho$$

imply $\rho\langle \varphi \rangle = \langle \varphi \rangle \rho$ by 6.31. Finally, we get

$$\rho R\langle \varphi \rangle Q = \langle \varphi \rangle \rho RQ = \langle \varphi \rangle L = L\langle\!\langle \varphi \rangle\!\rangle,$$
$$\rho\langle \varphi \rangle Q = \langle \varphi \rangle \rho Q = \langle \varphi \rangle QR, \langle\!\langle \varphi \rangle\!\rangle R = R\langle\!\langle \varphi \rangle\!\rangle,$$

hence $P\langle \varphi \rangle Q = \langle\!\langle \varphi \rangle\!\rangle$ by 6.31. The proof is complete.

Proposition 6.41. Let $D = \Delta(L^2, \langle R \rangle R)$. Then $\bar{n}D = \bar{n}\bar{n}$ for all n.

Proof.

$$\bar{n}D = L^2(\langle R \rangle R)^n = L^2\langle R \rangle^n R^n = LR^n LR^n.$$

Proposition 6.42. Let $\rho = C([I]L, R)$, $\sigma = D\langle \varphi \rangle \rho$ and $\tau = D\langle R\varphi \rangle \rho$. Then $\rho[\sigma] = ([\bar{0}\varphi], \rho[\tau])$.

Proof. Using the easy equalities

$$\sigma = (L\sigma, R\sigma) = (\bar{0}\varphi(\bar{0}, \bar{1}), \tau(\bar{0}, R^2)),$$

we get

$$(I, \bar{0}\varphi(\bar{0}, \bar{1})[\sigma]) = (I, \bar{0}\sigma[\sigma]) = (\bar{0}, \bar{1})[\sigma],$$

hence $[\bar{0}\varphi] \leq (\bar{0}, \bar{1})[\sigma]$. Similarly,

$$(I, \tau(\bar{0}, R^2)[\sigma]) = (I, R\sigma[\sigma]) = (\bar{0}, R^2)[\sigma]$$

implies $[\tau] \leq (\bar{0}, R^2)[\sigma]$.

On the other hand,

$$(I, \sigma(I, \bar{0}\varphi[\bar{0}\varphi], \tau[\tau])) = (I, \bar{0}\varphi(I, \bar{0}\varphi[\bar{0}\varphi]), \tau(I, \tau[\tau])) = (I, \bar{0}\varphi[\bar{0}\varphi], \tau[\tau])$$

implies $[\sigma] \leq (I, \bar{0}\varphi[\bar{0}\varphi], \tau[\tau])$, hence the above two inequalities are in fact equalities and

$$\rho[\sigma] = (L\rho[\sigma], R\rho[\sigma]) = ((\bar{0}, \bar{1})[\sigma], \rho(\bar{0}, R^2)[\sigma]) = ([\bar{0}\varphi], \rho[\tau]).$$

The proof is complete.

Proposition 6.43. Let $\rho = C([I]L, R)$ and $\Gamma = \lambda\theta.D\langle\theta\rangle\rho$. Then $\bar{n}\rho[\Gamma(\varphi)] = [\bar{n}\varphi]$ for all φ, n.

Proof. The previous statement gives that

$$\rho[\Gamma(\varphi)] = ([\bar{0}\varphi], \rho[\Gamma(R\varphi)])$$

for all φ, hence $\bar{0}\rho[\Gamma(\varphi)] = [\bar{0}\varphi]$. Supposing $\bar{n}\rho[\Gamma(\varphi)] = [\bar{n}\varphi]$ for all φ, we get

$$\overline{n+1}\rho[\Gamma(\varphi)] = \bar{n}\rho[\Gamma(R\varphi)] = [\bar{n}R\varphi] = [\overline{n+1}\varphi].$$

This completes the proof.

EXERCISES TO CHAPTER 6

Exercise 6.1. Prove the following equalities.

a. $A = LC[RL]\langle R\rangle[\langle L\rangle]$.
b. $(\varphi, \psi) = (RL, R)[\psi L^2][\varphi L]$.
c. $[\varphi] = \bar{1}[((\bar{0}, \varphi\bar{1}), \psi\bar{1})]$.
d. $\varphi[\psi] = R^2[((\bar{0}, \psi\bar{1}), \varphi\bar{1})]$.
e. $[\varphi]\psi[\chi] = R^2[(\chi(\bar{0}, \bar{1}), \psi(\bar{0}, \bar{1}), \varphi R^2)]$.
f. $(\varphi[\psi], \chi[\rho]) = (\varphi L, \chi R)[(L^2, \rho R)][\psi]$.
g. $[\varphi[\psi]] = R^2[((R^2, \psi\bar{1}), \bar{0}, \varphi\bar{1})]$.
h. $[\varphi[\psi]] = \bar{0}\bar{1}[(((\bar{0}, \varphi(R\bar{1}, R^2)), \bar{0}\bar{1}), \psi(R\bar{1}, R^2))]$.
i. $[\varphi[(\psi, \chi)]] = R^2\bar{1}[\sigma]$, where $\sigma = ((\psi\rho, R^2\bar{1}, \bar{0}, \varphi\rho), \chi\rho)$ and $\rho = (\bar{1}\bar{1}, \bar{0}\bar{1}, R^2)$.
j. $(\varphi, \psi) = \rho\Delta(R, \rho\Delta(\psi, L))\rho\Delta(I, \varphi)$, where $\rho = (\bar{0}, \bar{1})$.

Hint for assertion i. Show that

$$[(\psi, \chi)]R^2\bar{1}[\sigma] \leq \rho[\sigma], [\varphi[(\psi, \chi)]] \leq R^2\bar{1}[\sigma],$$
$$[\sigma] \leq (L, (\psi[(\psi, \chi)], I, I), \chi[(\psi, \chi)])[\varphi[(\psi, \chi)]].$$

Exercise 6.2 (Second Recursion Lemma, improved). Show that whenever $\psi\chi \leq \chi\chi_1$ and $\varphi(\chi, \tau\chi_1) \leq \tau$, then $(\mu\theta.\varphi(I, \theta\psi))\chi \leq \tau$.

Hint. Take σ as in 6.38, then $\mu\theta.\varphi(I, \theta\psi) = \varphi\bar{1}[\sigma]$ by 6.39. Show that $[\sigma]\chi \leq (\chi, \Delta((\chi, \tau\chi_1), \chi_1))$.

Exercise 6.3. $\Delta(\varphi, \psi) = \mu\theta.(\varphi, \varphi\psi, \theta\psi^2)$.

Hint. Writing θ_0 for $\mu\theta.(\varphi, \varphi\psi, \theta\psi^2)$, show that $\theta_0\psi \leq R\theta_0$ by exercise 6.2.

Exercise 6.4. Let $G = \Delta(\langle L\rangle, \langle R\rangle)$. Show that $\langle\!\langle I\rangle\!\rangle C = G(L, GR)$.

Hint. Show by 6.20 that $\langle L\rangle G = \langle I\rangle L$, hence $\langle L\rangle(L, GR) = L\langle\!\langle I\rangle\!\rangle C$;

similarly $\langle R\rangle G = GR$ and $\langle R\rangle(L, GR) = (L, GR)A^2$. The equality $\langle\!\langle I\rangle\!\rangle CA^2 = R\langle\!\langle I\rangle\!\rangle C$ completes the proof by 6.31.

Exercise 6.5. Establish the following assertions:

a. $P, Q \in cl(L, \langle\!\langle L\rangle\!\rangle, \langle\!\langle A\rangle\!\rangle/\circ, [\ \ ])$.
b. $P, Q \in cl(L, \langle\!\langle\!\langle L\rangle\!\rangle\!\rangle, \langle\!\langle\!\langle R\rangle\!\rangle\!\rangle/\circ, \Pi, [\ \ ])$.

Hint. Take $\sigma = \Delta(L^2, A)$, $\sigma_1 = \langle I\rangle\sigma\langle I\rangle$ and prove by exercises 6.3, 6.2 that $\sigma_1 = \sigma$, hence $Q = \mu\theta.\sigma_1(\theta R, L)$. Then show by 6.36–6.39 that

$$P, Q \in cl(L, \langle\!\langle I\rangle\!\rangle C, \langle\!\langle L\rangle\!\rangle, \langle\!\langle R\rangle\!\rangle, \langle\!\langle A\rangle\!\rangle/\circ, \Pi, [\ \ ]).$$

Make use of exercise 6.4, the equalities $\langle\!\langle R\rangle\!\rangle = \langle\!\langle L\rangle\!\rangle\langle\!\langle A\rangle\!\rangle$, $A = L^2\langle\!\langle A\rangle\!\rangle[L]^2$ and 6.9 to obtain assertion a.

Substitute $\langle I\rangle A$ for A in the above argument to show that

$$P, Q \in cl(L, \langle\!\langle I\rangle\!\rangle C, \langle\!\langle L\rangle\!\rangle, \langle\!\langle R\rangle\!\rangle, \langle\!\langle\!\langle I\rangle\!\rangle\!\rangle\langle\!\langle A\rangle\!\rangle/\circ, \Pi, [\ \ ]).$$

The equality $\langle A\rangle = C\langle I\rangle(\langle R\rangle, \langle RL\rangle)$ implies $\langle\!\langle\!\langle I\rangle\!\rangle\!\rangle\langle\!\langle A\rangle\!\rangle = \langle\!\langle\!\langle I\rangle\!\rangle\!\rangle\langle C\rangle\langle\!\langle I\rangle\!\rangle C(\langle\!\langle R\rangle\!\rangle, \langle\!\langle RL\rangle\!\rangle)$, hence

$$\langle\!\langle I\rangle\!\rangle C, \langle\!\langle\!\langle I\rangle\!\rangle\!\rangle\langle\!\langle A\rangle\!\rangle \in cl(L, \langle\!\langle\!\langle L\rangle\!\rangle\!\rangle, \langle\!\langle\!\langle R\rangle\!\rangle\!\rangle/\circ, \Pi, [\ \ ])$$

by

$$\begin{aligned}\langle\!\langle\!\langle I\rangle\!\rangle\!\rangle\langle C\rangle &= \langle G(L, GR)\rangle = \langle G\langle I\rangle(L, GR)\rangle\\ &= \langle G\rangle\langle\!\langle I\rangle\!\rangle C(\langle L\rangle, \langle G\rangle\langle R\rangle),\\ \langle G\rangle &= \langle\!\langle\!\langle L\rangle\!\rangle\!\rangle C[\langle\!\langle\!\langle R\rangle\!\rangle\!\rangle C].\end{aligned}$$

Exercise 6.6. Let $J = \lambda st.(2s+1)2^t$. Show that $\bar{s}\bar{t}P = \overline{J(s,t)}$ and $\overline{J(s,t)}Q = \bar{s}\bar{t}$ for all s, t.

Exercise 6.7.** Let $J{:}\omega^2 \to \omega$ and $\bar{s}\bar{t}\rho = \overline{J(s,t)}$, $\overline{J(s,t)}\sigma = \bar{s}\bar{t}$ for all s, t. Show that $\langle\!\langle\varphi\rangle\!\rangle = \langle\!\langle I\rangle\!\rangle\rho\langle\varphi\rangle\sigma$ for all φ.

Hint. Use 5.9**.

Axiom μA_1 ensures that the element $I^* = \mu\theta.(L, \langle\theta\rangle R)$ exists and has certain additional properties.

Exercise 6.8*. Prove that

$$\begin{aligned}(\pounds)^*\quad & (L, \langle I^*\rangle R) \le I^*,\\ & \langle I\rangle R\psi \le \langle\psi\rangle\psi_1 \,\&\, (L\psi, \langle\tau\rangle\psi_1) \le \tau \Rightarrow I^*\psi \le \tau,\\ & C(\langle L\rangle, P\tau Q\langle R\rangle) \le \tau \Rightarrow \langle I^*\rangle \le \tau.\end{aligned}$$

Hint. Supposing $\langle I\rangle R\psi \le \langle\psi\rangle\psi_1$ and $(L\psi, \langle\tau\rangle\psi_1) \le \tau$, show that the simple segment $\{\theta/\theta\psi \le \tau\}$ is closed under $\lambda\theta.(L, \langle\theta\rangle R)$. The element $\sigma = \mu\theta.C(\langle L\rangle, P\theta Q\langle R\rangle)$ exists by 6.39; show that the simple segment $\{\theta/\langle\theta\rangle \le \sigma\}$ is closed under $\lambda\theta.(L, \langle\theta\rangle R)$.

Exercise 6.9. Let I^* meet $(\pounds)^*$ and $\sigma = \mu\theta.C(\langle L\rangle, P\theta Q\langle R\rangle)$. Prove that $I^* = L\sigma(I, I)$.

Hint. It suffices to show that $\langle I^*\rangle = \sigma$. The inequality $C(\langle L\rangle,$

$P\langle I^*\rangle Q\langle R\rangle) \leq \langle I^*\rangle$ implies $\sigma \leq \langle I^*\rangle$ while, on the other hand, the last implication of (£)* ensures that $\langle I^*\rangle \leq \sigma$.

It is the case that $I^* = \langle I\rangle = I$ in example 4.1 and $I^* < \langle I\rangle < I$ in example 4.3.

Exercise 6.10. Let I^* meet (£)* and $\Gamma:\mathscr{F}^2 \to \mathscr{F}$ such that $\Gamma(\varphi,\psi) = (\varphi, \langle\Gamma(\varphi,\psi)\rangle\psi)$ for all φ,ψ. Show that the operation $\Delta^* = \lambda\varphi\psi.\mu\theta.(\varphi,\langle\theta\rangle\psi)$ exists and satisfies the following analogue to 6.30: if $\langle I\rangle\psi\chi \leq \langle\chi\rangle\chi_1$, $(\varphi\chi, \langle\tau\rangle\chi_1) \leq \tau$, then $\Delta^*(\varphi,\psi)\chi \leq \tau$.

Hint. Take $\Delta^* = \lambda\varphi\psi.I^*\Gamma(\varphi,\psi)$ by definition.

Exercise 6.11. Prove that the operation $\circ$ cannot be expressed in terms of Π, $\langle\ \ \rangle$, $[\ \ ]$ and members of $\mathscr{D}$, i.e. there is no expression $\mathscr{V}(\theta_1,\theta_2)$ constructed from θ_1,θ_2 and members of $\mathscr{D}$ using $\Pi, \langle\ \ \rangle, [\ \ ]$ such that $\varphi\psi = \mathscr{V}(\varphi,\psi)$ for all φ,ψ.

Hint. Suppose there is such an expression $\mathscr{V}$. Define $lh(I) = 0$, $lh(\alpha L) = lh(\alpha R) = lh(\alpha) + 1$. Let $\mathscr{D}_0$ be the finite set consisting of L and all the members of $\mathscr{D}$ to occur in $\mathscr{V}$. Take $n > lh(\alpha)$ for all $\alpha \in \mathscr{D}_0$. Show that if $\varphi \in cl(\mathscr{D}_0/\Pi, \langle\ \ \rangle, [\ \ ]) \cap \mathscr{D}$, then $lh(\varphi) < n$. Therefore, $L^n \notin cl(\mathscr{D}_0/\Pi, \langle\ \ \rangle, [\ \ ])$, which is a contradiction.

Exercise 6.12. Prove that the operation Δ cannot be expressed in terms of $\circ$, Π, $\langle\ \ \rangle$ and I, L, R.

Hint. Take $\mathscr{N} = \{x/x = \bar{k}_1 \ldots \bar{k}_n \& n, k_1, \ldots, k_n \in \omega\}$ $(= \mathscr{D} \setminus R\mathscr{D})$ and define $lh(I) = 0$, $lh(x\bar{n}) = lh(x) + 1$. Show that if $\varphi \in cl(I, L, R/\circ, \Pi, \langle\ \ \rangle)$, then $\exists n \forall x (lh(x) = n \Rightarrow x\varphi \in \mathscr{N})$. Therefore, $\Delta(I,(I,I)) \notin cl(I, L, R/\circ, \Pi, \langle\ \ \rangle)$.

Exercise 6.13. Show that the operation $[\ \ ]$ can not be expressed in terms of $\circ, \Pi, \Delta$ and I, L, R.

Hint. Take $\mathscr{N}$ as above and show that if $\varphi \in cl(I, L, R/\circ, \Pi, \Delta)$, then $\forall x \exists y (yx\varphi \in \mathscr{N})$. Therefore, $[R] \notin cl(I, L, R/\circ, \Pi, \Delta)$.

We give a detailed proof of the following Translation Independence Theorem since it introduces an important new technique called the *unwinding method.*

Exercise 6.14*.** Show that the operation $\langle\ \ \rangle$ can not be expressed in terms of $\circ, \Pi, [\ \ ]$ and I, L, R.

Hint. By way of contradiction, take $\alpha \in \mathscr{D} \setminus \{I\}$ and suppose $\langle\alpha\rangle \in cl(I, L, R/\circ, \Pi, [\ \ ])$. Propositions 6.8, 6.9, 6.15, 6.16 imply that $\langle\alpha\rangle = \bar{1}[\varphi]$ for a certain $\varphi \in cl(I, L, R/\circ, \Pi)$. Let $m, \varphi_0, \ldots, \varphi_m$ correspond to φ by 4.12 and $l > \max\{k/\exists i \leq m(\alpha\bar{k} = L\varphi_i[\varphi])\}$. (The last set is finite by 4.4.)

For each β such that $R^l\langle\alpha\rangle = \beta\varphi[\varphi]$ we construct a β_1 such that $\beta\varphi = \beta_1 R$ and $R^l\langle\alpha\rangle = \beta_1\varphi[\varphi]$. It follows from 4.12 that either $\beta\varphi \in \mathscr{D}$ or $\beta\varphi = \varphi_i$ for a certain $i \leq m$. The latter implies $R^l\langle\alpha\rangle = \beta\varphi[\varphi] = \varphi_i[\varphi]$; hence $\alpha\bar{l} = L\varphi_i[\varphi]$ contrary to the choice of l. Therefore, $\beta\varphi \in \mathscr{D}$. Supposing $\beta\varphi = I$, we get $\alpha\bar{l} = L\beta\varphi[\varphi] = I$ contrary to 4.4. Suppose that $\beta\varphi = \beta_1 L$. Then $R^l\langle\alpha\rangle = \beta_1 L[\varphi] = \beta_1$; hence $\langle\alpha\rangle \in \mathscr{D}$, which leads easily to a contradiction. Therefore,

$\beta\varphi = \beta_1 R$ for some β_1, namely $\beta_1 = \beta\varphi(I, I)$. It also follows that $R^l\langle\alpha\rangle = \beta\varphi[\varphi] = \beta_1 R[\varphi] = \beta_1\varphi[\varphi]$.

Take $\beta_0 = R^l L$. Then $\langle\alpha\rangle = \bar{1}[\varphi]$ implies $R^l\langle\alpha\rangle = \beta_0\varphi[\varphi]$ and we obtain by the construction suggested above an infinite sequence $\{\beta_n\}$ such that $\beta_n\varphi = \beta_{n+1}R$ for all n. The regular segment $\mathscr{E} = \{\theta/\forall n(\beta_n R\theta \leq O)\}$ is closed under $\lambda\theta.(I, \varphi\theta)$, hence $[\varphi]\in\mathscr{E}$. Therefore, $\alpha\bar{1} = L\beta_0 R[\varphi] \leq LO = O$, which is not the case. Thus the proof is complete.

It is worth mentioning that the above argument depends on the following corollary to the axiom μA_3. For any sequence $\{\alpha_n\}$,

$$(\pounds\pounds\pounds) \quad \forall n(\alpha_n\varphi = \alpha_{n+1}R) \Rightarrow \alpha_0 R[\varphi] = O.$$

CHAPTER 7

Relative recursiveness

In this chapter we introduce several abstract concepts of effective computability in IOS. The basic notion is that of relative recursiveness, although the others also play an important role. Some simple properties and characterizations of these notions are established. Both elements and mappings are studied.

Given an IOS $\mathscr{S} = (\mathscr{F}, I, \Pi, L, R)$, a subset $\mathscr{B}$ of $\mathscr{F}$ and an element $\varphi \in \mathscr{F}$, we say that

φ is *polynomial in* $\mathscr{B}$ iff
$\varphi \in cl(\{L, R\} \cup \mathscr{B}/\circ, \Pi)$,
φ is *primitive in* $\mathscr{B}$ iff
$\varphi \in cl(\{L, R\} \cup \mathscr{B}/\circ, \Pi, \langle\ \ \rangle)$,
φ is *primitive recursive in* $\mathscr{B}$ iff
$\varphi \in cl(\{L, R\} \cup \mathscr{B}/\circ, \Pi, \Delta)$,
prime φ is ~~*primitive*~~ *recursive in* $\mathscr{B}$ iff
$\varphi \in cl(\{L, R\} \cup \mathscr{B}/\circ, \Pi, [\ \ ])$,
φ is *recursive in* $\mathscr{B}$ iff
$\varphi \in cl(\{L, R\} \cup \mathscr{B}/\circ, \Pi, \langle\ \ \rangle, [\ \ ])$.

Our notion of prime recursiveness corresponds to the notion of recursiveness considered in Georgieva [1980].

The elements polynomial (primitive, primitive recursive, prime recursive, recursive) in $\varnothing$ are called *polynomial* (respectively *primitive*, *primitive recursive*, *prime recursive*, *recursive*). For example, the elements A, B introduced in the previous chapter are polynomial, while C, D, G, P are primitive recursive and I, O, Q are recursive.

Proposition 7.1. If φ is polynomial in $\mathscr{B}$, then φ is both primitive in $\mathscr{B}$ and prime recursive in $\mathscr{B}$. If φ is primitive in $\mathscr{B}$, then φ is primitive recursive in $\mathscr{B}$. If φ is primitive recursive in $\mathscr{B}$ or prime recursive in $\mathscr{B}$, then φ is recursive in $\mathscr{B}$.

This follows from the definitions, using 6.32.

Proposition 7.2. If φ is polynomial (respectively primitive, primitive recursive, prime recursive, recursive) in $\mathscr{B}$ and $\mathscr{B} \subseteq \mathscr{B}_1$, then φ is polynomial (primitive etc.) in $\mathscr{B}_1$.

This follows from the corresponding definitions.

Proposition 7.3. Let φ be polynomial (primitive etc.) in $\mathscr{B}$. If all the members of $\mathscr{B}$ are polynomial (primitive etc.) in $\mathscr{B}_1$, then so is φ.

This follows by a trivial induction on the construction of φ.

Proposition 7.4. If φ is polynomial (primitive etc.) in $\mathscr{B}$, then φ is polynomial (primitive etc.) in a finite subset of $\mathscr{B}$.

Proposition 7.5. An element φ is prime recursive in $\mathscr{B}$ iff

$$\varphi \in cl(\{L, A\} \cup \mathscr{B}/\circ, [\ \ ]),$$

while φ is recursive in $\mathscr{B}$ iff

$$\varphi \in cl(\{L, A\} \cup \mathscr{B}/\circ, \langle\ \ \rangle, [\ \ ]).$$

This follows from 6.9 and the equality $R = LA$.

Proposition 7.6. An element φ is primitive recursive in $\{I\} \cup \mathscr{B}$ iff $\varphi \in cl(\{R, B, I\} \cup \mathscr{B}/\circ, \Delta)$.

This follows from 6.29 and the equality $L = RB$.

Proposition 7.7. An element φ is recursive in $\mathscr{B}$ iff

$$\varphi \in cl(\{R, B\} \cup \mathscr{B}/\circ, \langle\ \ \rangle, [\ \ ]).$$

This follows from 5.12, 6.29.

Proposition 7.8. If φ is primitive in $\mathscr{B}$, then

$$\langle\varphi\rangle \in cl(\{C, P, Q, \langle L\rangle, \langle R\rangle\} \cup \langle\mathscr{B}\rangle/\circ, \Pi)$$

(where $\langle\mathscr{B}\rangle$ stands for $\{\langle\psi\rangle/\psi \in \mathscr{B}\}$).

Proof. By induction on the construction of φ.

If $\varphi \in \{L, R\} \cup \mathscr{B}$, then $\langle\varphi\rangle \in \{\langle L\rangle, \langle R\rangle\} \cup \langle\mathscr{B}\rangle$.

If φ, ψ have the required property, then so do $\varphi\psi$, (φ, ψ) and $\langle\varphi\rangle$ since $\langle\varphi\psi\rangle = \langle\varphi\rangle\langle\psi\rangle$, $\langle(\varphi, \psi)\rangle = C(\langle\varphi\rangle, \langle\psi\rangle)$ and $\langle\langle\varphi\rangle\rangle = P\langle\varphi\rangle Q$.

Proposition 7.9. If φ is primitive in $\{I\} \cup \mathscr{B}$, then φ is polynomial in $\{I, C, P, Q, \langle L\rangle, \langle R\rangle\} \cup \langle\mathscr{B}\rangle$.

This follows from 7.8 since $\varphi = L\langle\varphi\rangle(I, I)$.

Proposition 7.10. If φ is recursive in $\mathscr{B}$, then

$$\langle\varphi\rangle \in cl(\{C, P, Q, \langle L\rangle, \langle R\rangle\} \cup \langle\mathscr{B}\rangle/\circ, \Pi, [\ \ ]).$$

Proof. We complete the proof of 7.8 by considering one more case. Namely, whenever φ has the required property, then so does $[\varphi]$ since $\langle[\varphi]\rangle = C[\langle\varphi\rangle C]$.

Proposition 7.11 (Pull Back Theorem). The following are equivalent.

(1) φ is recursive in $\mathscr{B}$.

(2) $\varphi \in cl(\{L, C, P, Q, \langle L\rangle, \langle R\rangle\} \cup \langle\mathscr{B}\rangle/\circ, [\ \ ])$.

(3) $\varphi \in cl(\{L, \langle B \rangle, \langle\langle L \rangle\rangle, \langle\langle A \rangle\rangle\} \cup \langle \mathscr{B} \rangle / \circ, [\ \ ])$.
(4) $\varphi \in cl(\{L, \langle C \rangle, \langle\langle L \rangle\rangle, \langle\langle R \rangle\rangle\} \cup \langle \mathscr{B} \rangle / \circ, [\ \ ])$.

Proof. The implication (4)$\Rightarrow$(1) is immediate. We shall prove (1)$\Rightarrow$(2)$\Rightarrow$(3)$\Rightarrow$(4).

The implication (1)$\Rightarrow$(2) follows from 7.10 since $\varphi = L\langle \varphi \rangle [L]$; the operation Π is omitted by 6.9 and exercise 6.1a.

Assume (2). The equalities $C = \langle B \rangle^2 [\langle A \rangle^2]$, $\langle A \rangle = L \langle\langle A \rangle\rangle [L]$ and exercise 6.5a give

$$C, P, Q \in cl(L, \langle B \rangle, \langle\langle L \rangle\rangle, \langle\langle A \rangle\rangle / \circ, [\ \ ]),$$

which implies (3).

Suppose (3). We have

$$\langle B \rangle = \langle (LR, L) \rangle = C(\langle L \rangle \langle R \rangle, \langle L \rangle),$$
$$\langle A \rangle = \langle (R, RL) \rangle = C(\langle R \rangle, \langle RL \rangle).$$

Using 6.9, we get

$$\langle\langle A \rangle\rangle = \langle C \rangle C(\langle\langle R \rangle\rangle, \langle\langle RL \rangle\rangle) = \langle C \rangle CA[\langle\langle R \rangle\rangle L^2][\langle\langle RL \rangle\rangle L],$$

which implies (4) by exercise 6.1a. This completes the proof.

We now construct several primitive recursive elements to be used subsequently.

Proposition 7.12. Let $H = \Delta(\langle I \rangle, \langle I \rangle)$. Then $\bar{n}H = \langle I \rangle$ for all n.

Proof. H is primitive recursive since $\langle I \rangle = \Delta(L, R)$. It follows that $\bar{n}H = \langle I \rangle \langle I \rangle^n = \langle I \rangle$.

Proposition 7.13. $\bar{n}G = \langle \bar{n} \rangle$ for all n.

Proof.

$$\bar{n}G = \langle L \rangle \langle R \rangle^n = \langle LR^n \rangle = \langle \bar{n} \rangle.$$

Proposition 7.14. Let $T = \langle C \rangle C(H, GH)$. Then $\bar{m}\bar{n}T = (\bar{m}, \bar{n})$ for all m, n.

Proof.

$$\bar{m}\bar{n}T = \bar{m}C\bar{n}C(H, GH) = (\bar{m}\bar{n}H, \bar{m}\bar{n}GH) = (\bar{m}\langle I \rangle, \bar{m}\langle \bar{n} \rangle H)$$
$$= (\bar{m}, \bar{n}\langle I \rangle) = (\bar{m}, \bar{n}).$$

It is convenient to write $\langle \varphi \rangle_0$ for φ and $\langle \varphi \rangle_{n+1}$ for $\langle\langle \varphi \rangle\rangle_n$. Notice that whenever $m < n$, then $\langle R \rangle_m \langle \varphi \rangle_n = \langle \varphi \rangle_n \langle R \rangle_m$. Moreover, $\bar{s}_1 \ldots \bar{s}_n \langle \varphi \rangle_n = \varphi \bar{s}_1 \ldots \bar{s}_n$ for all $n, s_1, \ldots, s_n$.

Proposition 7.15. Let $G_0 = \langle I \rangle$ and $G_{n+1} = G_n \langle G \rangle_n$. Then $\bar{m}G_n = \langle \bar{m} \rangle_n$ for all m, n.

Proof. We have

$$\bar{m}G_0 = \bar{m} = \langle \bar{m} \rangle_0, \bar{m}G_{n+1} = \bar{m}G_n \langle G \rangle_n = \langle \bar{m} \rangle_n \langle G \rangle_n$$
$$= \langle \bar{m}G \rangle_n = \langle\langle \bar{m} \rangle\rangle_n = \langle \bar{m} \rangle_{n+1}.$$

Proposition 7.16. Let $D_1 = D$ and $D_{n+1} = \langle D_n \rangle D \langle G_n \rangle$ for $n > 0$. Then $\overline{s_1}\ldots\overline{s_n} D_n = \overline{s_1}\ldots\overline{s_n}\,\overline{s_1}\ldots\overline{s_n}$ for all $s_1, \ldots, s_n$.

Proof. Proposition 6.41 gives $\overline{s_1} D_1 = \overline{s_1}\,\overline{s_1}$, while $\overline{s_1}\ldots\overline{s_{n+1}} D_{n+1} = \overline{s_1}\ldots\overline{s_n}\,\overline{s_1}\ldots\overline{s_{n+1}} G_n \overline{s_{n+1}} = \overline{s_1}\ldots\overline{s_n}\,\overline{s_1}\ldots\overline{s_n} \langle \overline{s_{n+1}} \rangle_n \overline{s_{n+1}} = \overline{s_1}\ldots\overline{s_{n+1}}\,\overline{s_1}\ldots\overline{s_{n+1}}$.

Proposition 7.17. Let $C_1 = C$ and $C_{n+1} = \langle C_n \rangle C$ for $n > 0$. Then $\overline{s_1}\ldots\overline{s_n} C_n = (\overline{s_1}\ldots\overline{s_n} L, \overline{s_1}\ldots\overline{s_n} R)$ for all $s_1, \ldots, s_n$.

Proof. Proposition 6.35 gives $\overline{s_1} C_1 = (\overline{s_1} L, \overline{s_1} R)$, while $\overline{s_1}\ldots\overline{s_{n+1}} C_{n+1} = \overline{s_1}\ldots\overline{s_n} C_n \overline{s_{n+1}} C = (\overline{s_1}\ldots\overline{s_n} L, \overline{s_1}\ldots\overline{s_n} R)(\overline{s_{n+1}} L, \overline{s_{n+1}} R) = (\overline{s_1}\ldots\overline{s_{n+1}} L, \overline{s_1}\ldots\overline{s_{n+1}} R)$, completing the proof.

A notion of relative recursiveness for mappings is introduced by parametrizing the corresponding notion for elements. Namely, the unary mappings Γ *recursive in* $\mathscr{B}$ are defined inductively as follows.

1. The mappings $\Gamma = \lambda\theta.\theta$ and $\Gamma = \lambda\theta.\psi$, $\psi \in \{L, R\} \cup \mathscr{B}$, are recursive in $\mathscr{B}$.
2. If $\Gamma_1, \Gamma_2 : \mathscr{F} \to \mathscr{F}$ are recursive in $\mathscr{B}$, then so are $\Gamma = \lambda\theta.\Gamma_1(\theta)\Gamma_2(\theta)$, $\Gamma = \lambda\theta.(\Gamma_1(\theta), \Gamma_2(\theta))$, $\Gamma = \lambda\theta.\langle \Gamma_1(\theta) \rangle$ and $\Gamma = \lambda\theta.[\Gamma_1(\theta)]$. In other words, $\Gamma : \mathscr{F} \to \mathscr{F}$ is recursive in $\mathscr{B}$ iff for all θ the element $\Gamma(\theta)$ is uniformly recursive in $\{\theta\} \cup \mathscr{B}$.

To define the n-ary mappings $\Gamma : \mathscr{F}^n \to \mathscr{F}$ recursive in $\mathscr{B}$, just take $\Gamma = \lambda\theta_1\ldots\theta_n.\theta_i$, $1 \le i \le n$, and $\Gamma = \lambda\theta_1\ldots\theta_n.\psi$, $\psi \in \{L, R\} \cup \mathscr{B}$, in the first clause above.

Proposition 6.9 implies an equivalent definition with A substituted for R in the first clause and the case of Π dropped in the second one. Other equivalent definitions can be obtained by using 6.29 and (a parametrized version of) 7.11.

In particular, one may take the mappings $\Gamma = \lambda\theta_1\ldots\theta_n.\langle \theta_i \rangle$, $1 \le i \le n$, and $\Gamma = \lambda\theta_1\ldots\theta_n.\psi$, $\psi \in \mathscr{B}_0 \cup \langle \mathscr{B} \rangle$, in the first clause, omitting the cases of Π, $\langle\ \ \rangle$ in the second clause, where $\mathscr{B}_0$ is a set of primitive elements such that $cl(\mathscr{B}_0/\circ, [\ \ ])$ is the set of all recursive elements. (Therefore, $cl(\mathscr{B}_0 \cup \langle \mathscr{B} \rangle/\circ, [\ \ ])$ is exactly the set of all the elements recursive in $\mathscr{B}$.) For instance, $\mathscr{B}_0 = \{L, \langle B \rangle, \langle L \rangle_2, \langle A \rangle_2\}$ would do by 7.11. The set $\mathscr{B}_0$ may be assumed finite without loss of generality. Actually,

$$L, \langle B \rangle, \langle L \rangle_2, \langle A \rangle_2 \in cl(\mathscr{B}_0/\circ, [\ \ ]),$$

hence there is a finite subset $\mathscr{B}_1$ of $\mathscr{B}_0$ such that

$$L, \langle B \rangle, \langle L \rangle_2, \langle A \rangle_2 \in cl(\mathscr{B}_1/\circ, [\ \ ]).$$

Therefore, $\mathscr{B}_0$ may be replaced by $\mathscr{B}_1$.

Notions of mappings *polynomial, primitive, primitive recursive* and *prime recursive in* $\mathscr{B}$ are introduced in a similar way, while certain equivalent definitions are implied 7.5, 7.6. If $\mathscr{B} = \varnothing$, then one has absolute recursiveness, prime recursiveness etc.

The elementary properties 7.1–7.4 hold for mappings as well. Three properties germane to mappings follow.

Proposition 7.18. If $\Gamma_0:\mathscr{F}^m \to \mathscr{F}$ and $\Gamma_1,\dots,\Gamma_m:\mathscr{F}^n \to \mathscr{F}$ are recursive in $\mathscr{B}$, then so is

$$\Gamma = \lambda\theta_1\dots\theta_n.\Gamma_0(\Gamma_1(\theta_1,\dots,\theta_n),\dots,\Gamma_m(\theta_1,\dots,\theta_n)).$$

This follows by an easy induction on the construction of Γ_0.

Proposition 7.19. A n-ary mapping Γ is recursive in $\psi_1,\dots,\psi_m$ iff there is a $m+n$-ary recursive mapping Γ^* such that

$$\Gamma = \lambda\theta_1\dots\theta_n.\Gamma^*(\psi_1,\dots,\psi_m,\theta_1,\dots,\theta_n).$$

Proof. The 'if'-part follows by 7.18. The 'only if'-part is proved by a trivial induction on the construction of Γ.

Therefore, in studying n-ary mappings it suffices to confine attention to absolutely recursive mappings. On the other hand, one may consider only *unary* mappings when studying the relative recursiveness, because a pairing scheme Π, L, R for $\mathscr{F}$ is available.

Proposition 7.20. Let $\Gamma:\mathscr{F}^n \to \mathscr{F}$, $n > 1$, and

$$\Gamma^* = \lambda\theta.\Gamma(\bar{0}\theta,\dots,\overline{n-2}\theta, R^{n-1}\theta).$$

Then $\Gamma = \lambda\theta_1\dots\theta_n.\Gamma^*((\theta_1,\dots,\theta_n))$ and Γ is recursive in $\mathscr{B}$ iff so is Γ^*.

This follows from 7.18.

The following Transition Theorem shows that all the unary mappings recursive in $\mathscr{B}$ have the *transition property* established in 6.42 for [ ].

Proposition 7.21. If Γ is a unary mapping recursive in $\mathscr{B}$, then there is a mapping Γ^* recursive in $\mathscr{B}$ such that $\Gamma^*(\theta) = (\Gamma(L\theta), \Gamma^*(R\theta))$ for all θ. In particular, $\bar{n}\Gamma^*(\theta) = \Gamma(\bar{n}\theta)$ for all θ, n.

Proof. By induction on the construction of Γ.

Let $\Gamma = \lambda\theta.\langle\theta\rangle$. Taking $\Gamma^* = \lambda\theta.G\langle\theta\rangle$, we get

$$\Gamma^*(\theta) = (\langle L\rangle, G\langle R\rangle)\langle\theta\rangle = (\langle L\theta\rangle, G\langle R\theta\rangle) = (\Gamma(L\theta), \Gamma^*(R\theta)).$$

Let $\Gamma = \lambda\theta.\psi$, $\psi \in \mathscr{B}_0 \cup \langle\mathscr{B}\rangle$. Taking $\Gamma^* = \lambda\theta.\Delta(\psi, I)$, we get

$$\Gamma^*(\theta) = (\psi, \Delta(\psi, I)) = (\Gamma(L\theta), \Gamma^*(R\theta)).$$

Let Γ_1^*, Γ_2^* correspond to Γ_1, Γ_2.

If $\Gamma = \lambda\theta.\Gamma_1(\theta)\Gamma_2(\theta)$, then take $\Gamma^* = \lambda\theta.D\langle\Gamma_1^*(\theta)\rangle\Gamma_2^*(\theta)$. It follows that

$$\begin{aligned}\Gamma^*(\theta) &= (L^2, D\langle R\rangle R)\langle\Gamma_1^*(\theta)\rangle\Gamma_2^*(\theta)\\ &= (L\Gamma_1^*(\theta)L\Gamma_2^*(\theta), D\langle R\Gamma_1^*(\theta)\rangle R\Gamma_2^*(\theta))\\ &= (\Gamma_1(L\theta)\Gamma_2(L\theta), D\langle\Gamma_1^*(R\theta)\rangle\Gamma_2^*(R\theta)) = (\Gamma(L\theta), \Gamma^*(R\theta)).\end{aligned}$$

If $\Gamma = \lambda\theta.[\Gamma_1(\theta)]$, then take $\Gamma^* = \lambda\theta.\rho[D\langle\Gamma_1^*(\theta)\rangle\rho]$, where $\rho = C([I]L, R)$. Using 6.42, we get

$$\begin{aligned}\Gamma^*(\theta) &= ([L\Gamma_1^*(\theta)], \rho[D\langle R\Gamma_1^*(\theta)\rangle\rho]) = ([\Gamma_1(L\theta)], \rho[D\langle\Gamma_1^*(R\theta)\rangle\rho])\\ &= (\Gamma(L\theta), \Gamma^*(R\theta)),\end{aligned}$$

which completes the proof.

Notice that the mapping Γ^* constructed in 7.21 also satisfies the equality $\langle I\rangle\Gamma^*(\theta)=\Gamma^*(\theta)$ for all θ.

EXERCISES TO CHAPTER 7

Exercise 7.1. Let I^* satisfy (£)*. Prove that I^* is recursive.
Hint. Use exercise 6.9 and propositions 6.39, 6.38.

Exercise 7.2 (Modification Lemma). Let $W, L_1, R_1 \in \mathscr{F}$ and $L_1 W = L$, $R_1 W = R$. (In particular, whenever Π_1, L_1, R_1 is another pairing scheme for $\mathscr{F}$, then $W=(L,R)_1$ and L_1, R_1 satisfy these assumptions.) Taking $(\varphi,\psi)_1 = W(\varphi,\psi)$ show that $\mathscr{S}_1=(\mathscr{F}, I, \Pi_1, L_1, R_1)$ is an IOS and φ is recursive (prime recursive) in $\{L_1, R_1, (L,R)_1\}\cup\mathscr{B}$ iff φ is recursive$_1$ (prime recursive$_1$) in $\{L, R, (L_1, R_1)\}\cup\mathscr{B}$.
Hint. Show that $\langle\ \ \rangle_1=\lambda\varphi.\mu\theta.(\varphi L_1 L, R)_1(I,\theta R_1)$ satisfies (£) by exercise 6.2 and $[\ \ ]_1=\lambda\varphi.(L,R)_1[\varphi(L,R)_1]$ satisfies (££), then use exercise 4.6.

Exercise 7.3. Assuming $\langle I\rangle=I$ (cf. the comments to 6.22), show that the following are equivalent:

(1) φ is recursive in $\mathscr{B}$.
(2) $\varphi\in cl(\{L,P,Q,\langle L\rangle,\langle R\rangle\}\cup\langle\mathscr{B}\rangle/\circ,\Pi,[\ \ ])$.
(3) $\varphi\in cl(\{L,\langle L\rangle_2,\langle A\rangle_2\}\cup\langle\mathscr{B}\rangle/\circ,[\ \ ])$.
(4) $\varphi\in cl(\{L,\langle L\rangle_3,\langle R\rangle_3\}\cup\langle\mathscr{B}\rangle/\circ,\Pi,[\ \ ])$.

Hint. Use 7.11, exercise 6.4, the equality $\langle B\rangle=C(\langle LR\rangle,\langle L\rangle)$ and exercise 6.5b.

Exercise 7.4. Show that propositions 7.18–7.20 remain valid with 'prime recursive' (or 'primitive recursive', 'primitive', 'polynomial') substituted for 'recursive'.
The following exercise establishes a Parametrized Transition Theorem.

Exercise 7.5. Let Γ be a binary mapping recursive in $\mathscr{B}$. Prove that there is a mapping Γ^* recursive in $\mathscr{B}$ such that

$$\Gamma^*(\theta,\theta_1)=(\Gamma(\theta,L\theta_1),\Gamma^*(\theta,R\theta_1))\quad\text{for all }\theta,\theta_1.$$

Hint. Let $\Gamma_1=\lambda\theta.\Gamma(L\theta,R\theta)$ and Γ_1^* correspond to Γ_1 by 7.21. Then take $\Gamma^*=\lambda\theta\theta_1.\Gamma_1^*(C(\Delta(\theta,I),\theta_1))$.

Exercise 7.6. Construct a binary primitive recursive mapping Δ^* such that $\Delta^*(\theta_1,\theta_2)=(\Delta(L\theta_1,L\theta_2),\ \Delta^*(R\theta_1,R\theta_2))$ for all θ_1,θ_2. Using Δ^*, show that the Transition Theorem holds for relative primitive recursiveness, provided I is added to the initial elements.
Hint. Take $\Delta^*=\lambda\theta_1\theta_2.G(\theta_1,\Delta(D\langle\theta_1\rangle,D\langle\theta_2\rangle)\theta_2)$.

The next six exercises study 'multiple-valued' spaces (e.g. examples 4.8 but not 4.1, 4.3, 4.6, 4.7), in which L, R have an upper bound U. The element U will make it possible to discribe relative partial recursiveness and Friedman's computability by effective definitional schemes. Abstract partial recursiveness in spaces with elements U is studied in greater detail in Ivanov [1987].

Exercise 7.7. Let $L, R \le U$. Construct an element $\bar{\omega}$ recursive in U such that $\bar{n} \le \bar{\omega}$ for all n.

Hint. Take $\bar{\omega} = \bar{1}[\rho]$, where $\rho = \langle U \rangle C(L, R^2)$. Using the inequalities $\varphi = L(\varphi, \psi) \le U(\varphi, \psi)$ and $\psi \le U(\varphi, \psi)$, show that $\bar{n} \le \overline{n+1}\,[\rho]$ and $\overline{n+2}\,[\rho] \le \overline{n+1}\,[\rho]$, hence $\bar{n} \le \overline{n+1}\,[\rho] \le \bar{1}[\rho]$.

Exercise 7.8. Show that the following are equivalent.

(1) $L, R \le U$, $U(I, I) \le I$.
(2) $U(\varphi, \psi) = \sup\{\varphi, \psi\}$ for all φ, ψ.

Exercise 7.9*.** Let $L, R \le U$ and $U(I, I) \le I$. Construct an element $\bar{\omega}$ recursive in U such that $\bar{\omega}\varphi = \sup_n \bar{n}\varphi$ for all φ.

Hint. Take $\bar{\omega}$ as in exercise 7.7. Supposing $\bar{n}\varphi \le \tau$ for all n, show that $[\rho] \in \{\theta / L\theta \le I \,\&\, \forall n (\overline{n+1}\,\theta\varphi \le \tau)\}$.

Remark. This exercise ensures that whenever $\mathscr{U}$ is a subset of $\mathscr{F}$ of the form $\{\bar{n}\sigma / n \in \omega\}$, then $\bar{\omega}\sigma = \sup \mathscr{U}$, e.g. $\bar{\omega} = \sup_n \bar{n}$, $\bar{\omega}\Delta(I, \varphi) = \sup_n \varphi^n$, $\bar{\omega}\Delta(\varphi, \psi)\chi = \sup_n \varphi\psi^n\chi$ etc.

Exercise 7.10. Prove that the following are equivalent.

(1) $L, R \le U$, $U(I, I) \le I$, $LU \le UB^2, RU \le UA^2$.
(2) $\alpha U(\varphi, \psi) = \sup\{\alpha\varphi, \alpha\psi\}$ for all α, φ, ψ.

Notice that all the inequalities in (1) apart from $L, R \le U$ are in fact equalities.

Exercise 7.11. Let $\mathscr{S}$ be the IOS of example 4.8. Show that the element $U = \lambda s.\{L(s), R(s)\}$ satisfies condition (1) of exercise 7.10.

Hint. Show that whenever $\mathscr{H} \subseteq \mathscr{F}$ and $\varphi = \cup \mathscr{H}$, then $\psi\varphi\chi = \sup\{\psi\theta\chi / \theta \in \mathscr{H}\}$ for all ψ, χ. Therefore, condition (2) of exercise 7.10 is satisfied.

Exercise 7.12*.** Let U satisfy condition (1) of exercise 7.10. Construct an element $\bar{\omega}$ recursive in U such that $\alpha\bar{\omega}\varphi = \sup_n \alpha\bar{n}\varphi$ for all α, φ.

Hint. Take $\bar{\omega}$ as in exercise 7.7 or, as suggested by N. Georgieva, $\bar{\omega} = \bar{1}[U(L, R^2)]$.

While O is the greatest (and unique) lower bound of L, R in the examples considered so far, we shall see in chapters 25, 30 that sometimes these elements have a greatest lower bound V satisfying the additional assumptions considered below.

Exercise 7.13. Show that the following are equivalent.

(1) $V \leq L, R$, $I \leq V(I, I)$.
(2) $V(\varphi, \psi) = \inf\{\varphi, \psi\}$ for all φ, ψ.

The element O does not satisfy (1) since $O(I, I) < I$.

Exercise 7.14. Show that the following are equivalent.

(1) $V \leq L, R$, $I \leq V(I, I)$, $VB^2 \leq LV$, $VA^2 \leq RV$.
(2) $\alpha V(\varphi, \psi) = \inf\{\alpha\varphi, \alpha\psi\}$ for all α, φ, ψ.

As in the case of U, all the inequalities in (1) except $V \leq L, R$ are in fact equalities. However, there is no analogue to exercises 7.9***, 7.12***, so one can only construct greatest lower bounds of finite subsets of $\mathscr{F}$.

CHAPTER 8

Representation theorems

Three theorems on the representability of primitive recursive and partial recursive number theoretic functions are established in this chapter, providing a way of incorporating Ordinary Recursion Theory within the general axiomatic theory.

Let $f:\omega^n \dashrightarrow \omega$, $n>0$. An element φ *weakly represents* f iff $\bar{s_1}\ldots\bar{s_n}\varphi = \overline{f(s_1,\ldots,s_n)}$ whenever $f(s_1,\ldots,s_n)\downarrow$. An element φ *represents* f iff φ weakly represents f and whenever $f(s_1,\ldots,s_n)\uparrow$, then $\bar{s_1}\ldots\bar{s_n}\varphi = O$. We write simply $\bar{s_1}\ldots\bar{s_n}\varphi = \overline{f(s_1,\ldots,s_n)}$, meaning $\overline{f(s_1,\ldots,s_n)} = O$ whenever $f(s_1,\ldots,s_n)\uparrow$.

Notice that whenever f is an extension of g and φ weakly represents f, then φ weakly represents g. Obviously, whenever f is a total function, then φ weakly represents f iff φ represents f.

The following statement is called the Representation Theorem for primitive recursive functions.

Proposition 8.1. Let Ψ be a set of partial number theoretic functions; let $\mathscr{B} \subseteq \mathscr{F}$ and suppose that all the functions in Ψ are representable by members of $\mathscr{B}$. Let f be a function primitive recursive in Ψ. Then there is an element φ primitive recursive in $\mathscr{B}$ such that φ represents f. In particular, all primitive recursive functions are representable by primitive recursive elements.

Proof. By induction on the construction of f.

1. If $f = \lambda s.0$, then take $\varphi = \langle \bar{0} \rangle H$. It follows that $\bar{s}\varphi = \bar{0}\bar{s}H = \bar{0}\langle I \rangle = \bar{0}$ for all s.

2. If $f = \lambda s.s+1$, then take $\varphi = R$. It follows that $\bar{s}\varphi = \bar{s}R = \overline{s+1}$ for all s.

3. If $f = \lambda s_1 \ldots s_n.s_i$, $1 \le i \le n$, then take $\varphi = H^{n-i}G_{i-1}H^{i-1}$. It follows that

$$\bar{s_1}\ldots\bar{s_n}\varphi = \bar{s_1}\ldots\bar{s_i}G_{i-1}H^{i-1} = \bar{s_i}\bar{s_1}\ldots\overline{s_{i-1}}H^{i-1} = \bar{s_i}.$$

4. If $f \in \Psi$, then f is represented by a member of $\mathscr{B}$.

5. Let h be m-ary, $g_1,\ldots,g_m$ n-ary, their representing elements $\chi, \psi_1,\ldots,\psi_m$ and

$$f = \lambda s_1 \ldots s_n.h(g_1(s_1,\ldots,s_n),\ldots,g_m(s_1,\ldots,s_n)).$$

Then take $\varphi = D_n\langle\psi_1\rangle_n \ldots D_n\langle\psi_{m-1}\rangle_n\psi_m\chi$. The element φ is primitive recursive in $\chi, \psi_1,\ldots,\psi_m$ and

$$\bar{s_1}\ldots\bar{s_n}\varphi = \bar{s_1}\ldots\bar{s_n}\psi_1\ldots\bar{s_1}\ldots\bar{s_n}\psi_m\varphi,$$

hence φ represents f.

6. Let h be unary, g be ternary, their representing elements χ, ψ and

$$f(s,0)=h(s), \quad f(s,t+1)=g(s,t,f(s,t)).$$

Taking $\varphi = D^2\langle R\Delta(\langle I\rangle, (\langle I\rangle, D_2))\rangle\Delta(\chi,\psi)$, we get

$$\begin{aligned}\bar{s}\bar{0}\varphi &= \bar{s}\bar{0}\langle I\rangle(\langle I\rangle, D_2)\chi = \bar{s}\chi, \quad \bar{s}\overline{t+1}\varphi = \bar{s}\overline{t+1}(\langle I\rangle, D_2)^{t+2}\chi\psi^{t+1}\\ &= \bar{s}\bar{t}\bar{s}\bar{t}(\langle I\rangle, D_2)^{t+1}\chi\psi^t\psi = \bar{s}\bar{t}\bar{s}\bar{t}\varphi\psi.\end{aligned}$$

The element φ is primitive recursive in χ, ψ and an easy induction on t shows that φ represents f. This completes the proof.

Throughout proposition 8.1 and its proof 'weakly represent' can be substituted for 'represent'.

Two Representation Theorems for partial recursive functions follow.

Proposition 8.2. Let Ψ be a set of partial number theoretic functions, $\mathscr{B} \subseteq \mathscr{F}$ and suppose that all functions in Ψ are weakly representable by members of $\mathscr{B}$. Let f be a function μ-recursive in Ψ. Then there is an element φ recursive in $\mathscr{B}$ which weakly represents f. In particular, all the *general recursive functions* (i.e., total partial recursive functions) are representable by recursive elements.

Proof. We complete the proof of 8.1 (weak representability version) by adding one more case.

7. Let g be $n+1$-ary, let ψ weakly represent g and

$$f = \lambda s_1 \ldots s_n . \mu t(g(s_1, \ldots, s_n, t) = 0).$$

Taking $\psi_1 = \psi(L, [I]R)$, $\rho = G_n H^n$ (ρ represents $\lambda s_1 \ldots s_{n+1} . s_{n+1}$), $\psi_2 = D_{n+1} \circ \langle \psi_1 \rangle_{n+1} C_{n+1}(\rho L^2, R^2)$ and $\sigma = R[\psi_2]$, we get

$$\begin{aligned}\bar{s}_1 \ldots \bar{s}_n\bar{t}\psi_2 &= \bar{s}_1 \ldots \bar{s}_n\bar{t}\psi_1(\bar{s}_1 \ldots \bar{s}_n\bar{t}L, \bar{s}_1 \ldots \bar{s}_n\bar{t}R)(\rho L^2, R^2)\\ &= \bar{s}_1 \ldots \bar{s}_n\bar{t}\psi_1(\bar{t}L^2, \bar{s}_1 \ldots \bar{s}_n\overline{t+2});\end{aligned}$$

hence

$$\bar{s}_1 \ldots \bar{s}_n\bar{t}\sigma = \bar{s}_1 \ldots \bar{s}_n\bar{t}\psi_1(\bar{t}L, \bar{s}_1 \ldots \bar{s}_n\overline{t+1}\sigma).$$

Take $\varphi = \bar{1}[\sigma]$; this element is recursive in ψ. Suppose that $f(s_1, \ldots, s_n) = t$. Then $g(s_1, \ldots, s_n, r)\downarrow$ for all $r \leq t$, $g(s_1, \ldots, s_n, r) > 0$ for all $r < t$, while $g(s_1, \ldots, s_n, t) = 0$. Therefore, $\bar{s}_1 \ldots \bar{s}_n\bar{t}\psi_1 = L$ and $\bar{s}_1 \ldots \bar{s}_n\bar{r}\psi_1 = R$ for $r < t$; hence

$$\bar{s}_1 \ldots \bar{s}_n\bar{r}\sigma = \bar{s}_1 \ldots \bar{s}_n\overline{r+1}\,\sigma$$

for $r < t$. Finally, it follows that

$$\begin{aligned}\bar{s}_1 \ldots \bar{s}_n\varphi &= \bar{s}_1 \ldots \bar{s}_n\bar{0}\sigma[\sigma] = \bar{s}_1 \ldots \bar{s}_n\bar{t}\sigma[\sigma] = \bar{s}_1 \ldots \bar{s}_n\bar{t}\psi_1(\bar{t}L, \bar{s}_1 \ldots \bar{s}_n\overline{t+1}\sigma)[\sigma]\\ &= \bar{t}L[\sigma] = \bar{t},\end{aligned}$$

which completes the proof.

Proposition 8.3*.** Let Ψ be a set of partial number theoretic functions, let $\mathscr{B} \subseteq \mathscr{F}$ and suppose that all functions in Ψ are representable by members of $\mathscr{B}$. Let f be a function μ-recursive in Ψ. Then there is an element φ recursive in $\mathscr{B}$ which represents f. In particular, all partial recursive functions are representable by recursive elements.

Proof. Using the proofs of 8.1, 8.2, it suffices to complete point 7.

Let g, f, ψ_1, ψ_2, σ, φ be the same as in the previous proof and $f(s_1 \ldots, s_n)\uparrow$. We have to show that $\bar{s_1} \ldots \bar{s_n}\varphi = O$. There are two possibilities.

a. There is a t such that $g(s_1, \ldots, s_n, t)\uparrow$, while $g(s_1, \ldots, s_n, r) > 0$ for $r < t$. It follows that $\bar{s_1} \ldots \bar{s_n}\bar{t}\psi_1 = O$; hence

$$\bar{s_1} \ldots \bar{s_n}\varphi = \bar{s_1} \ldots \bar{s_n}\bar{t}\psi_1(\bar{t}L, \bar{s_1} \ldots \bar{s_n}\overline{t+1}\sigma)[\sigma] = O.$$

b. $g(s_1, \ldots, s_n, r) > 0$ for all r. Therefore, $\bar{s_1} \ldots \bar{s_n}\bar{r}\psi_1 = R$ for all r; hence

$$\bar{s_1} \ldots \bar{s_n}\bar{r}\psi_2 = \bar{s_1} \ldots \bar{s_n}\bar{r}\psi_1(\bar{r}L^2, \bar{s_1} \ldots \bar{s_n}\overline{r+2}) = \bar{s_1} \ldots \bar{s_n}\overline{r+1}R$$

for all r, which implies $\bar{s_1} \ldots \bar{s_1}\bar{0}R[\psi_2] = O$ by (£££). We get

$$\bar{s_1} \ldots \bar{s_n}\varphi = \bar{s_1} \ldots \bar{s_n}\bar{1}[\sigma] = \bar{s_1} \ldots \bar{s_n}\bar{0}\sigma[\sigma] = O[\sigma] = O,$$

which completes the proof.

Propositions 8.1, 8.3*** are reversed in the exercises to ensure that whenever a number theoretic function is representable by a (primitive) recursive element, then it is partial (respectively primitive) recursive.

While it is shown in the exercises below that the functions representable by prime recursive elements are quite trivial, all partial recursive functions are representable by elements prime recursive in several initial ones, such as $C, P, Q, \langle L \rangle, \langle R \rangle$. (The latter assertion follows by 7.11, 8.3***.) Moreover, an examination of the proofs of 8.1–8.3*** shows that the resources of IOS are only partially employed. The properties of $\Pi, [\quad]$ used are $\bar{0}(\varphi, \psi) = \varphi$, $\overline{n+1}(\varphi, \psi) = \bar{n}\psi$, $\bar{0}[\varphi] = I$ and $\overline{n+1}[\varphi] = \bar{n}\varphi[\varphi]$ together with (£££) in case. Instead of $\langle \varphi \rangle_n$ it suffices to have an element φ_n such that $\bar{s_1} \ldots \bar{s_n}\varphi_n = \varphi\bar{s_1} \ldots \bar{s_n}$ for all $s_1, \ldots, s_n$. (If $n > 1$, then $\varphi_n = P^{n-1}\varphi_1 Q^{n-1}$ would do.) Therefore, representability of partial recursive functions could be established in certain simpler algebraic systems. However, we shall not pursue further representability results since the theorems given above are sufficient for our present purposes.

EXERCISES TO CHAPTER 8

Exercise 8.1. Let $J: \omega^2 \to \omega$ be injective and partial recursive. Show that there are recursive elements ρ, σ which correspond to J in the sense of exercise 6.7**, i.e. $\bar{s}\bar{t}\rho = \overline{J(s,t)}$ and $\overline{J(s,t)}\sigma = \bar{s}\bar{t}$ for all s, t.

Hint. The existence of ρ follows by 8.2. Let $f_i = \lambda s.\psi_i(\mu t(J(\psi_0(t), \psi_1(t)) = s))$, $i = 0, 1$, where ψ_0, ψ_1 are the inverses of $\lambda st.2^s(2t+1) - 1$ considered in chapter 2. Take σ_0, σ_1 to weakly represent f_0, f_1 by 8.2, then take $\sigma = D\langle \sigma_0 \rangle \sigma_1$.

It should be mentioned that the existence of elements ρ, σ which correspond to the Cantor's pairing function was assumed as an axiom in an earlier version of IOS. However, D. Skordev proved that the existence of such elements follows from axiom μA_3.

Exercise 8.2*.** Let U satisfy condition (1) of exercise 7.8, Ψ be a set of partial number theoretic functions representable by members of $\mathscr{B} \subseteq \mathscr{F}$ and f be partial recursive in Ψ. Show that f is representable by an element recursive in $\{U\} \cup \mathscr{B}$.

Remark. By definition a n-ary function f is *partial recursive in* Ψ iff there is a $n+2$-ary function g primitive recursive in Ψ such that $f(s_1,\ldots,s_n)=t$ iff $\exists m(g(m,t,s_1,\ldots,s_n)=0)$. If Ψ consists of total functions, then f is partial recursive in Ψ iff f is μ-recursive in Ψ. In particular, f is partial recursive iff f is μ-recursive, which enabled us to introduce the partial recursive functions via Kleene's definition given at the beginning of chapter 2. However, in general the notion of relative partial recursiveness is broader than the notion of relative μ-recursiveness. (Cf. Myhill [1961], Skordev [1963].)

Hint. Let g be a $n+2$-ary function primitive recursive in Ψ and $f(s_1,\ldots, s_n)=t$ iff $\exists m(g(m,t,s_1,\ldots,s_n)=0)$. Take ψ to represent g by 8.1, a recursive element σ such that $\{\overline{k}\sigma/k\in\omega\} = \{\overline{t}\overline{m}/t, m\in\omega\}$ by exercise 8.1, and the element $\bar{\omega}$ of exercise 7.9***. Show that $\langle\bar{\omega}\sigma\langle D\rangle G\rangle_n\psi[R]$ represents f.

The next two exercises show that the unary functions representable by prime recursive elements are exactly those satisfying one of the following conditions.

(1) $\exists l \forall n(g(l+n)=g(l)+n)$.
(2) $\exists l \exists k>0 \forall n \geq l(g(n+k)=g(n))$,
i.e. $\exists l \exists k>0 \forall n(g(l+n)=g(l+\operatorname{rem}(n,k)))$.

Notice that while all such functions are primitive recursive in $\lambda s.\uparrow$, there are primitive recursive functions which satisfy neither (1) nor (2), e.g. $\lambda s.2s$.

Exercise 8.3. Show that if g satisfies (1) or (2), then it is representable by a prime recursive element ψ.

Hint. Let $g(l+n)=g(l)+n$ for all n. If $g(l)\downarrow$, then take $\psi=(\overline{g(0)},\ldots,\overline{g(l-1)}, R^{g(l)})$; if $g(l)\uparrow$, then take $\psi=(\overline{g(0)},\ldots,\overline{g(l-1)},O)$.

Let $g(l+n)=g(l+\operatorname{rem}(n,k))$ for a certain $k>0$ and all n. Then take $\psi=(\overline{g(0)},\ldots,\overline{g(l-1)}, R[(\overline{g(l)}L,\ldots,\overline{g(l+k-1)}L, R)])$.

Exercise 8.4*.** Prove that whenever a prime recursive element ψ represents $g:\omega \dashrightarrow \omega$, then g satisfies either (1) or (2).

Hint. We sketch a proof based on the unwinding method. There is a polynomial element φ such that $\psi=\bar{1}[\varphi]$. Let $m,\varphi_0,\ldots,\varphi_m$ correspond to φ by 4.12. If $\{l/\bar{l}\psi \neq O\}$ is finite, then g satisfies both (1) and (2). Otherwise, take $m+2$ distinct members $l_0,\ldots,l_{m+1}$ of this set. For each $i \leq m+1$ start constructing a sequence $\beta_{i,0},\beta_{i,1},\ldots$ such that $\beta_{i,0}=R^{l_i}L$ and $R^{l_i}\psi=\beta_{i,n}\varphi[\varphi]$. By proposition 4.12, either $\beta_{i,n}\varphi\in\mathscr{D}$ or $\beta_{i,n}\varphi=\varphi_{m_i}$ for a certain $m_i \leq m$. Supposing $\beta_{i,n}\varphi=I$, one gets $\bar{l}_i\psi=I$ contrary to the fact that ψ represents g. If $\beta_{i,n}\varphi=\beta R$, then take $\beta_{i,n+1}=\beta$. If $\beta_{i,n}\varphi=\beta L$ for certain i,n, then stop constructing all the sequences. If $\beta_{i,n}\varphi=\varphi_{m_i}$, then stop constructing the i-th one.

No sequence could be infinite since $\forall n(\beta_{i,n}\varphi = \beta_{i,n+1}R)$ would imply $\overline{l_i}\psi = O$ by (£££) contrary to the choice of l_i.

If $\beta_{i,n}\varphi = \beta L$ for certain i, n, then $R^{l_i}\psi = \beta$, hence $\overline{l_i}\psi = L\beta$. Therefore, $\beta = R^{g(l_i)}$ which implies $\overline{l_i + n}\,\psi = \bar{n}R^{g(l_i)} = \overline{g(l_i) + n}$ for all n, i.e. g satisfies (1).

Finally, one may get $m_0, \ldots, m_{m+1} \leq m$ such that $R^{l_i}\psi = \varphi_{m_i}[\varphi]$ for all $i \leq m+1$, in which case $m_i = m_j$ for certain $i \neq j \leq m+1$, hence $R^{l_i}\psi = R^{l_j}\psi$ and g satisfies (2). The proof is complete.

Similarly, the n-ary functions representable by prime recursive elements can be shown to compose the class Ψ introduced inductively as follows.

1. If g is unary and satisfies (1) or (2), then $g \in \Psi$.
2. If g is $n+1$-ary, $n > 0$, there are $k > 0$ and l such that $\lambda s_1 \ldots s_n . g(s_1, \ldots, s_n, i) \in \Psi$, $i \leq l + k - 1$, and $g(s_1, \ldots, s_n, s + k) = g(s_1, \ldots, s_n, s)$ for all $s \geq l$, then $g \in \Psi$.

Of course, all such functions are primitive recursive in $\lambda s. \uparrow$.

Exercise 8.5*.** Prove that whenever $f: \omega^n \dashrightarrow \omega$ is represented by a recursive element φ, then f is partial recursive. (If necessary, make use of the Normal Form Theorem 9.3 established in the next chapter.)

Hint. The unwinding method helps once again. Take a primitive element ψ such that $\varphi = \bar{1}[\psi]$ by 9.3. Fix a construction of ψ and for all α introduce a set $\mathrm{Sub}(\alpha\psi) \subseteq \mathscr{D}$ ('the sub elements of $\alpha\psi$') as follows.

$$\mathrm{Sub}(\alpha\beta) = \{\alpha\beta\},$$
$$\mathrm{Sub}(\alpha\psi_1\psi_2) = \cup\{\mathrm{Sub}(\beta\psi_2)/\beta \in \mathrm{Sub}(\alpha\psi_1)\},$$
$$\mathrm{Sub}(I(\psi_1, \psi_2)) = \mathrm{Sub}(\psi_1) \cup \mathrm{Sub}(\psi_2),$$
$$\mathrm{Sub}(\alpha L(\psi_1, \psi_2)) = \mathrm{Sub}(\alpha\psi_1),$$
$$\mathrm{Sub}(\alpha R(\psi_1, \psi_2)) = \mathrm{Sub}(\alpha\psi_2),$$
$$\mathrm{Sub}(I\langle\psi_1\rangle) = \{\beta\bar{n}/\beta \in \mathrm{Sub}(\psi_1) \,\&\, n \in \omega\},$$
$$\mathrm{Sub}(\alpha L\langle\psi_1\rangle) = \{\beta L/\beta \in \mathrm{Sub}(\alpha\psi_1)\},$$
$$\mathrm{Sub}(\alpha R\langle\psi_1\rangle) = \{\beta R/\beta \in \mathrm{Sub}(\alpha\langle\psi_1\rangle)\}$$

Notice that for all α and $\beta \in \mathrm{Sub}(\alpha\psi)$ there is a $\gamma \in \mathscr{D}$ such that $\gamma\alpha\psi = \beta$. Moreover,

$$\forall \beta \in \mathrm{Sub}(\alpha\psi)(\beta\rho \leq \beta\sigma) \Rightarrow \alpha\psi\rho \leq \alpha\psi\sigma.$$

Given $s_1, \ldots, s_n$, take $\alpha_0 = \bar{s}_1 \ldots \bar{s}_n \bar{1}$, $\mathscr{A}_0 = \{\alpha_0\}$, $\mathscr{A}_{i+1} = \cup\{\mathrm{Sub}(\alpha_i'\psi)/\alpha_i' R \in \mathscr{A}_i\}$ and $\mathscr{A} = \cup_i \mathscr{A}_i$. An easy induction on i shows that $\forall \alpha \in \mathscr{A} \exists \gamma(\gamma\alpha_0[\psi] = \alpha[\psi])$, which in particular gives $I \notin \mathscr{A}$ since $\gamma\alpha_0[\psi] = [\psi]$ would imply the false equality $L\gamma\bar{s}_1 \ldots \bar{s}_n\varphi = I$.

In order to compute $f(s_1, \ldots, s_n)$ start generating the members of $\mathscr{A}$ consecutively. There are two possibilities.

First, one may get $\alpha = \alpha' L$ for a certain $\alpha \in \mathscr{A}$. There is a γ such that $\gamma\alpha_0[\psi] = \alpha[\psi] = \alpha'$; hence $\bar{s}_1 \ldots \bar{s}_n\varphi \neq O$. Therefore, $f(s_1, \ldots, s_n)\downarrow$ and $t = f(s_1, \ldots, s_n)$ is effectively derived out of α since $\gamma\bar{t} = \alpha'$. The computation terminates.

Secondly, the process may fail to terminate if $\alpha=\alpha' R$ for all $\alpha\in\mathscr{A}$, in which case one expects that $f(s_1,\ldots,s_n)\uparrow$. Actually, consider the regular segment $\mathscr{E}=\{\theta/\forall\alpha\in\mathscr{A}(\alpha\theta L\leq\alpha[\psi])\}$. Let $\theta\in\mathscr{E}$ and $\alpha\in\mathscr{A}$. Then $\alpha(I,\psi\theta)L=\alpha' R(I,\psi\theta)L=\alpha'\psi\theta L$. It follows that $\alpha'\psi\theta L\leq\alpha'\psi[\psi]$ since $\beta\theta L\leq\beta[\psi]$ for all $\beta\in\mathrm{Sub}(\alpha'\psi)$. Therefore, $\alpha(I,\psi\theta)L\leq\alpha[\psi]$; hence $[\psi]\in\mathscr{E}$ by μA_3. In particular, $\alpha_0[\psi]L\leq\alpha_0[\psi]$ implies $\overline{s_1}\ldots\overline{s_n}\varphi\notin\mathscr{D}$, i.e. $f(s_1,\ldots,s_n)\uparrow$. This completes the proof.

The corresponding result for primitive recursiveness is to be found in the exercises to chapter 22.

CHAPTER 9

Basic theorems of recursion theory

Some standard theorems characteristic of Recursion Theory are established in this chapter: a Normal Form Theorem, First and Second Recursion Theorems, an Enumeration Theorem.

We begin with several normal form results throwing light on the structure of the recursive elements and mappings. In order to give more precise formulations we introduce the following special kinds of elements polynomial or primitive in a subset $\mathscr{B}$ of $\mathscr{F}$.

An element φ is *strictly polynomial in* $\mathscr{B}$ iff

$$\varphi \in cl(\{L, R\} \cup \mathscr{B} / \lambda\theta.\theta L, \lambda\theta.\theta R, \Pi),$$

which by the distributive law is equivalent to

$$\varphi \in cl(cl(\{L, R\} \cup \mathscr{B} / \lambda\theta.\theta L, \lambda\theta.\theta R) / \Pi).$$

Of course, all such elements are polynomial in $\mathscr{B}$, while 4.1, 4.12 imply that all polynomial elements are strictly polynomial (in $\varnothing$).

Let a finite set $\mathscr{B}_0$ of primitive elements be fixed as in chapter 7 such that $cl(\mathscr{B}_0/\circ, [\ \ ])$ contains all recursive elements. An element φ is *strictly primitive in* $\mathscr{B}$ iff φ is strictly polynomial in $\mathscr{B}_0 \cup \langle \mathscr{B} \rangle$. Notice that whenever φ is strictly primitive in $\mathscr{B}$, then so are φL, φR. All polynomial elements are strictly primitive (in $\varnothing$). It is also immediate that all the elements strictly primitive in $\mathscr{B}$ are primitive in $\mathscr{B}$. However, it can not be claimed that all the elements polynomial (primitive) in $\mathscr{B}$ are strictly polynomial (respectively strictly primitive) in $\mathscr{B}$ for all $\mathscr{B}$; relevant counterexamples will be given in the exercises to chapter 21.

Proposition 9.1. Let $\mathscr{C} \subseteq \mathscr{F}$ and $\varphi \in cl(\mathscr{C}/\circ, [\ \ ])$. Then there is an element σ strictly polynomial in $\mathscr{C}$ such that $\varphi = \bar{1}[\sigma]$.

Proof. We first prove by induction on the construction of φ that $\varphi = \rho[\sigma]$ with certain ρ, σ strictly polynomial in $\mathscr{C}$.

If $\varphi \in \mathscr{C}$, then $\varphi = R[\varphi L]$ by 6.8.

Let $\varphi = \rho[\sigma]$ and $\psi = \rho_1[\sigma_1]$. Then

$$\varphi\psi = \rho\bar{1}[((\rho_1 R^2, \sigma\bar{1}), \bar{0}, \sigma_1 R^2)]$$

by 6.15, while

$$[\varphi] = \bar{1}[((\bar{0}, \rho R^2), \bar{1}, \sigma R^2)]$$

by 6.16.

Now let $\varphi = \rho[\sigma]$ with ρ, σ strictly polynomial in $\mathscr{C}$. Then $\varphi = \bar{1}[(\rho R^2, \bar{0}, \sigma R^2)]$ by 6.14, which completes the proof.

The following is a slightly modified version of the Normal Form Theorem for elements prime recursive in $\mathscr{B}$ established in Georgieva [1980] and already used in the hints to exercises 6.14***, 8.4***.

Proposition 9.2. If φ is prime recursive in $\mathscr{B}$, then $\varphi = \bar{1}[\sigma]$ with a certain σ strictly polynomial in $\mathscr{B}$.

This follows from 7.5 and 9.1, taking $\mathscr{C} = \{L, A\} \cup \mathscr{B}$.

The following assertion is called the Normal Form Theorem for elements recursive in $\mathscr{B}$.

Proposition 9.3. If φ is recursive in $\mathscr{B}$, then $\varphi = \bar{1}[\sigma]$ for some σ strictly primitive in $\mathscr{B}$.

This follows from 9.1, taking $\mathscr{C} = \mathscr{B}_0 \cup \langle \mathscr{B} \rangle$.

We observe a curious phenomenon, namely, a strict stratification of the initial operations of an IOS. They are separately ordered as follows: translation, multiplication (right), pairing operation, iteration, multiplication (left). For, the Normal Form Theorem shows that the elements recursive in $\mathscr{B}$ can be constructed in a few stages, with different operations used in each of them. In the first stage members of $\mathscr{B}$ are taken and $\langle \quad \rangle$ is applied to each of them once. Secondly, elements from $\mathscr{B}_0$ are allowed and the operations $\lambda\theta.\theta L, \lambda\theta.\theta R$ are applied repeatedly. At the third stage Π is applied repeatedly. After that $[\quad]$ is applied just once. Finally, the elements so obtained are multiplied to the left by LR.

Certain modifications of 9.2, 9.3 can be obtained by applying exercise 4.9 to the element σ, stratifying the initial IOS-operations in another way.

Analogous normal form theorems in Ordinary Recursion Theory and the theory of Skordev combinatory spaces show that it is sufficient to use the least number operator, respectively the operation iteration just once.

The following Normal Form Theorem for mappings recursive in $\mathscr{B}$ is an analogue to that of Skordev [1978].

Proposition 9.4. If Γ is a n-ary mapping recursive in $\mathscr{B}$, then

$$\Gamma = \lambda\theta_1 \ldots \theta_n . \bar{1}[\varphi(I, \langle\theta_1\rangle, \ldots, \langle\theta_n\rangle)]$$

for a certain φ strictly primitive in $\mathscr{B}$.

Proof. Throughout this proof ρ will stand for $(I, \langle\theta_1\rangle, \ldots, \langle\theta_n\rangle)$. First we prove by induction on the construction of Γ that $\Gamma = \lambda\theta_1 \ldots \theta_n . \alpha[\psi\rho\tau]$ for certain ψ, τ recursive in $\mathscr{B}$.

Let $\Gamma = \lambda\theta_1 \ldots \theta_n . \langle\theta_i\rangle$, $1 \le i \le n$. If $i < n$, then $\Gamma = \lambda\theta_1 \ldots \theta_n . R[\bar{i}\rho L]$. If $i = n$, then $\Gamma = \lambda\theta_1 \ldots \theta_n . R[R^n \rho L]$.

Let $\Gamma = \lambda\theta_1 \ldots \theta_n . \psi$, $\psi \in \mathscr{B}_0 \cup \langle\mathscr{B}\rangle$. Then $\Gamma = \lambda\theta_1 \ldots \theta_n . R[\psi L^2 \rho]$.

Suppose that $\Gamma_1 = \lambda\theta_1 \ldots \theta_n . \alpha_1[\psi_1 \rho \tau_1]$ and $\Gamma_2 = \lambda\theta_1 \ldots \theta_n . \alpha_2 L[\psi_2 \rho \tau_2]$.

If $\Gamma = \lambda\theta_1 \ldots \theta_n . \Gamma_1(\theta_1, \ldots, \theta_n)\Gamma_2(\theta_1, \ldots, \theta_n)$, then 6.15 gives $\Gamma = \lambda\theta_1 \ldots \theta_n . \alpha_1 \bar{1}[\sigma]$, where

$$\begin{aligned}\sigma &= ((\alpha_2 R^2, \psi_1 \rho \tau_1 \bar{1}), \bar{0}, \psi_2 \rho \tau_2 R^2) = \psi_3(I, \langle\theta_1\rangle\bar{0}, \langle\theta_1\rangle\bar{1}, \ldots, \langle\theta_n\rangle\bar{0}, \langle\theta_n\rangle\bar{1})\tau_3 \\ &= \psi_3(I, \bar{0}\langle\theta_1\rangle_2, \bar{1}\langle\theta_1\rangle_2, \ldots, \bar{0}\langle\theta_n\rangle_2, \bar{1}\langle\theta_n\rangle_2)\tau_3 \\ &= \psi_4(I, \langle\theta_1\rangle_2, \langle\theta_2\rangle_2, \ldots, \langle\theta_n\rangle_2)\tau_3 \\ &= \psi_4(L, PR\langle\theta_1\rangle, \ldots, PR\langle\theta_n\rangle)(I, Q)\tau_3 = \psi\rho\tau\end{aligned}$$

with certain $\psi_3, \tau_3, \psi_4, \psi, \tau$ recursive in $\psi_1, \tau_1, \psi_2, \tau_2$.

If $\Gamma = \lambda\theta_1 \ldots \theta_n . [\Gamma_1(\theta_1, \ldots, \theta_n)]$, then 6.16 gives $\Gamma = \lambda\theta_1 \ldots \theta_n . \bar{1}[\sigma]$, where

$$\sigma = ((\bar{0}, \alpha_1 R^2), \bar{1}, \psi_1 \rho \tau_1 R^2) = \psi\rho\tau$$

with certain ψ, τ recursive in ψ_1, τ_1.

Now let $\Gamma = \lambda\theta_1 \ldots \theta_n . \alpha[\psi\rho\tau]$ with ψ, τ recursive in $\mathscr{B}$. There is by 9.3 an element φ_1 strictly primitive in $\mathscr{B}$ such that $\psi = \bar{1}[\varphi_1]$; hence

$$\psi\sigma\tau = \bar{1}[\varphi_1]\rho\tau = \bar{2}[(\rho\tau L, \varphi_1 R)]$$

by 6.13. It follows from 9.3 and exercise 6.1d that $\tau L = R^2[\varphi_2]$ with a certain φ_2 strictly primitive in $\mathscr{B}$. Therefore,

$$\psi\rho\tau = \bar{2}[(\rho R^2, \varphi_1 RL)[\varphi_2]] = \bar{2}\bar{1}[\sigma]$$

by 6.16, where

$$\sigma = ((\bar{0}, \rho R^4, \varphi_1 R\bar{2}), \bar{1}, \varphi_2 R^2) = ((\varphi_3, \rho R^4, \varphi_4), \varphi_5)$$

with $\varphi_3, \varphi_4, \varphi_5$ strictly primitive in $\mathscr{B}$. It follows from 6.16 that $\Gamma = \lambda\theta_1 \ldots \theta_n . \alpha\bar{1}[\sigma_1]$, where

$$\sigma_1 = ((\bar{0}, \bar{2}\bar{3}), \bar{1}, \sigma R^2) = (\varphi_6, \varphi_7, (\varphi_8, \rho R^6, \varphi_9), \varphi_{10})$$

with $\varphi_6, \ldots, \varphi_{10}$ strictly primitive in $\mathscr{B}$. Finally, we get $\Gamma = \lambda\theta_1 \ldots \theta_n . \bar{1}[\sigma_2]$ by 6.14, where

$$\sigma_2 = (\alpha\bar{3}, \bar{0}, \sigma_1 R^2) = \varphi\rho$$

with a certain φ strictly primitive in $\mathscr{B}$, using the equality $\langle\theta\rangle R = R\langle\theta\rangle$. This completes the proof.

In particular, one gets $\Gamma = \lambda\theta . \bar{1}[\varphi(I, \langle\theta\rangle)]$ whenever Γ is unary.

Three modified versions of the last normal form theorem follow.

Proposition 9.5. If Γ is a n-ary mapping recursive in $\mathscr{B}$, then

$$\Gamma = \lambda\theta_1 \ldots \theta_n . \bar{1}[\varphi(I, \langle\psi_1\rangle, \ldots, \langle\psi_m\rangle, \langle\theta_1\rangle, \ldots, \langle\theta_n\rangle)]$$

with certain $\psi_1, \ldots, \psi_m \in \mathscr{B}$ and a strictly primitive φ.

Proof. The mapping Γ is recursive in a finite subset $\{\psi_1, \ldots, \psi_m\}$ of $\mathscr{B}$. Let Γ^* correspond to Γ by 7.19. Applying 9.4 to the recursive mapping Γ^*, we get the desired normal form of Γ.

Proposition 9.6. If Γ is a n-ary mapping recursive in $\mathscr{B}$, then

$$\Gamma = \lambda\theta_1 \ldots \theta_n . \bar{1}[\varphi(I, \langle(\theta_1, \ldots, \theta_n)\rangle)]$$

with φ strictly primitive in $\mathscr{B}$.

Proof. Let Γ^* correspond to Γ by 7.20. Then we get the desired normal form by applying 9.4 to Γ^*.

Proposition 9.7. If Γ is a n-ary mapping recursive in $\mathscr{B}$, then

$$\Gamma = \lambda\theta_1 \ldots \theta_n . \bar{1}[\varphi(I, \langle(\psi_1, \ldots, \psi_m, \theta_1, \ldots, \theta_n)\rangle)]$$

with $\psi_1, \ldots, \psi_m \in \mathscr{B}$ and a strictly primitive φ.

This follows from 7.19, 9.6.

The next statement is an improved version of the Normal Form Theorem for unary mappings prime recursive in $\mathscr{B}$ established in Georgieva [1980]. First, however, we make a definition. To each such mapping Γ assign a natural number $c(\Gamma)$ viz., the least n such that Γ has a construction in which the clause with $\lambda\theta.\theta$ occurs n times. Such a 'complexity' could also be introduced for unary mappings recursive, primitive recursive etc. in $\mathscr{B}$. Notice that in the case of recursiveness $c(\Gamma) \leq 1$ by 9.4.

Proposition 9.8. Let Γ be a unary mapping prime recursive in $\mathscr{B}$ and $c(\Gamma) = n$. Then

$$\Gamma = \lambda\theta . \bar{1}[\varphi(I, \theta\bar{4}, \ldots, \theta\overline{n+3})]$$

with a certain φ strictly polynomial in $\mathscr{B}$.

Proof. To begin with, we prove by induction on the construction of Γ (the clause $\lambda\theta.\theta$ assumed to appear n times) that

$$\Gamma = \lambda\theta . \alpha[\psi(I, \theta\alpha_1, \ldots, \theta\alpha_n)]$$

with a certain ψ polynomial in $\mathscr{B}$.

If $\Gamma = \lambda\theta.\theta$, then $\Gamma = \lambda\theta.R[R(I, \theta L)]$.

If $\Gamma = \lambda\theta.\psi$, $\psi \in \{L, A\} \cup \mathscr{B}$, then $\Gamma = \lambda\theta.R[\psi L]$.

Let $\Gamma_1 = \lambda\theta.\alpha[\psi_1(I, \theta\alpha_1, \ldots, \theta\alpha_k)]$, $\Gamma_2 = \lambda\theta.\beta[\psi_2(I, \theta\beta_1, \ldots, \theta\beta_l)]$ and $\Gamma = \lambda\theta.\Gamma_1(\theta)\Gamma_2(\theta)$.

Using 6.15, we obtain

$$\Gamma = \lambda\theta.\alpha'[\psi(I, \theta\alpha'_1, \ldots, \theta\alpha'_{k+l})]$$

with ψ polynomial in ψ_1, ψ_2.

The case of iteration is treated by means of 6.16.

Now let $\Gamma = \lambda\theta.\alpha[\psi(I, \theta\alpha_1, \ldots, \theta\alpha_n)]$, where ψ is polynomial in $\mathscr{B}$. It follows from 9.2 that $\psi = \bar{1}[\varphi_1]$ with φ_1 strictly polynomial in $\mathscr{B}$, hence

$$\psi(I, \theta\alpha_1, \ldots, \theta\alpha_n) = \bar{1}[\varphi_1](I, \theta\alpha_1, \ldots, \theta\alpha_n) = \bar{2}[((L, \theta\alpha_1 L, \ldots, \theta\alpha_n L), \varphi_1 R)]$$

by 6.13. We get $\Gamma = \lambda\theta.\alpha\bar{1}[\sigma]$ by 6.16, where

$$\sigma = ((\bar{0}, \bar{4}), \bar{1}, (\bar{2}, \theta\alpha_1\bar{2}, \ldots, \theta\alpha_n\bar{2}), \varphi_1 R^3) = (\varphi_2, \varphi_3, (\varphi_4, \theta\beta_1, \ldots, \theta\beta_n), \varphi_5)$$

with $\varphi_2, \ldots, \varphi_5$ strictly polynomial in $\mathscr{B}$. Let $\tau = (\beta_1, \ldots, \beta_n, I)$. Then

$$\sigma = (\varphi_2 R^n, \varphi_3 R^n, (\varphi_4 R^n, \theta\bar{0}, \ldots, \theta\overline{n-1}), \varphi_5 R^n)\tau.$$

Substituting $R[\tau L]$ for τ and using exercise 6.1h, 6.14, we get the desired normal form. The proof is complete.

In particular, one gets $\Gamma = \lambda\theta.\bar{1}[\varphi(I, \theta\bar{4})]$ whenever $c(\Gamma) = 1$. A normal form $\Gamma = \lambda\theta.\bar{1}[\varphi(I, \theta R^4)]$ can also be obtained in this case.

The assumption $c(\Gamma) = n$ above can obviously be replaced by $c(\Gamma) \leq n$. Moreover, 9.8 can be generalized to n-ary mappings, $n > 1$.

It is also worth mentioning that the proof of 9.4 produces an element φ such that for all $\theta_1, \ldots, \theta_n$ the element $\varphi(I, \langle\theta_1\rangle, \ldots, \langle\theta_n\rangle)$ is uniformly strictly primitive in $\{\theta_1, \ldots, \theta_n\} \cup \mathscr{B}$, while in the case of 9.8 $\varphi(I, \theta\bar{4}, \ldots, \theta\overline{n+3})$ is uniformly strictly polynomial in $\{\theta\} \cup \mathscr{B}$. The elements φ in 9.4–9.8 can be also modified by means of exercise 4.9.

While the natural concept of inductiveness was introduced at the beginning of chapter 5, the notion of recursiveness was taken as fundamental and has been regarded as central so far. We are now going to justify this by showing that the class of all inductive mappings is identical with that of all recursive ones. The following statement is the easy half of such a justification.

Proposition 9.9. If a mapping Γ is recursive, then it is inductive.

The proof is by induction on the construction of Γ, making use of the simple fact that composition of mappings preserves their inductiveness.

It should be mentioned that the desire to obtain 9.9 was the only reason for taking $\lambda\theta_1 \ldots \theta_n.\psi$, $\psi \in \{I, L, R\}$ among the initial inductive mappings, for otherwise it would only follow that for every n-ary recursive mapping Γ there is a $n+3$-ary inductive mapping Γ_1 such that $\Gamma = \lambda\theta_1 \ldots \theta_n.\Gamma_1(I, L, R, \theta_1, \ldots, \theta_n)$.

On the other hand, all inductive mappings turn out to be recursive. To ensure this one should prove that the μ-operation preserves recursiveness, i.e. whenever $\Gamma : \mathscr{F}^{n+1} \to \mathscr{F}$ is recursive, then so is $\lambda\theta_1 \ldots \theta_n.\mu\theta.\Gamma(\theta_1, \ldots, \theta_n, \theta)$. In other words, one needs the First Recursion Theorem. In fact, propositions 6.38, 6.39 were the first steps toward such a theorem and we are now able to complete its proof in two final steps, using axiom μA_1 in the latter.

Proposition 9.10 (Third Recursion Lemma). The element $\mu\theta.\varphi[\psi(I, \theta\chi)]$ exists and is recursive in φ, ψ, χ for all φ, ψ, χ.

Proof. We show that the element in question equals $\varphi R^2\bar{1}[\sigma]$, where $\sigma = \Delta(\sigma_1, \sigma_2)$, $\sigma_1 = (\chi\bar{1}\bar{1}, R^2\bar{1}, \bar{0}, \psi(\bar{1}\bar{1}, \varphi R^2\bar{2}))$, $\sigma_2 = (\bar{0}\bar{1}, R^2)$.

Using 6.13 and exercise 6.1i it follows that

$$\varphi[\psi(I, \varphi R^2\bar{1}[\sigma]\chi)] = \varphi[\psi(\bar{0}, \varphi R^2\bar{2})[(\chi\bar{0}, \sigma R)]] = \varphi R^2\bar{1}[(\sigma_1, \sigma\sigma_2)]$$
$$= \varphi R^2\bar{1}[\sigma].$$

Suppose that $\varphi[\psi(I,\tau\chi)]\leq\tau$. Take $\rho=[\psi(I,\tau\chi)]$, $\rho_1=\Delta((\chi\rho,\rho,\rho),\chi\rho)$. Then

$$\begin{aligned}\sigma_1(I,\rho_1)&=(\chi\rho,\rho,I,\psi(\rho,\varphi\rho\chi\rho))=(\chi\rho,\rho,I,\psi(I,\varphi\rho\chi)\rho)\\&\leq(\chi\rho,\rho,I,\psi(I,\tau\chi)\rho)=(\chi\rho,\rho,\rho)=L\rho_1,\\\sigma_2(I,\rho_1)&=(\chi\rho,\rho_1\chi\rho)=(I,\rho_1)\chi\rho\end{aligned}$$

and $\rho_1\chi\rho=R\rho_1$; hence $\sigma(I,\rho_1)\leq\rho_1$ by 6.30. Therefore, $R[\sigma]\leq\rho_1$ by 6.11, which implies

$$\varphi R^2\bar{1}[\sigma]\leq\varphi R^2\bar{0}\rho_1=\varphi\rho\leq\tau.$$

This completes the proof.

Proposition 9.11. If Γ is a unary mapping prime recursive in $\mathscr{B}$ and $c(\Gamma)\leq 1$, then the element $\mu\theta.\Gamma(\theta)$ exists and is recursive in $\mathscr{B}$.

This follows from 9.8, 9.10.

Proposition 9.12* (Fourth Recursion Lemma). The element $\mu\theta.\varphi[\psi(I,\langle\theta\rangle)]$ is recursive in φ,ψ for all φ,ψ.

Proof. The element $\theta_0=\mu\theta.\varphi[\psi(I,\langle\theta\rangle)]$ exists by 5.2*. It remains to show that it is recursive in φ,ψ.

Take $\varphi_1=\langle\varphi\rangle C$, $\psi_1=\langle\psi\rangle C(CL,PR)$ and $\chi_1=QC$. The element $\theta_1=\mu\theta.\varphi_1[\psi_1(I,\theta\chi_1)]$ is recursive in φ_1,ψ_1,χ_1 by 9.10, hence it is recursive in φ,ψ. Using 6.37, 6.40, we get

$$\begin{aligned}\langle\varphi[\psi(I,\langle\theta\rangle)]\rangle&=\langle\varphi\rangle C[\langle\psi\rangle C(\langle I\rangle,\langle\theta\rangle_2)C]=\varphi_1[\langle\psi\rangle C(C,P\langle\theta\rangle QC)]\\&=\varphi_1[\psi_1(I,\langle\theta\rangle\chi_1)]\end{aligned}$$

for all θ. It follows in particular that

$$\langle\theta_0\rangle=\langle\varphi[\psi(I,\langle\theta_0\rangle)]\rangle=\varphi_1[\psi_1(I,\langle\theta_0\rangle\chi_1)];$$

hence $\theta_1\leq\langle\theta_0\rangle$.

On the other hand, consider the simple segment

$$\mathscr{E}=\{\theta/\langle\theta\rangle\leq\theta_1\}=\{\theta/\langle\theta\rangle\leq\langle I\rangle\theta_1\}.$$

If $\theta\in\mathscr{E}$, then

$$\langle\varphi[\psi(I,\langle\theta\rangle)]\rangle=\varphi_1[\psi_1(I,\langle\theta\rangle\chi_1)]\leq\varphi_1[\psi_1(I,\theta_1\chi_1)]=\theta_1,$$

hence $\theta_0\in\mathscr{E}$ by μA_1. Therefore, $\langle\theta_0\rangle=\theta_1$, hence $\theta_0=L\theta_1(I,I)$, which completes the proof.

We call the technique used above the *translation method* since in order to solve the equality $\varphi[\psi(I,\langle\theta\rangle)]=\theta$ we have transformed ('translated') it by means of $\langle\ \ \rangle$ into the more tractable form $\varphi_1[\psi_1(I,\theta\chi_1)]=\theta$.

The following statement is called the First Recursion Theorem, being actually an analogue of theorem XXVI of Kleene [1952] and the theorem proved in Skordev [1979].

Proposition 9.13*. If Γ is a unary mapping recursive in $\mathscr{B}$, then the element $\mu\theta.\Gamma(\theta)$ is recursive in $\mathscr{B}$.

This follows from 9.12* since Γ has a normal form $\lambda\theta.\varphi[\psi(I,\langle\theta\rangle)]$ with φ,ψ recursive in $\mathscr{B}$ by 9.4. Next, we give a parametrized version of the First Recursion Theorem.

Proposition 9.14*. If Γ is a $n+1$-ary mapping recursive in $\mathscr{B}$, then the mapping $\Gamma_1=\lambda\theta_1\ldots\theta_n.\mu\theta.\Gamma(\theta_1,\ldots,\theta_n,\theta)$ is recursive in $\mathscr{B}$.

Proof. We recall that the existence of Γ_1 is ensured by 5.2*. The element θ_0 in the proof of 9.12* is constructed uniformly in φ,ψ; in essence, we have obtained a recursive binary mapping Γ_0 such that $\Gamma_0(\varphi,\psi)=\mu\theta.\varphi[\psi(I,\langle\theta\rangle)]$ for all φ,ψ. On the other hand, it follows immediately from 9.4 that

$$\Gamma=\lambda\theta_1\ldots\theta_{n+1}.\bar{1}[\Gamma_2(\theta_1,\ldots,\theta_n)(I,\langle\theta_{n+1}\rangle)]$$

with a certain mapping Γ_2 recursive in $\mathscr{B}$. Therefore,

$$\Gamma_1=\lambda\theta_1\ldots\theta_n.\Gamma_0(\bar{1},\Gamma_2(\theta_1,\ldots,\theta_n));$$

hence Γ_1 is recursive in $\mathscr{B}$ by 7.18.

Proposition 9.15*. If a mapping Γ is inductive, then it is recursive.

This follows from 9.14*.

The First Recursion Theorem makes it possible to solve systems of inequalities as well.

Proposition 9.16*. Let a system of inequalities

(1) $$\Gamma_i(\theta_0,\ldots,\theta_n)\le\theta_i,\ i\le n,$$

be given, where $n>0$ and $\Gamma_0,\ldots,\Gamma_n$ are recursive in $\mathscr{B}$. Then there are elements $\varphi_0,\ldots,\varphi_n$ recursive in $\mathscr{B}$ such that $\varphi_0,\ldots,\varphi_n$ satisfy (1) and whenever $\tau_0,\ldots,\tau_n$ satisfy (1), then $\varphi_i\le\tau_i$ for all $i\le n$.

Proof. The system (1) may be reduced to a single inequality in a manner similar to that of Skordev [1979]. Namely, the mapping

$$\Gamma=\lambda\theta.(\Gamma_0(\bar{0}\theta,\ldots,\overline{n-1}\,\theta,R^n\theta)),\ldots,\Gamma_n(\bar{0}\theta,\ldots,\overline{n-1}\,\theta,R^n(\theta))$$

is recursive in $\mathscr{B}$, hence so is $\theta_0=\mu\theta.\Gamma(\theta)$ by 9.13*. Take $\varphi_i=\bar{\imath}\theta_0$ for $i<n$ universal element (mapping).

$$\Gamma_i(\varphi_0,\ldots,\varphi_n)=\bar{\imath}\Gamma(\theta_0)=\bar{\imath}\theta_0=\varphi_i,\ i<n,$$
$$\Gamma_n(\varphi_0,\ldots,\varphi_n)=R^n\Gamma(\theta_0)=R^n\theta_0=\varphi_n;$$

hence $\varphi_0,\ldots,\varphi_n$ satisfy (1).

Let $\tau_0,\ldots,\tau_n$ be another solution to (1). Then

$$\Gamma((\tau_0,\ldots,\tau_n))=(\Gamma_0(\tau_0,\ldots,\tau_n),\ldots,\Gamma_n(\tau_0,\ldots,\tau_n))\le(\tau_0,\ldots,\tau_n),$$

hence $\theta_0\le(\tau_0,\ldots,\tau_n)$, which implies $\varphi_i\le\tau_i$ for all $i\le n$. This completes the proof.

In reality, the mappings $\Gamma_0,\ldots,\Gamma_n$ in (1) may be of various arities $m\le n$ since any such mapping Γ_i may be replaced by the n-ary mapping $\lambda\theta_1\ldots\theta_n.\Gamma_i(\theta_1,\ldots,\theta_m)$.

The remainder of this chapter is devoted to universal elements and mappings.

Let $\mathscr{U} \subseteq \mathscr{F}$ and let $\mathscr{M}$ be a set of unary mappings over $\mathscr{F}$. An element σ is *universal for* $\mathscr{U}$ iff $\mathscr{U} \subseteq \{\bar{n}\sigma / n \in \omega\}$, while a unary mapping Σ is *universal for* $\mathscr{M}$ iff $\mathscr{M} \subseteq \{\lambda\theta.\bar{n}\Sigma(\theta)/n \in \omega\}$. We consider unary mappings since the relevant definitions and results for n-ary mappings are obtained in an obvious way from 7.20. Notice that $\mathscr{U}$ (respectively $\mathscr{M}$) is countable whenever it admits a universal element (mapping).

Proposition 9.17. Enumeration Lemma. Let $\mathscr{B}$ be finite. Then there is an element τ recursive in $\mathscr{B}$ and universal for the elements polynomial in $\mathscr{B}$.

Proof. Let $\{L, R\} \cup \mathscr{B} = \{\psi_0, \ldots, \psi_m\}$ and $\mathscr{U} = cl(\psi_0, \ldots, \psi_m / \circ, \Pi)$. It follows by 4.13 that

$$\mathscr{U} = cl(\psi_0, \ldots, \psi_m / \lambda\theta.\psi_i\theta, i \leq m, \Pi).$$

Take $\psi = (\psi_0, \ldots, \psi_m, L)$, $\chi = (\langle\psi_0\rangle, \ldots, \langle\psi_m\rangle, L)$ and $\tau = R[Q(\psi L, Q\chi R, QTR, L)]$. To show that τ is universal for $\mathscr{U}$, we shall use the equality $\tau = Q(\psi, Q\chi\tau, QT\tau, I)$ and the fact that $\overline{J(s,t)}Q = \bar{s}\bar{t}$ by exercise 6.6, where $J = \lambda st.(2s+1)2^t$.

If $i \leq m$, then

$$\overline{J(i,0)}\tau = \bar{i}\bar{0}(\psi, Q\chi\tau, QT\tau, I) = \bar{i}\psi = \psi_i.$$

Let $i \leq m$ and $\varphi = \bar{k}\tau$. Then

$$\overline{J(J(k,i),1)}\tau = \overline{J(k,i)}Q\chi\tau = \bar{k}\langle\psi_i\rangle\tau = \psi_i\bar{k}\tau = \psi_i\varphi.$$

Let $\varphi = \bar{k}\tau$, $\psi = \bar{l}\tau$. Then

$$\overline{J(J(k,l),2)}\tau = \bar{k}\bar{l}T\tau = (\bar{k}\tau, \bar{l}\tau) = (\varphi, \psi)$$

by 7.14. We are done.

The following Enumeration Theorem is an analog to theorem XXII of Kleene [1952] and the corresponding statement of Skordev [1980].

Proposition 9.18. Let $\mathscr{B}$ be finite and $\mathscr{U}$ be the set of all elements recursive in $\mathscr{B}$. Then there is an element $\sigma \in \mathscr{U}$ which is universal for $\mathscr{U}$.

Proof. There exists by 9.17 an element $\tau \in \mathscr{U}$ universal for the elements polynomial in $\mathscr{B}_0 \cup \langle\mathscr{B}\rangle$. Take $\rho = C([I]L, R)$ and $\sigma = \langle\bar{1}\rangle\rho[D\langle\tau\rangle\rho]$. It follows that $\sigma \in \mathscr{U}$ and $\bar{n}\sigma = \bar{1}[\bar{n}\tau]$ for all n by 6.43.

Let $\varphi \in \mathscr{U}$. Then 9.3 implies that $\varphi = \bar{1}[\psi]$ with a certain ψ strictly primitive in $\mathscr{B}$, hence polynomial in $\mathscr{B}_0 \cup \langle\mathscr{B}\rangle$. There is a number n such that $\psi = \bar{n}\tau$; hence

$$\varphi = \bar{1}[\psi] = \bar{1}[\bar{n}\tau] = \bar{n}\sigma.$$

Therefore, σ is universal for $\mathscr{U}$. We are done.

Alternatively, one may use in the above proof the equivalence (1)⇔(2) of 7.11 and 9.2 instead of 9.3.

The following statement is called the Parametrized Enumeration Theorem.

Proposition 9.19. Let $\mathscr{B}$ be finite and $\mathscr{M}$ be the set of all unary mappings recursive in $\mathscr{B}$. Then there is a mapping $\Sigma \in \mathscr{M}$ which is universal for $\mathscr{M}$.

The proof follows that of 9.18. Just take

$$\Sigma = \lambda\theta.\langle \bar{1} \rangle \rho[D\langle \tau(I, \langle \theta \rangle)\rangle \rho]$$

and make use of the Normal Form Theorem 9.4.

Assume from now on that a subset $\mathscr{B}$ of $\mathscr{F}$ is fixed, $\mathscr{U}$ is the set of all elements recursive in $\mathscr{B}$ and $\mathscr{M}$ is the set of all unary mappings recursive in $\mathscr{B}$. As in the proof of 9.17, J will stand for the pairing function $\lambda st.(2s+1)2^t$.

An element ψ is *natural* iff for all s there is a t such that $\bar{s}\psi = \bar{t}$. In other words, natural elements are those which represent total unary number theoretic functions.

An element σ is *principal universal for* $\mathscr{U}$ iff for all $\varphi \in \mathscr{U}$ there is a natural primitive recursive element ψ such that $\bar{n}\varphi = \bar{n}\psi\sigma$ for all n. A mapping Σ is *principal universal for* $\mathscr{M}$ iff for all $\Gamma \in \mathscr{M}$ there is a natual primitive recursive element ψ such that $\bar{n}\Gamma(\theta) = \bar{n}\psi\Sigma(\theta)$ for all n, θ. We shall concentrate on principal universal elements since the corresponding statements for principal universal mappings may be obtained by an immediate parametrization.

Proposition 9.20. If σ is principal universal for $\mathscr{U}$, then σ is universal for $\mathscr{U}$.

Proof. Let $\varphi \in \mathscr{U}$. Then $[I]\varphi \in \mathscr{U}$; hence there is a natural element ψ such that $\bar{n}[I]\varphi = \bar{n}\psi\sigma$ for all n. It follows that $\bar{0}\psi = \bar{k}$ for a certain k, hence

$$\varphi = \bar{0}[I]\varphi = \bar{0}\psi\sigma = \bar{k}\sigma.$$

This completes the proof.

Proposition 9.21. If σ is universal for $\mathscr{U}$, then $Q\sigma$ is principal universal for $\mathscr{U}$.

Proof. Let $\varphi \in \mathscr{U}$. Then $\varphi = \bar{k}\sigma$ with a certain k. Taking $\psi = \bar{k}P$, we get $\bar{n}\psi = \overline{J(n, k)}$ for all n; hence ψ is a natural primitive recursive element. It follows that

$$\bar{n}\varphi = \bar{n}\bar{k}\sigma = \overline{J(n, k)}Q\sigma = \bar{n}\bar{k}PQ\sigma = \bar{n}\psi Q\sigma$$

for all n, hence $Q\sigma$ is principal universal for $\mathscr{U}$. The proof is complete.

The following statement is called the Second Recursion Theorem. It is an analogue to theorem XXVII of Kleene [1952] and the corresponding theorem of Skordev [1980].

Proposition 9.22. If $\sigma \in \mathscr{U}$ is principal universal for $\mathscr{U}$ and $\varphi \in \mathscr{U}$, then there is an n such that $\bar{n}\varphi = \bar{n}\sigma$.

Proof. We have $Q\sigma \in \mathscr{U}$; hence there is a natural primitive recursive element ψ such that $\bar{n}Q\sigma = \bar{n}\psi\sigma$ for all n. Proposition 9.20 implies that $DP\psi\varphi = \bar{m}\sigma$ for a certain m since $DP\psi\varphi \in \mathscr{U}$. Taking $\bar{n} = \overline{J(m, m)}\psi$, we get

$$\bar{n}\varphi = \bar{m}\bar{m}P\psi\varphi = \bar{m}DP\psi\varphi = \bar{m}\bar{m}\sigma = \overline{J(m, m)}Q\sigma = \overline{J(m, m)}\psi\sigma = \bar{n}\sigma,$$

which completes the proof.

A parametrized version of the Second Recursion Theorem states that whenever $\Sigma \in \mathscr{M}$ is principal universal for $\mathscr{M}$ and $\Gamma \in \mathscr{M}$ then there is an n such that $\bar{n}\Gamma(\theta) = \bar{n}\Sigma(\theta)$ for all θ.

The Enumeration Theorem and the Second Recursion Theorem imply a fixed point result called the Recursion Theorem.

Proposition 9.23. If Γ is a uniary mapping recursive in $\mathscr{B}$, then it has a fixed point φ recursive in $\mathscr{B}$.

Proof. The mapping Γ is recursive in a finite subset $\mathscr{B}_1$ of $\mathscr{B}$. Let Γ^* correspond to Γ by 7.21 and let σ be recursive in $\mathscr{B}_1$ and principal universal for the elements recursive in $\mathscr{B}_1$; such an element σ exists by 9.18, 9.21.

The element $\Gamma^*(\sigma)$ is recursive in $\mathscr{B}_1$; hence there is an n such that $\bar{n}\Gamma^*(\sigma) = \bar{n}\sigma$ by 9.22. Taking $\varphi = \bar{n}\sigma$, we get

$$\Gamma(\varphi) = \Gamma(\bar{n}\sigma) = \bar{n}\Gamma^*(\sigma) = \bar{n}\sigma = \varphi,$$

which completes the proof.

The element φ so constructed is not necessarily a least fixed point of Γ; hence 9.23 is not a First Recursion Theorem. However, a recursive fixed point of a given mapping is often all one needs. The advantage of the above Recursion Theorem is that it depends on no μ-axiom stronger than μA_0.

A parametrized version of 9.23 states that whenever Γ is a $n+1$-ary mapping recursive in $\mathscr{B}$, then there is a n-ary mapping Γ_1 recursive in $\mathscr{B}$ such that

$$\Gamma(\theta_1, \ldots, \theta_n, \Gamma_1(\theta_1, \ldots, \theta_n)) = \Gamma_1(\theta_1, \ldots, \theta_n)$$

for all $\theta_1, \ldots, \theta_n$. Systems of equalities may also be solved.

The following statement is an analogue to a theorem of Rice [1953].

Proposition 9.24. Let $\sigma \in \mathscr{U}$ be principal universal for $\mathscr{U}$ and $\varnothing \subset \mathscr{E} \subset \mathscr{U}$. Then there is no natural element $\psi \in \mathscr{U}$ such that $\bar{n}\sigma \in \mathscr{E}$ iff $\bar{n}\psi = \bar{0}$.

Proof. Suppose there is such an element $\psi \in \mathscr{U}$. Take k, l such that $\bar{k}\sigma \in \mathscr{E}$, $\bar{l}\sigma \in \mathscr{U} \setminus \mathscr{E}$, then take $\tau = \psi(\bar{l}, [I]\bar{k})$. It follows that $\tau\sigma \in \mathscr{U}$, hence $\bar{n}\tau\sigma = \bar{n}\sigma$ for a certain n by the Second Recursion Theorem.

If $\bar{n}\psi = \bar{0}$, then $\bar{n}\sigma \in \mathscr{E}$, $\bar{n}\tau = \bar{l}$, hence

$$\bar{l}\sigma = \bar{n}\tau\sigma = \bar{n}\sigma \in \mathscr{E},$$

which is not the case.

If $\bar{n}\psi \neq \bar{0}$, then $\bar{n}\sigma \notin \mathscr{E}$, $\bar{n}\tau = \bar{k}$, hence $\bar{k}\sigma = \bar{n}\tau\sigma = \bar{n}\sigma \notin \mathscr{E}$, which again is not the case. The proof is complete.

We close our study of universal elements and mappings by a discussion of the Theory of Numberings. Our reasons for doing this are first to avoid any overwork and, secondly, the proofs in the general theory turn out to be adaptations of the original proofs of Ordinary Recursion Theory, see 9.22, 9.24 for example.

To each element $\sigma \in \mathscr{U}$ universal for $\mathscr{U}$ assign a *numbering* σ^* of $\mathscr{U}$ by the equality

$$\sigma^* n = \bar{n}\sigma, n \in \omega.$$

Take the set Ψ_0 of all unary partial number theoretic functions weakly representable by recursive elements. Proposition 8.2 implies that all unary partial recursive functions are in Ψ_0 and if f is a unary function μ-recursive in Ψ_0, then $f \in \Psi_0$. Obviously, there are nonrecursive functions in Ψ_0.

Proposition 9.25. Let $\sigma \in \mathscr{U}$ be principal universal for $\mathscr{U}$. Then σ^* is a *precomplete numbering* of $\mathscr{U}$ in the sense of Maltsev [1961], where the last notion is relativized by considering members of Ψ_0 instead of unary partial recursive functions.

Proof. Let $f \in \Psi_0$. Then f is weakly represented by a recursive element φ. It follows that $\varphi\sigma \in \mathscr{U}$; hence there is a natural primitive recursive element ψ such that $\bar{n}\varphi\sigma = \bar{n}\psi\sigma$ for all n. The element ψ represents a total function $g \in \Psi_0$ and if $f(n)\downarrow$, then

$$\sigma^* g(n) = \overline{g(n)}\sigma = \bar{n}\psi\sigma = \bar{n}\varphi\sigma = \overline{f(n)}\sigma = \sigma^* f(n).$$

It now follows from a statement of Ershov [1977] that σ^* is a precomplete numbering. This completes the proof.

It is worth mentioning that the arguments in Ershov's book concerning precomplete numberings depend on the closure properties of the class of all unary partial recursive functions, so they remain valid for the above relativized notion of precomplete numbering.

Proposition 9.26*.** If $\sigma \in \mathscr{U}$ is principal universal for $\mathscr{U}$, then σ^* is a *complete numbering* in the sense of Maltsev [1965].

Proof. Let f be a unary partial number theoretic function. Then f is represented by a recursive element φ by 8.3***. It follows again that $\varphi\sigma \in \mathscr{U}$; hence there is a natural primitive recursive element ψ such that $\bar{n}\varphi\sigma = \bar{n}\psi\sigma$ for all n. The element ψ represents a general recursive function g by exercise 8.5***.

If $f(n)\downarrow$, then $\sigma^* g(n) = \sigma^* f(n)$ as in the proof of 9.25.

We have $O \in \mathscr{U}$ and whenever $f(n)\uparrow$, then $\bar{n}\varphi = O$; hence

$$\sigma^* g(n) = \overline{g(n)}\sigma = \bar{n}\psi\sigma = \bar{n}\varphi\sigma = O\sigma = O.$$

Therefore, σ^* is a complete numbering. The proof is complete.

As a corollary to 9.25 we get the following analogue to a theorem of Rogers [1967].

Proposition 9.27. Let $\sigma, \sigma_1 \in \mathscr{U}$ be principal universal for $\mathscr{U}$. Then there is a natural recursive element φ such that $\bar{n}\sigma = \bar{n}\varphi\sigma_1$ for all n and φ represents a permutation of ω, i.e. $\bar{m}\varphi \neq \bar{n}\varphi$ whenever $m \neq n$, and for all n there is an m such that $\bar{m}\varphi = \bar{n}$.

Proof. Since $\sigma \in \mathscr{U}$, there is a natural primitive recursive element ψ such that $\bar{n}\sigma = \bar{n}\psi\sigma_1$ for all n. The element ψ represents a total function $g \in \Psi_0$ and it follows that

$$\sigma^* n = \bar{n}\sigma = \bar{n}\psi\sigma_1 = \overline{g(n)}\sigma_1 = \sigma_1^* g(n)$$

for all n, hence g reduces σ^* to σ_1^*. Similarly, the numbering σ_1^* is reducible to σ^* by a total function from Ψ_0. Using a (relativized) theorem of Maltsev [1961], we conclude that σ^* and σ_1^* are recursively in Ψ_0 isomorphic, i.e., there is a μ-recursive in Ψ_0 function f such that $\sigma^* n = \sigma_1^* f(n)$ for all n and f is a permutation of ω. Then $f \in \Psi_0$; hence there is a recursive element φ to weakly represent f. However, f is total, hence φ represents f and

$$\bar{n}\sigma = \sigma^* n = \sigma_1^* f(n) = \overline{f(n)}\sigma_1 = \bar{n}\varphi\sigma_1$$

for all n. This completes the proof.

Proposition 9.27 may also be proved directly by making use of the Recursion Theorem 9.23. A parametrized version can be established, too.

EXERCISES TO CHAPTER 9

Exercise 9.1. Give an alternative proof of 9.10 based on 6.16 and exercise 6.2 instead of exercise 6.1i and proposition 6.30.

Hint. Show that $\mu\theta.\varphi[\psi(I,\theta\chi)] = \varphi\bar{1}[\sigma]$, where $\sigma = \mu\theta.((\bar{0}, \psi(\bar{2}, \varphi\bar{4})), \bar{1}, \chi\bar{2}, \theta R^3)$. To that end, supposing $\varphi[\psi(I,\tau\chi)] \leq \tau$ and writing ρ for $[\psi(I,\tau\chi)]$, prove that $R[\sigma] \leq \mu\theta.(\rho, \rho, \chi\rho, \theta\chi\rho)$. (If μA_1 is allowed, then one does not need exercise 6.2 to get the last inequality.)

Exercise 9.2. Let $I^* \in \mathscr{F}$ satisfy condition (£)* of exercise 6.8*. Using the Parametrized Recursion Theorem and exercises 6.1h, 6.10, give an alternative proof of 9.12*.

Hint. The operation Δ^* of exercise 6.10 exists and is recursive by the Parametrized Recursion Theorem and exercise 7.1. Take

$$\sigma = \Delta^*(((\bar{0}, \psi(R\bar{1}, \langle \varphi\bar{0}\bar{1} \rangle C(R\bar{1}, R^2))), \bar{0}\bar{1}), C(R\bar{1}, R^2)).$$

Using exercise 6.1h, show that $\varphi\bar{0}\bar{1}[\sigma]$ is a solution of $\varphi[\psi(I, \langle\theta\rangle)] = \theta$. Supposing $\varphi[\psi(I,\langle\tau\rangle)] \leq \tau$, take $\rho = [\psi(I,\langle\tau\rangle)]$ and prove that $R[\sigma] \leq \Delta^*((\rho,\rho),\rho)$.

Combining 9.4 and exercise 9.2 one gets a second proof of the First Recursion Theorem which depends on μA_0 plus (£)*. Another proof given in Ivanov [1980] assumes that $\langle\ \ \rangle$, $[\ \ ]$ satisfy the requirements of 5.13; there is also a proof due to J. Zashev which depends on similar assumptions and uses an ordinary coding of terms.

The following exercise establishes a parametrized version of 9.16*.

Exercise 9.3*. Let a system of inequalities

$$\Gamma_i(\theta_1,\ldots,\theta_{n+m}) \leq \theta_{n+i}, \quad 1 \leq i \leq m,$$

be given, and let $\Gamma_1,\ldots,\Gamma_m$ be recursive in $\mathscr{B}$. Show that there are n-ary mappings $\Gamma_1^*,\ldots,\Gamma_m^*$ recursive in $\mathscr{B}$ such that

$$\Gamma_i(\theta_1,\ldots,\theta_n, \Gamma_1^*(\theta_1,\ldots,\theta_n),\ldots,\Gamma_m^*(\theta_1,\ldots,\theta_n)) = \Gamma_i^*(\theta_1,\ldots,\theta_n)$$

for $1 \leq i \leq m$ and all $\theta_1,\ldots,\theta_n$, and if $\tau_1,\ldots,\tau_{n+m}$ satisfy the above system, then $\Gamma_i^*(\tau_1,\ldots,\tau_n) \leq \tau_{n+i}$, $1 \leq i \leq m$.

If the element U of exercise 7.8 is available, then even more general systems of inequalities can be solved. The idea of using least upper bounds is due to J. Zashev.

Exercise 9.4. Let U satisfy condition (1) of exercise 7.8 and suppose given a system of inequalities

$$\Gamma_j(\theta_0,\ldots,\theta_n) \leq \theta_{i_j}, \quad j \leq m,$$

with $i_0,\ldots,i_m \leq n$ and suppose that $\Gamma_0,\ldots,\Gamma_m$ are recursive in $\{U\} \cup \mathscr{B}$. Transform this system into an equivalent one introduced by mappings recursive in $\{U\} \cup \mathscr{B}$ and such that each θ_i appears at most once on the right.

Hint. Show that each pair of inequalities

$$\Gamma_k(\theta_0,\ldots,\theta_n) \leq \theta_i, \quad \Gamma_l(\theta_0,\ldots,\theta_n) \leq \theta_i$$

is equivalent to the single inequality

$$U(\Gamma_k(\theta_0,\ldots,\theta_n), \quad \Gamma_l(\theta_0,\ldots,\theta_n)) \leq \theta_i.$$

It is worth mentionning that the least solution turns an inequality $\Gamma_j(\theta_0,\ldots,\theta_n) \leq \theta_{i_j}$ into an equality, provided θ_{i_j} occurs just once on the right side of the given system.

Let $\mathscr{B} \subseteq \mathscr{F}$ and $\mathscr{U}$ (respectively $\mathscr{M}$, stand for the set of all the elements (unary mappings) recursive in $\mathscr{B}$.

Exercise 9.5. Following the proof of 9.17, prove the Enumeration Theorem by means of the Recursion Theorem.

Hint. Let $\mathscr{B} = \{\psi_1,\ldots,\psi_m\}$, $\psi = (L, R, \psi_1,\ldots,\psi_m, L)$ and $\rho = C([I]L, R)$. The mapping

$$\Gamma = \lambda\theta . Q(\psi, Q\langle\theta\rangle\theta, QT\theta, G\langle\theta\rangle, \rho[D\langle\theta\rangle\rho], L)$$

is recursive in $\mathscr{B}$, hence it has a fixed point σ recursive in $\mathscr{B}$. Show that σ is universal for $\mathscr{U}$.

Exercise 9.6. Show that whenever Σ is universal for $\mathscr{M}$, then $\Sigma(\psi)$ is universal for the elements recursive in $\{\psi\} \cup \mathscr{B}$. In particular, $\Sigma(I)$ is universal for $\mathscr{U}$.

Exercise 9.7. Show that $\mathscr{U}$ (respectively $\mathscr{M}$) has a universal element $\sigma \in \mathscr{U}$ (mapping $\Sigma \in \mathscr{M}$) iff the set $\mathscr{B}$ is finitely generated, i.e. all the members of $\mathscr{B}$ are recursive in a finite subset $\mathscr{B}_1$ of $\mathscr{B}$.

Both the Recursion Theorem 9.23 and its parametrized version can be reduced by 7.20 to the following assertion: Whenever Γ_1 is a unary mapping recursive in $\mathscr{B}$, then there is a unary mapping Γ recursive in $\mathscr{B}$ such that $\Gamma_1((\theta, \Gamma(\theta))) = \Gamma(\theta)$ for all θ. One may put the question whether more complicated equalities can be solved. For instance, is there a mapping Γ recursive in $\mathscr{B}$ such that $\Gamma_1((\theta, \Gamma(\Gamma_2(\theta)))) = \Gamma(\theta)$ for all θ, provided Γ_1, Γ_2 are recursive in $\mathscr{B}$?

Speaking somewhat informally, let $\Gamma_1,\ldots,\Gamma_m$ be unary mappings over $\mathscr{F}$, let Γ stand for a unary mapping and let $\mathscr{V}(\theta, \Gamma)$ be an expression constructed by composing $\Pi, \Gamma, \Gamma_1, \ldots, \Gamma_m$. (The pairing operation is added to compensate for the unarity of the mappings.) The following exercise establishes a Generalized Recursion Theorem.

Exercise 9.8. Let $\Gamma_1, \ldots, \Gamma_m$ be recursive in $\mathscr{B}$. Prove that there is a mapping Γ recursive in $\mathscr{B}$ such that $\mathscr{V}(\theta, \Gamma) = \Gamma(\theta)$ for all θ.

Hint. The mappings $\Gamma_1, \ldots, \Gamma_m$ are recursive in a finite subset $\mathscr{B}_1$ of $\mathscr{B}$. Take a mapping Σ recursive in $\mathscr{B}_1$ and principal universal for the unary mappings recursive in $\mathscr{B}_1$, and consider the expression $\mathscr{V}_1(\theta, \theta_1, \Gamma)$ obtained from $\mathscr{V}(\theta, \Gamma)$ by substituting $\theta_1 \Gamma$ for Γ. The mapping $\Gamma^* = \lambda \theta \theta_1 . \mathscr{V}_1(\theta, \theta_1, \Sigma)$ is recursive in $\mathscr{B}_1$; hence there is by the Parametrized Transition Theorem a mapping Γ^{**} recursive in $\mathscr{B}_1$ such that $\bar{n}\Gamma^{**}(\theta, \theta_1) = \Gamma^*(\theta, \bar{n}\theta_1)$ for all n, θ, θ_1. There is n such that $\bar{n}\Gamma^{**}(\theta, I) = \bar{n}\Sigma(\theta)$ for all θ by the Parametrized Second Recursion Theorem. Taking $\Gamma = \lambda\theta.\bar{n}\Sigma(\theta)$, one gets

$$\mathscr{V}(\theta, \Gamma) = \mathscr{V}_1(\theta, \bar{n}, \Sigma) = \Gamma^*(\theta, \bar{n}) = \bar{n}\Gamma^{**}(\theta, I) = \bar{n}\Sigma(\theta) = \Gamma(\theta)$$

for all θ, which completes the proof.

The above theorem can obviously be restated for n-ary mappings as follows. If $\Gamma_1, \ldots, \Gamma_m$ are mappings over $\mathscr{F}$ recursive in $\mathscr{B}$, Γ stands for a n-ary mapping and the expression $\mathscr{V}(\theta_1, \ldots, \theta_n, \Gamma)$ is constructed by composing $\Gamma, \Gamma_1, \ldots, \Gamma_m$, then there is a mapping Γ recursive in $\mathscr{B}$ such that $\mathscr{V}(\theta_1, \ldots, \theta_n, \Gamma) = \Gamma(\theta_1, \ldots, \theta_n)$ for all $\theta_1, \ldots, \theta_n$.

A generalized First Recursion Theorem established in the exercises to chapter 12 will show that moreover, the equalities considered above have solutions Γ recursive in $\mathscr{B}$ which are least with respect to the pointwise partial order.

Exercise 9.9. Let $\Gamma: \mathscr{F} \to \mathscr{F}$ be recursive in $\mathscr{B}$. Construct a mapping Γ^* recursive in $\mathscr{B}$ such that $\bar{n}\Gamma^*(\theta) = \Gamma^n(\theta)$ for all n, θ.

Hint. Use exercise 9.8 to obtain a mapping Γ^* recursive in $\mathscr{B}$ such that $\Gamma^*(\theta) = (\theta, \Gamma^*(\Gamma(\theta)))$ for all θ.

The following exercise strengthens both the Second Recursion Theorem and the Recursion Theorem.

Exercise 9.10. Let $\sigma \in \mathscr{U}$ be principal universal for $\mathscr{U}$. Prove that for every binary mapping Γ recursive in $\mathscr{B}$ there is a natural primitive recursive element ψ such that $\Gamma(\bar{n}\psi, \bar{n}) = \bar{n}\psi\sigma$ for all n.

Hint. Applying the Parametrized Transition Theorem twice, obtain a $\rho \in \mathscr{U}$ such that $\Gamma(\bar{m}, \bar{n}) = \bar{m}\bar{n}\rho$ for all m, n. Take natural primitive recursive elements ψ_0, ψ_1 such that $\bar{n}Q\sigma = \bar{n}\psi_0\sigma, \bar{n}\langle DP\psi_0 \rangle \rho = \bar{n}\psi_1\sigma$ for all n, then take $\psi = \psi_1 DP\psi_0$ and show that $\bar{n}\psi\bar{n}\rho = \bar{n}\psi\sigma$ for all n.

PART C

Diversified recursion theory on operative spaces

CHAPTER 10

Translation-like operations

An IOS $\mathscr{S}=(\mathscr{F}, I, \Pi, L, R)$ has four initial operations, namely $\circ$, Π, $\langle\ \ \rangle$, $[\ \ ]$. Since our intention when choosing them in chapter 2 was to get a minimal collection, it may happen in some spaces that there exist wider collections of initial operations which also support a sensible recursion theory.

On the other hand, the role played so far by the operation $\langle\ \ \rangle$ suggests that similar operations are 'nice' from the recursion theoretic viewpoint. Roughly speaking, such a translation-like operation $\langle\!\langle\ \ \rangle\!\rangle$ maps $\mathscr{F}$ injectively into itself, transforming $\circ$, Π, $\langle\ \ \rangle$, $[\ \ ]$, $\langle\!\langle\ \ \rangle\!\rangle$ into $\langle\!\langle\ \ \rangle\!\rangle$-free operations, e.g. $\langle\!\langle \varphi\psi \rangle\!\rangle = \Gamma(\langle\!\langle \varphi \rangle\!\rangle, \langle\!\langle \psi \rangle\!\rangle)$ for a certain mapping Γ recursive in $\mathscr{F}$.

The present chapter shows that a good deal of the theory developed in chapter 9 can actually be reestablished in this more complicated situation. Of particular interest is the axiomatically defined operation *St*. An example of a translation-like operation is given in the exercises below, while others are considered in chapters 21, 22, 24, 25, 28, 30. The concept of translation-like operation is important for the considerations in chapters 14, 15 and crucial for the study of Skordev combinatory spaces in chapter 27.

Formally, assume that a monotonic operation $\langle\!\langle\ \ \rangle\!\rangle:\mathscr{F}\to\mathscr{F}$ and fixed elements $K_0,\ldots,K_m \in \mathscr{F}$ are given together with mappings Σ_0, $\Sigma_1:\mathscr{F}^2\to\mathscr{F}$, $\Sigma_2-\Sigma_5:\mathscr{F}\to\mathscr{F}$ recursive in $\mathscr{B}_0=\{K_0,\ldots,K_m, \langle\!\langle K_0\rangle\!\rangle,\ldots,\langle\!\langle K_m\rangle\!\rangle\}$ such that

(0) $\langle\!\langle \varphi\psi \rangle\!\rangle = \Sigma_0(\langle\!\langle \varphi \rangle\!\rangle, \langle\!\langle \psi \rangle\!\rangle)$,

(1) $\langle\!\langle (\varphi,\psi) \rangle\!\rangle = \Sigma_1(\langle\!\langle \varphi \rangle\!\rangle, \langle\!\langle \psi \rangle\!\rangle)$,

(2) $\langle\!\langle \langle \varphi \rangle \rangle\!\rangle = \Sigma_2(\langle\!\langle \varphi \rangle\!\rangle)$,

(3) $\langle\!\langle [\varphi] \rangle\!\rangle = \Sigma_3(\langle\!\langle \varphi \rangle\!\rangle)$,

(4) $\langle\!\langle \langle\!\langle \varphi \rangle\!\rangle \rangle\!\rangle = \Sigma_4(\langle\!\langle \varphi \rangle\!\rangle)$,

(5) $\varphi = \Sigma_5(\langle\!\langle \varphi \rangle\!\rangle)$

for all φ, ψ. Under these conditions $\langle\!\langle\ \ \rangle\!\rangle$ is said to be a t-*operation*. It is convenient to include I, L, R in the list $K_0,\ldots,K_m$. On the other hand, members of $\mathscr{B}_0$ will sometimes be dropped, provided they can be obtained from remaining ones by means of the basic IOS-operations.

The operation $\langle\ \ \rangle$ is itself a t-operation with mappings $\Sigma_0 = \lambda\theta_1\theta_2.\theta_1\theta_2$, $\Sigma_1=\lambda\theta_1\theta_2.C(\theta_1,\theta_2)$, $\Sigma_2=\Sigma_4=\lambda\theta.P\theta Q$, $\Sigma_3=\lambda\theta.C[\theta C]$, $\Sigma_5=\lambda\theta.L\theta[L]$ which are prime recursive in C, P, Q.

It would appear that the basic features of t-operations are (0), (1), (4).

Actually, whenever $\langle\!\langle\ \ \rangle\!\rangle$ satisfies (0)–(4), then it is easily seen that $\langle\!\langle\ \ \rangle\!\rangle_1 = \lambda\theta.(\theta, \langle\!\langle\theta\rangle\!\rangle)$ satisfies (0)–(5), while it will be shown below that (0), (1) imply (2), (3) by means of a modified μ-axiom.

An element φ is t-*recursive in* $\mathscr{B} \subseteq \mathscr{F}$ if

$$\varphi \in cl(\mathscr{B}_0 \cup \mathscr{B}/\circ, \Pi, \langle\ \ \rangle, [\ \ ], \langle\!\langle\ \ \rangle\!\rangle)$$

and similarly for mappings. The notions of prime recursiveness etc. are also extended in this way.

The following Parametrized Pull Back Theorem plays a key role here by ensuring that all t-recursive mappings have the property given in (0)–(4) for the initial operations $\circ$, Π, $\langle\ \ \rangle$, $[\ \ ]$, $\langle\!\langle\ \ \rangle\!\rangle$.

Proposition 10.1. If Γ is a unary mapping t-recursive in $\mathscr{B}$, then there is a mapping Γ^* recursive in $\mathscr{B}_0 \cup \langle\!\langle\mathscr{B}\rangle\!\rangle$ such that $\Gamma = \lambda\theta.\Gamma^*(\langle\!\langle\theta\rangle\!\rangle)$.

Proof. It suffices to construct a mapping Γ^{**} recursive in $\mathscr{B}_0 \cup \langle\!\langle\mathscr{B}\rangle\!\rangle$ such that $\langle\!\langle\Gamma(\theta)\rangle\!\rangle = \Gamma^{**}(\langle\!\langle\theta\rangle\!\rangle)$ for all θ and then take $\Gamma^* = \lambda\theta.\Sigma_5(\Gamma^{**}(\theta))$.

Take $(\lambda\theta.\theta)^{**} = \lambda\theta.\theta$, $(\lambda\theta.\psi)^{**} = \lambda\theta.\langle\!\langle\psi\rangle\!\rangle$ for all $\psi \in \{K_0, \dots, K_m\} \cup \mathscr{B}$, $(\lambda\theta.\langle\!\langle\psi\rangle\!\rangle)^{**} = \lambda\theta.\Sigma_4(\langle\!\langle\psi\rangle\!\rangle)$ for all $\psi \in \{K_0, \dots, K_m\}$.

Let Γ_1^{**}, Γ_2^{**} correspond respectively to Γ_1, Γ_2. Then take

$$\begin{aligned}
(\lambda\theta.\Gamma_1(\theta)\Gamma_2(\theta))^{**} &= \lambda\theta.\Sigma_0(\Gamma_1^{**}(\theta), \Gamma_2^{**}(\theta)),\\
(\lambda\theta.(\Gamma_1(\theta), \Gamma_2(\theta)))^{**} &= \lambda\theta.\Sigma_1(\Gamma_1^{**}(\theta), \Gamma_2^{**}(\theta)),\\
(\lambda\theta.\langle\Gamma_1(\theta)\rangle)^{**} &= \lambda\theta.\Sigma_2(\Gamma_1^{**}(\theta)),\\
(\lambda\theta.[\Gamma_1(\theta)])^{**} &= \lambda\theta.\Sigma_3(\Gamma_1^{**}(\theta)),\\
(\lambda\theta.\langle\!\langle\Gamma_1(\theta)\rangle\!\rangle)^{**} &= \lambda\theta.\Sigma_4(\Gamma_1^{**}(\theta)).
\end{aligned}$$

This completes the proof.

As an easy corollary we get a Pull Back Theorem for elements.

Proposition 10.2. An element φ is t-recursive in $\mathscr{B}$ iff it is recursive in $\mathscr{B}_0 \cup \langle\!\langle\mathscr{B}\rangle\!\rangle$. In particular, φ is t-recursive iff it is recursive in $\mathscr{B}_0$.

Proof. If φ is t-recursive in $\mathscr{B}$, then so is the mapping $\Gamma = \lambda\theta.\varphi$. Take Γ^* to correspond to Γ by 10.1. Then $\varphi = \Gamma^*(\langle\!\langle I\rangle\!\rangle)$; hence φ is recursive in $\mathscr{B}_0 \cup \langle\!\langle\mathscr{B}\rangle\!\rangle$, which completes the proof.

Relative t-recursiveness possesses the ordinary properties stated at the beginning of chapter 7. Certain specific characterizations may be obtained by 7.11, 10.2. Analogs to 7.18–7.20 for mappings t-recursive in $\mathscr{B}$ may also be established.

The Pull Back Theorems and the normal form results 9.3–9.7 imply corresponding Normal t-Form Theorems.

Proposition 10.3. If φ is t-recursive in $\mathscr{B}$, then $\varphi = \bar{1}[\sigma]$ with a certain σ strictly primitive in $\mathscr{B}_0 \cup \langle\!\langle\mathscr{B}\rangle\!\rangle$. This follows from 10.2, 9.3.

Proposition 10.4. If Γ is a n-ary mapping t-recursive in $\mathscr{B}$, then

$$\Gamma = \lambda\theta_1 \dots \theta_n.\bar{1}[\varphi(I, \langle\langle\!\langle\theta_1\rangle\!\rangle\rangle, \dots, \langle\langle\!\langle\theta_n\rangle\!\rangle\rangle)]$$

with a certain φ strictly primitive in $\mathscr{B}_0 \cup \langle\!\langle\mathscr{B}\rangle\!\rangle$.

Proof. Proposition 10.1 and an analogue to 7.20 imply that there is an n-ary mapping Γ^* recursive in $\mathscr{B}_0 \cup \langle \mathscr{B} \rangle$ such that $\Gamma = \lambda\theta_1 \ldots \theta_n.\ \Gamma^*(\langle \theta_1 \rangle, \ldots, \langle \theta_n \rangle)$. Applying 9.4 to Γ^* we get the desired normal form.

Two t-Enumeration Theorems follow.

Proposition 10.5. If $\mathscr{B}$ is finite, then there is a unary mapping Σ t-recursive in $\mathscr{B}$ and universal for the unary mappings t-recursive in $\mathscr{B}$.

Proof. Take a unary mapping Σ^* recursive in $\mathscr{B}_0 \cup \langle \mathscr{B} \rangle$ and universal for the unary mappings recursive in $\mathscr{B}_0 \cup \langle \mathscr{B} \rangle$ by 9.19. Then, using 10.1 we see that $\Sigma = \lambda\theta.\Sigma^*(\langle \theta \rangle)$ is universal for the unary mappings t-recursive in $\mathscr{B}$.

Proposition 10.6. If $\mathscr{B}$ is finite, then there is an element σ t-recursive in $\mathscr{B}$ and universal for the elements t-recursive in $\mathscr{B}$.

Proof. Take the mapping Σ of the previous statement. The element $\sigma = \Sigma(I)$ is t-recursive in $\mathscr{B}$ and whenever φ is t-recursive in $\mathscr{B}$, then so is $\Gamma = \lambda\theta.\varphi$; hence there is n such that $\Gamma = \lambda\theta.\bar{n}\Sigma(\theta)$, which implies $\varphi = \bar{n}\sigma$. This completes the proof.

Let $\mathscr{B}$ be a fixed subset of $\mathscr{F}$, let $\mathscr{U}$ stand for the set of all elements t-recursive in $\mathscr{B}$ and $\mathscr{M}$ for the set of all unary mappings t-recursive in $\mathscr{B}$. Notions of element and mapping principal universal respectively for $\mathscr{U}$, $\mathscr{M}$ are introduced by the same definitions as in chapter 9. All the further statements and proofs of that chapter then remain valid in the present context. However, the analog to the Recursion Theorem 9.23 requires the following t-Transition Theorem.

Proposition 10.7. Let Γ be a unary mapping t-recursive in $\mathscr{B}$. Then there is a mapping Γ^* t-recursive in $\mathscr{B}$ such that $\Gamma^*(\theta) = (\Gamma(L\theta), \Gamma^*(R\theta))$ for all θ. In particular, $\bar{n}\Gamma^*(\theta) = \Gamma(\bar{n}\theta)$ for all n, θ.

Proof. It suffices to consider the mapping $\langle\ \ \rangle$. Let a t-recursive mapping $\langle\ \ \rangle^*$ correspond to $\langle\ \ \rangle$ such that $\langle \theta \rangle^* = (\langle L\theta \rangle, \langle R\theta \rangle^*)$ for all θ. Then, given a unary mapping Γ t-recursive in $\mathscr{B}$, take Γ_1 to correspond to Γ by 10.1, Γ_1^* to correspond to Γ_1 by 7.21 and finally take $\Gamma^* = \lambda\theta.\Gamma_1^*(\langle \theta \rangle^*)$. It follows that

$$\begin{aligned}\Gamma^*(\theta) &= (\Gamma_1(L\langle \theta \rangle^*), \Gamma_1^*(R\langle \theta \rangle^*)) \\ &= (\Gamma_1(\langle L\theta \rangle), \Gamma_1^*(\langle R\theta \rangle^*)) = (\Gamma(L\theta), \Gamma^*(R\theta))\end{aligned}$$

for all θ; hence Γ^* corresponds to Γ.

The mappings $\Gamma_1 = \lambda\theta.\Sigma_0(\langle L \rangle, \theta)$, $\Gamma_2 = \lambda\theta.\Sigma_0(\langle R \rangle, \theta)$ are recursive in $\mathscr{B}_0$. Take a mapping Γ_1^* to correspond to Γ_1 by 7.21 and a mapping Γ_3 recursive in $\mathscr{B}_0$ such that $\Gamma_3(\theta) = (\theta, \Gamma_3(\Gamma_2(\theta)))$ for all θ by exercise 9.8, then take $\langle\ \ \rangle^* = \lambda\theta.\Gamma_1^*(\Gamma_3(\langle \theta \rangle))$. It follows that

$$\begin{aligned}\langle \theta \rangle^* = \Gamma_1^*(\Gamma_3(\langle \theta \rangle)) &= (\Gamma_1(L\Gamma_3(\langle \theta \rangle)), \Gamma_1^*(R\Gamma_3(\langle \theta \rangle))) \\ &= (\Gamma_1(\langle \theta \rangle), \Gamma_1^*(\Gamma_3(\Gamma_2(\langle \theta \rangle)))) \\ &= (\langle L\theta \rangle, \Gamma_1^*(\Gamma_3(\langle R\theta \rangle))) \\ &= (\langle L\theta \rangle, \langle R\theta \rangle^*)\end{aligned}$$

for all θ, which completes the proof.

A Generalized t-Recursion Theorem may also be established by adapting the hint to exercise 9.8. To obtain a First t-Recursion Theorem, however, one has to employ a stronger μ-axiom.

The notion of t-*inductive mapping* is introduced by modifying the definition given in chapter 5. Namely, replace $\{I, L, R\}$ by $\mathscr{B}_0$ in the first clause and add to the second clause the following case of $\langle\!\langle\ \rangle\!\rangle$:

If $\Gamma_1 : \mathscr{F}^n \to \mathscr{F}$ is t-inductive, then so is $\Gamma = \lambda\theta_1 \ldots \theta_n . \langle\!\langle \Gamma_1(\theta_1, \ldots, \theta_n) \rangle\!\rangle$.

A t-*simple segment* is a subset of $\mathscr{F}$ of the form $\{\theta / \langle\!\langle \theta \rangle\!\rangle \leq \Sigma_6(\tau)\}$, where Σ_6 is a given mapping recursive in $\mathscr{B}_0$ such that $\Sigma_6(\langle\!\langle \theta \rangle\!\rangle) = \langle\!\langle \theta \rangle\!\rangle$ for all θ.

Axiom tμA. For any $n + 1$-ary t-inductive mapping Γ and any $\theta_1, \ldots, \theta_n$ the inequality $\Gamma(\theta_1, \ldots, \theta_n, \theta) \leq \theta$ has a solution which is a member of all t-simple segments closed under the mapping $\lambda\theta . \Gamma(\theta_1, \ldots, \theta_n, \theta)$.

The axioms μA_1, μA_2, μA_3 are now completed by axiom tμA,

$$\mathrm{t}\mu A_i = \mu A_i + \mathrm{t}\mu A, \quad i = 1, 2, 3.$$

The use of a modified μ-axiom will be indicated by the corresponding number of asterisks.

We now prove the First t-Recursion Theorem.

Proposition 10.8*. If Γ is a unary mapping t-recursive in $\mathscr{B}$, then the element $\mu\theta . \Gamma(\theta)$ exists and is t-recursive in $\mathscr{B}$.

Proof. We modify the translation method of the proof of 9.12* so that the role of $\langle\ \rangle$ is played by $\langle\!\langle\ \rangle\!\rangle$.

Let θ_0 be the solution to $\Gamma(\theta) \leq \theta$ assumed by tμA, let Γ^* correspond to $\lambda\theta . \langle\!\langle \Gamma(\theta) \rangle\!\rangle$ by 10.1 and let $\Gamma_1 = \lambda\theta . \Sigma_6(\Gamma^*(\theta))$. Then Γ_1 is recursive in $\mathscr{B}_0 \cup \langle\!\langle \mathscr{B} \rangle\!\rangle$, hence $\theta_1 = \mu\theta . \Gamma_1(\theta)$ exists and is recursive in $\mathscr{B}_0 \cup \langle\!\langle \mathscr{B} \rangle\!\rangle$ by 9.13*.

It follows that

$$\langle\!\langle \Gamma(\theta) \rangle\!\rangle = \Sigma_6(\langle\!\langle \Gamma(\theta) \rangle\!\rangle) = \Sigma_6(\Gamma^*(\langle\!\langle \theta \rangle\!\rangle)) = \Gamma_1(\langle\!\langle \theta \rangle\!\rangle)$$

for all θ. In particular, $\Gamma_1(\langle\!\langle \theta_0 \rangle\!\rangle) = \langle\!\langle \Gamma(\theta_0) \rangle\!\rangle \leq \langle\!\langle \theta_0 \rangle\!\rangle$ implies $\theta_1 \leq \langle\!\langle \theta_0 \rangle\!\rangle$. On the other hand, consider the t-simple segment

$$\mathscr{E} = \{\theta / \langle\!\langle \theta \rangle\!\rangle \leq \theta_1\} = \{\theta / \langle\!\langle \theta \rangle\!\rangle \leq \Sigma_6(\Gamma^*(\theta_1))\}.$$

If $\theta \in \mathscr{E}$, then

$$\langle\!\langle \Gamma(\theta) \rangle\!\rangle = \Gamma_1(\langle\!\langle \theta \rangle\!\rangle) \leq \Gamma_1(\theta_1) = \theta_1,$$

hence $\langle\!\langle \theta_0 \rangle\!\rangle \in \mathscr{E}$ by tμA. Therefore, $\langle\!\langle \theta_0 \rangle\!\rangle = \theta_1$, hence $\theta_0 = \Sigma_5(\theta_1)$ and θ_0 is t-recursive in $\mathscr{B}$.

We still have to show that $\theta_0 = \mu\theta . \Gamma(\theta)$. If $\Gamma(\tau) \leq \tau$, we get $\Gamma_1(\langle\!\langle \tau \rangle\!\rangle) \leq \langle\!\langle \tau \rangle\!\rangle$, hence $\theta_1 \leq \langle\!\langle \tau \rangle\!\rangle$ which implies $\theta_0 = \Sigma_5(\theta_1) \leq \Sigma_5(\langle\!\langle \tau \rangle\!\rangle) = \tau$. This completes the proof.

Analogues to 9.14*–9.16* and exercises 9.3*, 9.4 may also be established for t-recursiveness. In particular, it follows from the analogue to 9.15* and the proof of 10.8* that the solution to the inequality $\Gamma(\theta_1, \ldots, \theta_n, \theta) \leq \theta$ assumed in tμA is exactly $\mu\theta . \Gamma(\theta_1, \ldots, \theta_n, \theta)$. (This fact will be used in the proofs of 10.9*, 10.10*.)

The following two statements show that the existence of mappings Σ_0, Σ_1 implies the existence of Σ_2, Σ_3 by tμA$_1$.

Proposition 10.9*. Let Σ_0, Σ_1 satisfy (0), (1). Then there is a unary mapping Σ_2 recursive in $\mathscr{B}_0$ satisfying (2).

Proof. Take

$$\Gamma = \lambda\theta\theta_1.\Sigma_6(\Sigma_1(\Sigma_0(\theta, \langle\!\langle L\rangle\!\rangle), \Sigma_0(\theta_1, \langle\!\langle R\rangle\!\rangle)))$$

and $\Sigma_2 = \lambda\theta.\mu\theta_1.\Gamma(\theta, \theta_1)$. The mapping Γ is recursive in $\mathscr{B}_0$, hence so is Σ_2 by 9.14*. It follows that

$$\Gamma(\langle\!\langle\varphi\rangle\!\rangle, \langle\!\langle\langle\varphi\rangle\rangle\!\rangle) = \Sigma_6(\langle\!\langle(\varphi L, \langle\varphi\rangle R)\rangle\!\rangle) = \Sigma_6(\langle\!\langle\langle\varphi\rangle\rangle\!\rangle) = \langle\!\langle\langle\varphi\rangle\rangle\!\rangle,$$

hence $\Sigma_2(\langle\!\langle\varphi\rangle\!\rangle) \leq \langle\!\langle\langle\varphi\rangle\rangle\!\rangle$. On the other hand, consider the t-simple segment $\mathscr{E} = \{\theta/\langle\!\langle\theta\rangle\!\rangle \leq \Sigma_2(\langle\!\langle\varphi\rangle\!\rangle)\}$. If $\theta \in \mathscr{E}$, then

$$\langle\!\langle(\varphi L, \theta R)\rangle\!\rangle = \Gamma(\langle\!\langle\varphi\rangle\!\rangle, \langle\!\langle\theta\rangle\!\rangle) \leq \Gamma(\langle\!\langle\varphi\rangle\!\rangle, \Sigma_2(\langle\!\langle\varphi\rangle\!\rangle)) = \Sigma_2(\langle\!\langle\varphi\rangle\!\rangle),$$

hence $\langle\!\langle\langle\varphi\rangle\rangle\!\rangle \leq \Sigma_2(\langle\!\langle\varphi\rangle\!\rangle)$ by tμA. This completed the proof.

Proposition 10.10*. Let Σ_0, Σ_1 satisfy (0), (1). Then there is a mapping Σ_3 recursive in $\mathscr{B}_0$ satisfying (3).

The proof repeats that of 10.9* with

$$\Gamma = \lambda\theta\theta_1.\Sigma_6(\Sigma_1(\langle\!\langle I\rangle\!\rangle, \Sigma_0(\theta, \theta_1)))$$

and $\Sigma_3 = \lambda\theta.\mu\theta_1.\Gamma(\theta, \theta_1)$.

We bring this chapter to a close by introducing a specific t-operation which subsumes almost all the t-operations studied in the book.

Let $\mathscr{L}$ be a nonempty subset of $\mathscr{F}$, $St: \mathscr{F} \to \mathscr{F}$ (*storing operation*), K_0, K_1, $K_2 \in \mathscr{F}$ and suppose that the following axioms be satisfied with x, y ranging over $\mathscr{L}$.

$$\text{sA}_1.\ xK_0(L, R) = (xL, xR).$$

$$\text{sA}_2.\ xyK_1 \in \mathscr{L}, \quad xyK_1K_2 = xy.$$

$$(\$)\ xSt(\varphi) = \varphi x,$$

$$\forall x(\varphi x\psi \leq \rho x\sigma) \Rightarrow St(\varphi)\psi \leq St(\rho)\sigma.$$

One may assume without loss of generality that $K_0 = St(I)K_0$. For otherwise $St(I)K_0$ satisfies sA$_1$ and $St(I)K_0 = St(I)St(I)K_0$ by (\$), hence K_0 can be replaced by $St(I)K_0$. Similarly, we assume $K_1 = St(St(I))K_1$.

Propositions 6.35, 5.6, 5.7** and exercise 6.6 imply that $\langle\ \ \rangle$ is a storing operation with $\mathscr{L} = \{\bar{n}/n \in \omega\}$, $K_0 = C$, $K_1 = P$ and $K_2 = Q$.

We now establish some properties of the storing operation.

Proposition 10.11. *St* is monotonic.

Proof. If $\varphi \leq \psi$, then $\varphi x \leq \psi x$ for all x; hence $St(\varphi) \leq St(\psi)$ by (\$).

Proposition 10.12. $St(\varphi\psi) = St(\varphi)St(\psi)$.

Proof. We have $\psi x = xSt(\psi)$; hence $\varphi\psi x = \varphi x St(\psi)$ for all x, which implies $St(\varphi\psi) = St(\varphi)St(\psi)$ by (\$).

Proposition 10.13. $St((\varphi,\psi))=K_0(St(\varphi),St(\psi))$.

Proof. The equalities

$$(\varphi,\psi)x=(\varphi x,\psi x)=(xSt(\varphi),xSt(\psi))=xK_0(St(\varphi),St(\psi))$$

imply $St((\varphi,\psi))=K_0(St(\varphi),St(\psi))$ by ($).

Proposition 10.14. There is a mapping Σ_2 recursive in $K_0, St(L), St(R)$ such that $St(\langle\varphi\rangle)=\Sigma_2(St(\varphi))$ for all φ.

Proof. The mapping

$$\Sigma_2=\lambda\theta.\mu\theta_1.K_0(\theta St(L),\theta_1 St(R))$$

exists and is recursive in K_0, $St(L)$, $St(R)$ by 6.38, 6.39. It follows that

$$K_0(St(\varphi)St(L),St(\langle\varphi\rangle)St(R))=St((\varphi L,\langle\varphi\rangle R))=St(\langle\varphi\rangle),$$

hence $\Sigma_2(St(\varphi))\le St(\langle\varphi\rangle)$. On the other hand, $Rx=xSt(R)$ and

$$(\varphi Lx,x\Sigma_2(St(\varphi))St(R))=xK_0(St(\varphi)St(L),\Sigma_2(St(\varphi))St(R))=x\Sigma_2(St(\varphi))$$

imply $\langle\varphi\rangle x\le x\Sigma_2(St(\varphi))$ for all x by (£); hence

$$St(\langle\varphi\rangle)\le St(I)\Sigma_2(St(\varphi))=\Sigma_2(St(\varphi))$$

by ($). The proof is complete.

Proposition 10.15. $St([\varphi])=K_0[St(\varphi)K_0]St(I)$.

Proof. We have

$$K_0(St(I),St(\varphi)St([\varphi]))=St((I,\varphi[\varphi]))=St([\varphi]),$$

hence $K_0[St(\varphi)K_0]St(I)\le St([\varphi])$ by 6.10. On the other hand,

$$\begin{aligned}(x,\varphi xK_0[St(\varphi)K_0]St(I))&=(xSt(I),xSt(\varphi)K_0[St(\varphi)K_0]St(I))\\&=xK_0(I,St(\varphi)K_0[St(\varphi)K_0])St(I)\\&=xK_0[St(\varphi)K_0]St(I);\end{aligned}$$

hence $[\varphi]x\le xK_0[St(\varphi)K_0]St(I)$ for all x by (££). Therefore, $St([\varphi])\le K_0[St(\varphi)K_0]St(I)$ by ($), which completes the proof.

Proposition 10.16. $St(St(\varphi))=K_1St(\varphi)K_2$.

Proof. We have

$$\varphi xy=\varphi xyK_1K_2=xyK_1St(\varphi)K_2$$

for all x, y; hence $St(\varphi)y=St(I)yK_1St(\varphi)K_2$ for all y, which implies

$$St(St(\varphi))=St(St(I))K_1St(\varphi)K_2=K_1St(\varphi)K_2.$$

Proposition 10.17 (Storing Operation Theorem). The operation $\langle\ \rangle=\lambda\varphi.(\varphi,St(\varphi))$ is a t-operation with functional elements K_0, K_1, K_2. (So the corresponding set $\mathscr{B}_0$ consists of L, R, K_0, K_1, K_2, $St(I)$, $St(L)$, $St(R)$, $St(K_0)$, $St(K_1)$, $St(K_2)$.)

This follows from 10.12–10.16.

Proposition 10.18. Let $I \in \mathscr{L} \circ \mathscr{F}$. Then St is a t-operation and taking $\Sigma_6 = \lambda\theta . St(I)\theta$, all t-simple segments are normal.

Proof. Let $I = K_3 K_4$ for certain $K_3 \in \mathscr{L}$, $K_4 \in \mathscr{F}$. Then $\varphi = \varphi K_3 K_4 = K_3 St(\varphi) K_4$ for all φ, hence St is a t-operation with functional elements $K_0 - K_4$ by 10.12–10.16. Every t-simple segment $\mathscr{E} = \{\theta / St(\theta) \leq St(I)\tau\}$ is normal since $\mathscr{E} = \{\theta / \forall x(\theta x \leq x\tau)\}$ by ($). The proof is complete.

Proposition 10.19.** Let $\psi_0, \psi_1 \in \mathscr{F}$ such that $\bar{n}\psi_0 \in \mathscr{L}$ and $\bar{n}\psi_0\psi_1 = \bar{n}$ for all n. Then the operation $\langle \quad \rangle$ can be expressed in terms of St and the elements ψ_0, ψ_1. (Assuming without loss of generality that $\psi_0 = \langle I \rangle \psi_0$.)

Since $\langle \quad \rangle$ is a storing operation, 10.19** is a particular instance of the following more general statement.

Proposition 10.20. Let St, St^* be storing operations with corresponding sets $\mathscr{L}$, $\mathscr{L}^*$ and let $\psi_0, \psi_1 \in \mathscr{F}$ such that $\mathscr{L}\psi_0 \subseteq \mathscr{L}^*$, $\psi_0\psi_1 = St(I)$. Then St can be expressed in terms of St^* and ψ_0, ψ_1. If $\mathscr{L}\psi_0 = \mathscr{L}^*$ also holds, then St^* can also be expressed in terms of St, ψ_0 and ψ_1. (Assuming without loss of generality that $\psi_0 = St(I)\psi_0$, $\psi_1 = St^*(I)\psi_1$.)

Proof. We have

$$\varphi x = \varphi x \psi_0 \psi_1 = x \psi_0 St^*(\varphi) \psi_1$$

for all $x \in \mathscr{L}$, hence

$$St(\varphi) = St(I)\psi_0 St^*(\varphi)\psi_1 = \psi_0 St^*(\varphi)\psi_1$$

by ($). If $\mathscr{L}\psi_0 = \mathscr{L}^*$, then for all $x^* \in \mathscr{L}^*$ there is a $x \in \mathscr{L}$ such that $x^* = x\psi_0$, hence

$$\varphi x^* = \varphi x \psi_0 = x St(\varphi)\psi_0 = x\psi_0\psi_1 St(\varphi)\psi_0 = x^*\psi_1 St(\varphi)\psi_0$$

which implies $St^*(\varphi) = \psi_1 St(\varphi)\psi_0$ by ($). This completes the proof.

EXERCISES TO CHAPTER 10

Exercise 10.1. Let $\mathscr{S}$ be the IOS of example 4.7 or 4.8 and $J: M^2 \to M$ be injective. Define $St(\varphi)$ by $St(\varphi)(J(s,t)) = J(s, \varphi(t))$ and $St(\varphi)(s)\uparrow$ otherwise. Prove that St is a t-operation and $\mathscr{S}$, St satisfy the axiom $t\mu A_3$.

Hint. Take $\mathscr{L} = \{\tilde{s}/s \in M\}$, where $\tilde{s} = \lambda t . J(s,t)$. Specify elements K_0, K_1, K_2 to satisfy sA_1, sA_2. Use exercise 5.3 and 10.18.

In spaces with t-operations the operation $\langle \quad \rangle$ is alsmost always expressible in terms of $\langle\!\langle \quad \rangle\!\rangle$, i.e. $\langle \quad \rangle$ is prime t-recursive in certain elements which can be added to $\mathscr{B}_0$. A useful criterion for this is provided by the proof of 5.13: Whenever the axiom μA_3 holds for $\lambda\theta_1\theta.(I, \theta_1\theta)$ and there is a prime t-recursive mapping $\langle \quad \rangle_1$ such that $\bar{n}\langle\varphi\rangle_1 = \varphi\bar{n}$ for all n, φ, then the operation $\langle \quad \rangle$ is prime t-recursive. Let us, accordingly, discuss this situation.

In the next four exercises the operation $\langle \quad \rangle$ is assumed to be prime t-recursive and the mappings $\Sigma_0, \Sigma_1, \Sigma_3 - \Sigma_5$ to be prime recursive in $\mathscr{B}_0$. (Σ_2 is no longer necessary.) It follows immediately that relative t-recursiveness and relative prime t-recursiveness are equivalent, both for elements and mappings.

Exercise 10.2. Prove that 10.1, 10.2 take place with 'prime recursive' substituted for 'recursive'. In particular, there is a mapping Σ_7 prime recursive in $\mathscr{B}_0$ such that $\langle\varphi\rangle = \Sigma_7(\langle\!\langle\varphi\rangle\!\rangle)$ for all φ.

The normal Form Theorems 10.3, 10.4 can be further specified.

Exercise 10.3. Prove that whenever φ is t-recursive in $\mathscr{B}$, then $\varphi = \bar{1}[\sigma]$ for a certain σ strictly polynomial in $\mathscr{B}_0 \cup \langle\!\langle\mathscr{B}\rangle\!\rangle$.

Hint. Use 9.2.

Exercise 10.4. Prove that whenever a unary mapping Γ is t-recursive in $\mathscr{B}$, then

$$\Gamma = \lambda\theta.\bar{1}[\varphi(I, \langle\!\langle\theta\rangle\!\rangle\bar{4}, \ldots, \langle\!\langle\theta\rangle\!\rangle\overline{n+3})]$$

with φ strictly polynomial in $\mathscr{B}_0 \cup \langle\!\langle\mathscr{B}\rangle\!\rangle$ and $n = c(\Sigma_4)c(\Sigma_7)$. In particular, $\Gamma = \lambda\theta.\bar{1}[\varphi(I, \langle\!\langle\theta\rangle\!\rangle\bar{4})]$ provided $c(\Sigma_4) = c(\Sigma_7) = 1$.

Hint. Use 10.4, 9.8.

Under certain assumptions, a First t-Recursion Theorem can be proved using only μA_0 and $t\mu A$.

Exercise 10.5. Let Σ_6 be prime recursive in $\mathscr{B}_0$ and suppose that $c(\Sigma_4) = c(\Sigma_6) = c(\Sigma_7) = 1$. Using $t\mu A$, prove that whenever a unary mapping Γ is t-recursive in $\mathscr{B}$, then the element $\mu\theta.\Gamma(\theta)$ exists and is t-recursive in $\mathscr{B}$.

Hint. The mapping $\lambda\theta.\langle\!\langle\Gamma(\theta)\rangle\!\rangle$ has a normal form $\lambda\theta.\bar{1}[\varphi(I, \langle\!\langle\theta\rangle\!\rangle\bar{4})]$ with φ t-recursive in $\mathscr{B}$ by exercise 10.4. The mapping $\Gamma_1 = \lambda\theta.\Sigma_6(\bar{1}[\varphi(I, \theta\bar{4})])$ is prime recursive in $\{\varphi\} \cup \mathscr{B}_0$ and $c(\Gamma_1) = 1$; hence the element $\theta_1 = \mu\theta.\Gamma_1(\theta)$ exists and is recursive in $\{\varphi\} \cup \mathscr{B}_0$ by 9.11. Following the proof of 10 8*, show that $\Sigma_5(\theta_1) = \mu\theta.\Gamma(\theta)$.

What happens if there is a second t-operation $\langle\!\langle\ \rangle\!\rangle^*$ over $\mathscr{F}$? Certainly, one would like to consider t*, t-recursiveness, i.e. to add both $\langle\!\langle\ \rangle\!\rangle, \langle\!\langle\ \rangle\!\rangle^*$ to the initial IOS-operations. One way of doing this is to proceed as in the present chapter with t-recursiveness playing the role of recursiveness and $\langle\!\langle\ \rangle\!\rangle^*$ playing the role of $\langle\!\langle\ \rangle\!\rangle$. There is also another option, namely to combine $\langle\!\langle\ \rangle\!\rangle, \langle\!\langle\ \rangle\!\rangle^*$ into a single t-operation, which allows a direct application of the theory already developed.

Exercise 10.6. Let $\langle\!\langle\ \rangle\!\rangle^*$ satisfy (0)–(5) with corresponding mappings $\Sigma_0^* - \Sigma_5^*$ t-recursive in a set $\mathscr{B}_0^*$ and suppose that there exists a mapping Σ^* t-recursive in $\mathscr{B}_0^*$ such that $\langle\!\langle\langle\!\langle\varphi\rangle\!\rangle\rangle\!\rangle^* = \Sigma^*(\langle\!\langle\varphi\rangle\!\rangle^*)$ for all φ. Show that $\langle\!\langle\ \rangle\!\rangle^{**} = \lambda\varphi.\langle\!\langle\langle\!\langle\varphi\rangle\!\rangle^*\rangle\!\rangle$ is a t-operation and there are mappings Γ, Γ^* recursive in $\mathscr{B}_0 \cup \langle\!\langle\mathscr{B}_0^*\rangle\!\rangle$ such that $\langle\!\langle\varphi\rangle\!\rangle = \Gamma(\langle\!\langle\varphi\rangle\!\rangle^{**})$, $\langle\!\langle\varphi\rangle\!\rangle^* = \Gamma^*(\langle\!\langle\varphi\rangle\!\rangle^{**})$ for all φ. (Therefore, t*, t-recursiveness and t**-recursiveness are equivalent.)

One drawback of the above construction is that if both $\langle\!\langle\ \rangle\!\rangle, \langle\!\langle\ \rangle\!\rangle^*$ satisfy corresponding $t\mu$-axioms, then no such axiom for $\langle\!\langle\ \rangle\!\rangle^{**}$ seems to be automatically implied.

Exercise 10.7. Show that the implication in ($) can be replaced by its particular instance $\forall x(x\varphi \leq x\psi) \Rightarrow St(I)\varphi \leq St(I)\psi$, taking the equality $St(I)St(\varphi) = St(\varphi)$ as an axiom.

Exercise 10.8.** Let $\psi_0, \psi_1 \in \mathscr{F}$ satisfy $\bar{n}\psi_0 \in \mathscr{L}$, $\bar{n}\psi_0\psi_1 = \bar{n}$ for all n. Show that the operation $\langle\ \ \rangle = \lambda\varphi.(\varphi, St(\varphi))$ satisfies the assumptions of exercises 10.2–10.5 and so does the operation St, provided $I \in \mathscr{L} \circ \mathscr{F}$.

The following exercise establishes a 'boldface' t-Transition Theorem. As opposed to the '*lightface*' results, their '*boldface*' versions are concerned with a situation in which some or all members of $\mathscr{L}$ are taken as initial elements and used as enumeration indices. In other words, $\mathscr{L}$ plays the role of $\{\bar{n}/n \in \omega\}$.

Exercise 10.9. Let K_4, K_5, $K_6 \in \mathscr{F}$, $xK_4 = I$, $xK_5 = xx$, $xyK_6 = yx$ for all x, $y \in \mathscr{L}$ and allow in tμA segments of the form $\{\theta/x\theta \leq \tau\}$. Take $\langle\ \ \rangle = St$, an arbitrary member of $\mathscr{L}$ for K_3 and add $K_3 - K_6$ to the initial elements. Show that for every binary mapping Γ t-recursive in $\mathscr{B}$ there is a mapping Γ^* t-recursive in $\mathscr{B}$ such that $x\Gamma^*(\theta) = \Gamma(x, \theta)$ for all x, θ.

Hint. By induction on the construction of Γ.

Remark. These additional assumptions are satisfied by the operation St of exercise 10.1 and others to be considered later. One may use exercise 10.9 to get 'boldface' Normal Form, Enumeration, Second Recursion and Rice Theorems. For instance, given a finite $\mathscr{B}$ and not necessarily finite $\mathscr{L}_0 \subseteq \mathscr{L}$, there exists a σ t-recursive in $\mathscr{B}$ such that, if φ is t-recursive in $\mathscr{L}_0 \cup \mathscr{B}$, then $\varphi = x\bar{n}\sigma$ for certain $x \in cl(\mathscr{L}_0/\lambda xy.xyK_1)$, $n \in \omega$; similarly for mappings. (A related Enumeration Theorem is suggested by Skordev [1982a].) 'Boldface' representability results can also be obtained, with prime computability over $\mathscr{L}$ playing the role of μ-recursiveness. The First Recursion Theorem is not affected by the 'lightface–boldface' division of the theory. As for a 'boldface' Rogers Theorem, that is quite a different matter. And we can not refer to a 'boldface' Theory of Numberings, because there is no such theory available.

Exercise 10.10. Let $\mathscr{L}$ be a nonempty set of mappings $X:\mathscr{F} \to \mathscr{F}$, $St:\mathscr{F} \to \mathscr{F}$, K_0, K_1, $K_2 \in \mathscr{F}$ and suppose that the following axioms are satisfied.

sA$_1$. $X(K_0(\varphi, \psi)) = (X(\varphi), X(\psi))$.

sA$_2$. For all X, Y there is a Z such that

$$X(Y(K_1\varphi)) = Z(\varphi) \quad \text{and} \quad Z(K_2) = X(Y(I)).$$

($) $X(St(\varphi)\psi) = \varphi X(\psi), RX(I) = X(I)St(R)$,

$$\forall X(\varphi X(\psi) \leq \rho X(\sigma)) \Rightarrow St(\varphi)\psi \leq St(\rho)\sigma.$$

Show that $\langle\ \ \rangle = \lambda\varphi.(\varphi, St(\varphi))$ is a t-operation with functional elements K_0, K_1, K_2.

Hint. Follow the proofs of 10.10–10.16.

The storing operation introduced in this way generalizes the former one which corresponds to a set $\mathscr{L}$ of mappings X such that $X(\varphi) = X(I)\varphi$ for all φ. Another example will be given in the exercises to chapter 21.

CHAPTER 11

The collection operation

This chapter is devoted to an infinitary pairing operation called *collection*. For instance, in example 3.1 the operation Co in question brings the members of a sequence $\{\varphi_n\}$ together in a single element $\varphi = Co(\{\varphi_n\})$ such that (in the notations of chapter 3) $\varphi(s01^n) = \varphi_n(s)$ for all n, s. Although this operation is elementary, being infinitary, it is classically noneffective.

One can reasonably argue that the present study of effective computability should have nothing to do with noneffective operations. However, collection has some interesting aspects. It is natural and also has effective applications. In particular, one may introduce the concept of a sequence $\{\varphi_n\}$ recursive in $\mathscr{B}$ as a sequence for which the element $Co(\{\varphi_n\})$ is recursive in $\mathscr{B}$. More important is the fact that essential parts of the recursion theory on IOS admit an adaptation with Co added to the initial operations; we are interested in such extensions anyway. Finally, while in chapter 2 the operations $\langle\ \ \rangle$, $[\ \ ]$ were intuitively interpreted by expressions $(\varphi\bar{0}, \varphi\bar{1}, \varphi\bar{2}, \ldots)$, $(I, \varphi(I, \varphi(I, \ldots)))$, collection enables one to formally treat infinite expressions constructed by means of $\circ$, Π. In this respect, there is a clear parallel with the so called infinite diagrams in Scott [1971].

Given an IOS $\mathscr{S} = (\mathscr{F}, I, \Pi, L, R)$, we say that $Co: \mathscr{F}^\omega \to \mathscr{F}$ is a *collection operation* iff the following three axioms are satisfied.

cA_1. Co is monotonic.

cA_2. $Co\{\varphi_n\}\psi = Co\{\varphi_n\psi\}$.

cA_3. $Co\{\varphi_n\} = (\varphi_0, Co\{\varphi_{n+1}\})$,

where $\{\varphi_{n+1}\}$ is the sequence $\varphi_1, \varphi_2, \ldots$ and $Co\{\varphi_n\}$ is written for $Co(\{\varphi_n\})$.

The axioms of Co are obviously infinitary versions of those of Π. A simple sufficient condition which ensures the existence of a collection operation will be given in the exercises, together with examples of IOS which satisfy it.

Some properties of collection follow, concerning in particular its interconnections with the initial IOS-operations.

Proposition 11.1. $\bar{n}Co\{\varphi_n\} = \varphi_n$ for all n.
This follows from axiom cA_3.

11.1 shows that the element $\sigma = Co\{\varphi_n\}$ is universal for $\{\varphi_n / n \in \omega\}$. Therefore, all countable subsets $\mathscr{B}$ of $\mathscr{F}$ have universal elements, though not necessarily in $\mathscr{B}$. Indeed, whenever σ is universal for $\{\bar{n}/n \in \omega\}$, then 4.4 easily implies $\sigma \neq \bar{n}$ for all n.

Proposition 11.2. $Co\{\varphi_n\bar{n}\}Co\{\psi_n\} = Co\{\varphi_n\psi_n\}$. In particular, $Co\{\bar{n}\}Co\{\varphi_n\} = Co\{\varphi_n\}$.

Proof. Using cA_2 and 11.1, we get

$$Co\{\varphi_n\bar{n}\}Co\{\psi_n\} = Co\{\varphi_n\bar{n}Co\{\psi_n\}\} = Co\{\varphi_n\psi_n\}.$$

Proposition 11.3. $Co\{(\varphi_n, \psi_n)\} = Co\{(\bar{n}L, \bar{n}R)\}(Co\{\varphi_n\}, Co\{\psi_n\})$.

Proof. It follows that

$$(\varphi_n, \psi_n) = (\bar{n}Co\{\varphi_n\}, \bar{n}Co\{\psi_n\}) = (\bar{n}L, \bar{n}R)(Co\{\varphi_n\}, Co\{\psi_n\})$$

for all n, which completes the proof by cA_2.

Proposition 11.4. $Co\{[\varphi_n]\} = Co\{(L, \overline{n+1})\}\ [\sigma]$, where $\sigma = Co\{\varphi_n(L, \overline{n+1})\}$.

Proof. Since $Co\{(L, \overline{n+1})\}[\sigma] = Co\{(I, \overline{n+1}[\sigma])\}$ and $Co\{[\varphi_n]\} = Co\{(I, R[\varphi_n])\}$, it suffices to show that $R[\varphi_n] = \overline{n+1}\ [\sigma]$ for all n.

The equality

$$\sigma(I, Co\{R[\varphi_n]\}) = Co\{\varphi_n(I, R[\varphi_n])\} = Co\{R[\varphi_n]\}$$

implies $R[\sigma] \leq Co\{R[\varphi_n]\}$ by 6.11, hence $\overline{n+1}\ [\sigma] \leq R[\varphi_n]$ for all n. On the other hand,

$$\varphi_n(I, \overline{n+1}\ [\sigma]) = \varphi_n(L, \overline{n+1})\,[\sigma] = \overline{n+1}\ [\sigma],$$

hence $R[\varphi_n] \leq \overline{n+1}\ [\sigma]$ for all n by 6.11. This completes the proof.

The above properties indicate that collection behaves to some extent as a t-operation. The following statement shows that consecutive applications of Co can be reduced to a single one.

Proposition 11.5. Let $J{:}\omega^2 \to \omega$ be a bijection, $\rho = Co\{Co\{\overline{J(m,n)}\}_n\}_m$ and $\varphi_{J(m,n)} = \varphi_{m,n}$ for all m, n. Then $Co\{Co\{\varphi_{m,n}\}_n\}_m = \rho Co\{\varphi_k\}$.

Proof.

$$Co\{Co\{\varphi_{m,n}\}_n\}_m = Co\{Co\{\varphi_{J(m,n)}\}_n\}_m = Co\{Co\{\overline{J(m,n)}Co\{\varphi_k\}\}_n\}_m = \rho Co\{\varphi_k\}.$$

Proposition 11.6. $\Delta(\varphi, \psi) = \bar{1}[\sigma]$, where $\sigma = Co\{(\varphi\psi^n L, \overline{n+2})\}$.

Proof. We have

$$\begin{aligned}\sigma(I, Co\{\Delta(\varphi, \psi)\psi^n\}) &= Co\{(\varphi\psi^n, \Delta(\varphi, \psi)\psi^{n+1})\} = Co\{(\varphi, \Delta(\varphi, \psi)\psi)\psi^n\} \\ &= Co\{\Delta(\varphi, \psi)\psi^n\},\end{aligned}$$

hence $R[\sigma] \leq Co\{\Delta(\varphi, \psi)\psi^n\}$ by 6.11. In particular, $\bar{1}[\sigma] \leq \Delta(\varphi, \psi)$. On the other hand,

$$\sigma(\psi, R^2[\sigma]) = \sigma(\psi L, R^2)[\sigma] = R\sigma[\sigma] = R^2[\sigma]$$

implies $R[\sigma]\psi \leq R^2[\sigma]$ by 6.11 again. Therefore,

$$(\varphi, \bar{1}[\sigma]\psi) \leq (\varphi, \bar{2}[\sigma]) = (\varphi\bar{0}, \bar{2})[\sigma] = \bar{0}\sigma[\sigma] = \bar{1}[\sigma],$$

hence $\Delta(\varphi, \psi) \leq \bar{1}[\sigma]$. The proof is complete.

Primitive recursion can also be expressed by the equality $\Delta(\varphi, \psi) = \langle I \rangle Co\{\varphi\psi^n\}$ which follows easily from 6.34. Conversely, $Co\{\varphi\psi^n\} = Co\{\bar{n}\}\Delta(\varphi, \psi)$ follows by making use of cA_2.

The notion of relative recursiveness is naturally extended by adding *Co* to the initial IOS-operations. Namely, φ is *c-recursive in* $\mathscr{B}$ iff

$$\varphi \in cl(\{L, R\} \cup \mathscr{B}/\circ, \Pi, \langle \quad \rangle, [\quad], Co).$$

Proposition 6.9, 11.6 imply that, equivalently, φ is c-recursive in $\mathscr{B}$ iff $\varphi \in cl(\{L, A\} \cup \mathscr{B}/\circ, [\quad], Co)$. Notions of elements *prime* c-*recursive*, *primitive* c-*recursive*, c-*primitive* and c-*polynomial in* $\mathscr{B}$ are introduced in a similar way. It follows from 11.6 that the relative c-recursiveness and prime c-recursiveness are equivalent, while the equality $\Delta(\varphi, \psi) = \langle I \rangle Co\{\varphi\psi^n\}$ implies that φ is primitive c-recursive in $\mathscr{B}$ iff it is c-polynomial in $\{\langle I \rangle\} \cup \mathscr{B}$. The ordinary properties from the beginning of chapter 7 arise again with 7.4 modified essentially as follows.

Proposition 11.7. If φ is c-recursive (c-polynomial) in $\mathscr{B}$, then φ is c-recursive (respectively, c-polynomial) in a countable subset of $\mathscr{B}$.

It is worth mentioning that, while there are $\max\{\mathrm{Card}(\mathscr{B}), \omega\}$ elements recursive in $\mathscr{B}$, there are $\mathrm{Card}(\{L, R\} \cup \mathscr{B})^\omega$ elements c-recursive in $\mathscr{B}$ since $Co\{\varphi_n\} = Co\{\psi_n\}$ iff $\forall n(\varphi_n = \psi_n)$.

The classically noneffective nature of *Co* finds its expression in the fact that all partial number theoretic functions are representable in terms of *Co*.

Proposition 11.8. Every partial number theoretic function f is represented by an element $\varphi \in cl(\mathscr{D} \cup \{O\}/Co)$.

Proof. By induction on the arity of f.

If f is unary, then $Co\{\overline{f(n)}\}$ represents it. (Recall our convention that $\overline{f(n)} = O$ whenever $f(n)\uparrow$.)

Let f be $n+1$-ary and for all m an element φ_m represent $\lambda s_1 \ldots s_n.\, f(s_1, \ldots, s_n, m)$ by the induction clause. Then $\varphi = Co\{\varphi_m\}$ represents f since

$$\bar{s}_1 \ldots \bar{s}_n \bar{m} \varphi = \bar{s}_1 \ldots \bar{s}_n \varphi_m = \overline{f(s_1, \ldots, s_n, m)}$$

for all $s_1, \ldots, s_n$, m.

Passing from elements to mappings, one sees that collection makes it possible to consider ω-ary mappings. A notion of mapping $\Gamma: \mathscr{F}^\omega \to \mathscr{F}$ *c-recursive in* $\mathscr{B}$ is introduced inductively as follows, taking in view of 6.9, 11.6 the element A initial and omitting the operations Π, $\langle \quad \rangle$.

1. The mappings $\Gamma = \lambda\{\theta_n\}.\, \theta_i$, $i \geq 0$ and $\Gamma = \lambda\{\theta_n\}.\, \psi$, $\psi \in \{L, A\} \cup \mathscr{B}$, are c-recursive in $\mathscr{B}$
2. If $\Gamma_m: \mathscr{F}^\omega \to \mathscr{F}$ are c-recursive in $\mathscr{B}$ for all m, then so are

$$\Gamma = \lambda\{\theta_n\}.\Gamma_0\{\theta_n\}\Gamma_1\{\theta_n\}, \quad \Gamma = \lambda\{\theta_n\}.[\Gamma_0\{\theta_n\}],$$
$$\Gamma = \lambda\{\theta_n\}.Co\{\Gamma_m\{\theta_n\}\}.$$

Of course, n-ary mappings c-recursive in $\mathscr{B}$ may be considered as well: an ω-ary mapping with the clause $\lambda\{\theta_n\}.\, \theta_i$ used for finitely many i in its construction has only finitely many genuine arguments.

The properties of the collection operation given above allow us to establish a Normal c-Form Theorem.

Proposition 11.9. If $\Gamma:\mathscr{F}^{\omega}\to\mathscr{F}$ is c-recursive in $\mathscr{B}$, then

$$\Gamma=\lambda\{\theta_n\}.\bar{1}[\varphi(I,Co\{\theta_{i_n}\bar{n}\})]$$

with a certain sequence $\{i_n\}$ and φ c-polynomial in $\mathscr{B}$. Another normal form is $\Gamma=\lambda\{\theta_n\}.\bar{1}[\varphi(I,Co\{\langle\theta_n\rangle\})]$.

Proof. We shall prove first that

$$\Gamma=\lambda\{\theta_n\}.\psi[\varphi(I,Co\{\theta_{i_n}\bar{n}\}\chi)]$$

with certain ψ, φ, χ c-polynomial in $\mathscr{B}$.

If $\Gamma=\lambda\{\theta_n\}.\theta_i$, then

$$\Gamma=\lambda\{\theta_n\}.R[\overline{i+1}(I,Co\{\theta_n\bar{n}\}Co\{L\})],$$

$\{L\}$ standing for the sequence $L, L,\ldots.$

If $\Gamma=\lambda\{\theta_n\}.\psi$, $\psi\in\{L,A\}\cup\mathscr{B}$, then $\Gamma=\lambda\{\theta_n\}.R[\psi L^2(I,Co\{\theta_n\bar{n}\})]$.

Let $\Gamma_m=\lambda\{\theta_n\}.\psi_m[\varphi_m(I,Co\{\theta_{i_{m,n}}\bar{n}\}_n\chi_m)]$ and ψ_m,φ_m,χ_m be c-polynomial in $\mathscr{B}$ for all m.

If $\Gamma=\lambda\{\theta_n\}.\Gamma_0\{\theta_n\}\Gamma_1\{\theta_n\}$, then we get by 6.15 elements ψ, φ', χ', χ'' c-polynomial in $\mathscr{B}$ such that

$$\Gamma=\lambda\{\theta_n\}.\psi[\varphi'(I,Co\{\theta_{i_{0,n}}\bar{n}\}\chi',Co\{\theta_{i_{1,n}}\bar{n}\}\chi'')].$$

Taking $\varphi=\varphi'(L,Co\{\overline{2n+1}\},\ Co\{\overline{2n+2}\})$, $\chi=Co(\bar{0}\chi',\bar{0}\chi'',\bar{1}\chi',\bar{1}\chi'',\ldots)$ and $j_n=i_{\mathrm{rem}(n,2),n}$ for all n, we get

$$\Gamma=\lambda\{\theta_n\}.\psi[\varphi(I,Co\{\theta_{j_n}\bar{n}\}\chi)].$$

The case of iteration is a bit simpler, using 6.16 instead of 6.15.

If $\Gamma=\lambda\{\theta_n\}.Co\{\Gamma_m\{\theta_n\}\}$, then we get by 11.4 elements ψ, φ', $\chi_{m,n}$ c-polynomial in $\mathscr{B}$ for all m, n such that $\Gamma=\lambda\{\theta_n\}.\psi[\varphi'(I,Co\{Co\{\theta_{i_{m,n}}\chi_{m,n}\}_n\}_m)]$. Take J, ρ as in 11.5, $j_{J(m,n)}=i_{m,n}$, $\sigma_{J(m,n)}=\chi_{m,n}$ for all m, n and $\chi=Co\{\sigma_k\}$. It follows by 11.5 that

$$Co\{Co\{\theta_{i_{m,n}}\chi_{m,n}\}_n\}_m=\rho Co\{\theta_{j_k}\sigma_k\}_k=\rho Co\{\theta_{j_k}\bar{k}\}\chi$$

for all $\{\theta_n\}$. Therefore, $\Gamma=\lambda\{\theta_n\}.\psi[\varphi(I,Co\{\theta_{j_k}\bar{k}\}\chi)]$ with $\varphi=\varphi'(L,\rho R)$.

Now let $\Gamma=\lambda\{\theta_n\}.\psi[\varphi(I,Co\{\theta_{i_n}\bar{n}\}\chi)]$ with ψ, φ, χ c-polynomial in $\mathscr{B}$. Substituting $\varphi(L,Co\{\theta_{i_n}\overline{n+1}\})[\chi L]$ for $\varphi(I,Co\{\theta_{i_n}\bar{n}\}\chi)$ and using 6.16, 6.14, we almost get the desired normal form except that $Co\{\theta_{i_n}\overline{n+5}\}$ appears instead of $Co\{\theta_{i_n}\bar{n}\}$. However, $Co\{\theta_{i_n}\overline{n+5}\}=R^5Co\{\theta_{j_n}\bar{n}\}$, provided $j_{n+5}=i_n$ for all n.

The alternative normal form is obtained immediately by

$$Co\{\theta_{i_n}\bar{n}\}=Co\{\bar{n}\bar{i}_n\}Co\{\langle\theta_n\rangle\}.$$

This completes the proof.

A finitary version of 11.9 gives normal forms

$$\Gamma=\lambda\theta_1\ldots\theta_n.\bar{1}[\varphi(I,Co\{\theta_{i_m}\bar{m}\})],$$

$1 \le i_m \le n$ for all m, or

$$\Gamma = \lambda\theta_1 \ldots \theta_n.\bar{1}[\varphi(I, \langle\theta_1\rangle, \ldots, \langle\theta_n\rangle)]$$

with φ c-polynomial in $\mathscr{B}$. The elements c-recursive in $\mathscr{B}$ have a normal form $\bar{1}[\sigma]$ with σ c-polynomial in $\mathscr{B}$.

A First c-Recursion Theorem follows.

Proposition 11.10*. If $\Gamma: \mathscr{F}^\omega \to \mathscr{F}$ is c-recursive in $\mathscr{B}$, then there is a mapping $\Gamma_0: \mathscr{F}^\omega \to \mathscr{F}$ c-recursive in $\mathscr{B}$ such that $\Gamma_0 = \lambda\{\theta_n\}.\mu\theta.\Gamma(\theta, \{\theta_n\})$. (That is, for all $\{\theta_n\}$ $\Gamma(\Gamma_0\{\theta_n\}, \{\theta_n\}) = \Gamma_0\{\theta_n\}$ and whenever $\Gamma(\tau, \{\theta_n\}) \le \tau$, then $\Gamma_0\{\theta_n\} \le \tau$.)

Proof. The Normal c-Form Theorem gives

$$\Gamma = \lambda\{\theta_n\}.\bar{1}[\varphi(I, \langle\theta_0\rangle, Co\{\langle\theta_{n+1}\rangle\})]$$

for a certain φ c-recursive in $\mathscr{B}$. The mapping $\Gamma^* = \lambda\theta\theta_1.\bar{1}[\varphi(I, \langle\theta\rangle, \theta_1)]$ is recursive in φ, hence so is $\Gamma^{**} = \lambda\theta_1.\mu\theta.\Gamma^*(\theta, \theta_1)$ by 9.14*. Therefore, the mapping $\Gamma_0 = \lambda\{\theta_n\}.\ \Gamma^{**}(Co\{\langle\theta_n\rangle\})$ is c-recursive in $\mathscr{B}$. It follows that

$$\begin{aligned} \Gamma(\Gamma_0\{\theta_n\}, \{\theta_n\}) &= \Gamma^*(\Gamma_0\{\theta_n\}, Co\{\langle\theta_n\rangle\}) \\ &= \Gamma^*(\Gamma^{**}(Co\{\langle\theta_n\rangle\}), Co\{\langle\theta_n\rangle\}) \\ &= \Gamma^{**}(Co\{\langle\theta_n\rangle\}) = \Gamma_0\{\theta_n\}. \end{aligned}$$

Whenever $\Gamma(\tau, \{\theta_n\}) \le \tau$, then

$$\Gamma^*(\tau, Co\{\langle\theta_n\rangle\}) = \Gamma(\tau, \{\theta_n\}) \le \tau,$$

hence $\Gamma^{**}(Co\{\langle\theta_n\rangle\}) \le \tau$, i.e., $\Gamma_0\{\theta_n\} \le \tau$. The proof is complete.

A finitary version of 11.10* states that whenever $\Gamma: \mathscr{F}^{n+1} \to \mathscr{F}$ is c-recursive in $\mathscr{B}$, then the mapping $\lambda\theta_1 \ldots \theta_n.\mu\theta.\Gamma(\theta_1, \ldots, \theta_n, \theta)$ exists and is c-recursive in $\mathscr{B}$. In particular, the element $\mu\theta.\Gamma(\theta)$ exists and is c-recursive in $\mathscr{B}$, provided $n = 0$.

The collection operation also enables us to solve infinite systems of inequalities.

Proposition 11.11*. Let $\Gamma_m: \mathscr{F}^\omega \to \mathscr{F}$ be c-recursive in $\mathscr{B}$ for all m. Then there is a sequence $\{\varphi_n\}$ of elements c-recursive in $\mathscr{B}$ which is the least solution of the system $\Gamma_m\{\theta_n\} \le \theta_m$, $m \ge 0$.

Proof. The mapping $\Gamma = \lambda\theta.Co\{\Gamma_m\{\bar{n}\theta\}\}$ is c-recursive in $\mathscr{B}$; hence so is the element $\varphi = \mu\theta.\Gamma(\theta)$ by 11.10*. Taking $\varphi_n = \bar{n}\varphi$, we get

$$\Gamma_m\{\varphi_n\} = \Gamma_m\{\bar{n}\varphi\} = \bar{m}\Gamma(\varphi) = \bar{m}\varphi = \varphi_m$$

for all m.

If $\Gamma_m\{\tau_n\} \le \tau_m$ for all m, then

$$\Gamma(Co\{\tau_n\}) = Co\{\Gamma_m\{\tau_n\}\} \le Co\{\tau_m\},$$

hence $\varphi \le Co\{\tau_n\}$, which implies $\varphi_n \le \tau_n$ for all n. (Notice that this is the first and only time cA_1 was used. Without this axiom one would get that $\{\varphi_n\}$ is a least solution to the corresponding system of equalities.) The proof is complete.

The mappings Γ_m may have finitely many arguments. For instance, a simple infinite system of inequalities is $\Gamma_m(\theta_{m+1}) \leq \theta_m$, $m \geq 0$. A parametrized version of 11.11* concerns systems of the form $\Gamma_{2m}\{\theta_n\} \leq \theta_{2m}$, $m \geq 0$.

This concludes our discussion of the collection operation. As compared with similar considerations in the previous two chapters, there is a noticeable omission. Despite having a First c-Recursion Theorem, no c-Enumeration Theorem has been established. In chapter 9 an element σ was universal for a countable set $\mathscr{U}$ iff $\mathscr{U} \subseteq \{\bar{n}\sigma/n\in\omega\}$; given a finite set $\mathscr{B}$, there were countably many elements recursive in $\mathscr{B}$. Now, given a countable set $\mathscr{B}$, there are 2^ω elements c-recursive in $\mathscr{B}$; hence the elements $\bar{n}$, $n\in\omega$ are inadequate as enumeration indices. We shall return to this problem in the exercises.

As mentioned in the introduction to this chapter, *infinite expressions* can be given a formal interpretation by using collection operation. We sketch the idea and give several examples.

Given a sequence $\{\varphi_n\}$, what value, if any, should the infinite expression $(I, \varphi_0(I, \varphi_1(I, \ldots)))$ have? One would like it to be a member of $\mathscr{F}$, of course. Suppose for a moment that it is σ_0 and similarly $\sigma_{n+1} = (I, \varphi_{n+1}(I, \varphi_{n+2}(I, \ldots)))$ for all n. Then it follows immediately that the sequence $\{\sigma_n\}$ satisfies the system

(1) $$(I, \varphi_n\theta_{n+1}) = \theta_n, \quad n \geq 0.$$

However, this system of equalities actually does have solutions. Proposition 11.11* ensures that (1) has a least solution which is a sequence consisting of elements c-recursive in $\{\varphi_n\}$. Now one can return to formally define σ_n for all n. Namely, take $\{\sigma_n\}$ to be the least solution of (1); then σ_0 is the value of the expression we are interested in. The correctness of this definition needs proof since equivalent systems of equalities can be attached to an expression in various ways.

Therefore, arbitrary infinite expressions can be interpreted by means of systems of equalities and vice versa, single equalities and systems of equalities yield infinite expressions by consecutively replacing right sides of equalities by their left ones. For instance, given the equality $(I, \varphi\theta) = \theta$, we get $(I, \varphi(I, \varphi(I, \ldots)))$ by consecutively substituting $(I, \varphi\theta)$ for θ. Other basic operations besides $\circ$, Π may also be used.

If $\varphi_n = \varphi$ for all n and $\{\sigma_n\}$ is the least solution of (1), then it is quite immediate that $\sigma_n = \sigma_0$ for all n. Therefore, σ_0 is the least solution to $(I, \varphi\theta) = \theta$, i.e. $\sigma_0 = [\varphi]$. So one may write $[\varphi] = (I, \varphi(I, \varphi(I, \ldots)))$. The equality $\langle\varphi\rangle = (\varphi\bar{0}, \varphi\bar{1}, \ldots)$ is verified in exercise 11.6 below.

Let $\{\sigma_n\}$ be the least solution of the system $(\varphi_n, \theta_{n+1}) = \theta_n$, $n \geq 0$. Then $\sigma_0 = (\varphi_0, \varphi_1, \ldots)$ and it easily follows that $Co\{\varphi_n\} = Co\{\bar{n}\}\sigma_0$, hence collection itself is expressed via an infinite expression.

EXERCISES TO CHAPTER 11

Exercise 11.1. Let $\mathscr{S}$ be an OS to satisfy the following condition.

($c*$) The semigroup $\mathscr{F}$ has a least element O and for any increasing sequence $\{\theta_n\}$ there is a θ such that $\theta\psi = \sup_n(\theta_n\psi)$ and $(\psi, \theta) = \sup_n(\psi, \theta_n)$ for all ψ.

Determine an operation Co satisfying the axioms $cA_1 - cA_3$.

Hint. Given a sequence $\{\varphi_n\}$, define $Co\{\varphi_n\}$ as $\sup_n \theta_n$, where $\theta_0 = O$ and $\theta_{n+1} = (\varphi_0, \varphi_1, \ldots, \varphi_n, O)$.

Notice that $Co\{\varphi_n\} = (\varphi_0, \varphi_1, \ldots)$ and $\langle I \rangle = Co\{\bar{n}\}$ in such a case. It follows by what we know about the IOS of examples 4.7, 4.8, 3.2 that those spaces satisfy $(c*)$.

Exercise 11.2. Show that $\Gamma: \mathscr{F}^\omega \to \mathscr{F}$ is c-recursive in $\{\psi_n\}$ iff there is a c-recursive mapping $\Gamma^*: \mathscr{F}^\omega \to \mathscr{F}$ such that

$$\Gamma = \lambda\{\theta_n\}.\Gamma^*(\psi_0, \theta_0, \psi_1, \theta_1, \ldots).$$

Exercise 11.3. Show that $(\varphi_0, \varphi_1, \ldots) = \bar{1}[\rho]Co\{\varphi_n\}$, where $\rho = Co\{(\bar{n}L, \overline{n+2})\}$.

Hint. By definition $(\varphi_0, \varphi_1, \ldots) = \sigma_0$, where $\{\sigma_n\}$ is the least solution to the system $(\varphi_n, \theta_{n+1}) = \theta_n$, $n \geq 0$. Following the proof of 11.11*, show that $\sigma_n = \bar{n}\sigma$ for all n, where $\sigma = \mu\theta.Co\{(\varphi_n, \overline{n+1}\,\theta)\} = \mu\theta.\rho(Co\{\varphi_n\}, \theta) = R[\rho]Co\{\varphi_n\}$.

Exercise 11.4. Show that $(\varphi_0, \varphi_1, \ldots)\psi = (\varphi_0\psi, \varphi_1\psi, \ldots)$.

Hint. Make use of the previous exercise.

Exercise 11.5. Show that $(I, \varphi_0(I, \varphi_1(I, \ldots))) = \bar{1}[\rho]$, where $\rho = Co\{(L, \varphi_n \overline{n+2})\}$.

Hint. See the hint to exercise 11.3.

Exercise 11.6. Show that $\Delta(\varphi, \psi) = (\varphi, \varphi\psi, \varphi\psi^2, \ldots)$. In particular, $\langle \varphi \rangle = (\varphi\bar{0}, \varphi\bar{1}, \ldots)$.

Hint. The sequence $\{\Delta(\varphi, \psi)\psi^n\}$ is a solution of the system $(\varphi\psi^n, \theta_{n+1}) = \theta_n$, $n \geq 0$; hence $\sigma_0 = (\varphi, \varphi\psi, \ldots) \leq \Delta(\varphi, \psi)$. Use exercise 11.4 to show that $(\varphi, \sigma_0\psi) = \sigma_0$, which implies $\Delta(\varphi, \psi) \leq \sigma_0$.

Further to the enumeration problem for c-recursiveness, let $\mathscr{S}$ be augmented by a storing operation St satisfying the assumptions of exercise 10.9. Let $K_7 \in \mathscr{F}$, $\varnothing \subset \mathscr{L}_0 \subseteq \mathscr{L}$ and suppose that for every sequence $\{x_n \bar{k}_n\}_n$, $x_n \in \mathscr{L}_0$, there exist $x \in \mathscr{L}_0$, $k \in \omega$ such that $x\bar{k}K_7 = Co\{x_n \bar{k}_n\}$. (This imposes additional constraints, e.g. $\mathrm{Card}(M) \geq 2^\omega$ in examples 4.7, 4.8.) Assume also that $\mathscr{L}_0^2 K_1 \subseteq \mathscr{L}_0$. Adding $K_0 - K_7$ to the initial elements and Co, St to the initial operations, the notion of recursiveness is then extended to t,c-recursiveness for which the following 'boldface' t,c-Enumeration Theorem can be established.

Exercise 11.7. Show under the above assumptions that for any countable $\mathscr{B} \subseteq \mathscr{F}$ there is an element σ t,c-recursive in $\mathscr{B}$ such that whenever φ is t,c-recursive in $\mathscr{L}_0 \cup \mathscr{B}$, then $\varphi = x\bar{n}\sigma$ for certain $x \in \mathscr{L}_0$, $n \in \omega$.

Hint. Modify the proof of 9.18, using an analogue to 10.2. Alternatively, follow the idea of exercise 9.5.

As in the case of t-recursiveness discussed after exercise 10.9, 'boldface' Second Recursion and Rice Theorems can also be established.

CHAPTER 12

Consecutive spaces

Given an OS $\mathscr{S} = (\mathscr{F}, I, \Pi, L, R)$, we have so far studied elements 'computable' from certain elements by means of the initial operations $\circ, \Pi$ plus $\langle \quad \rangle, [\quad]$, provided $\mathscr{S}$ is iterative, and possibly $\langle\!\langle \quad \rangle\!\rangle$, *Co*. To accommodate notions such as Kleene's [1955] concept of hyperarithmetical function, however, one should be able to consider elements 'computable' from certain operations, particularly operations involving quantifiers.

This leads to the idea of trying to construct another OS $\mathscr{S}' = (\mathscr{F}', I', \Pi', L', R')$ the carrier $\mathscr{F}'$ of which would consist of mappings over $\mathscr{F}$. It is an important feature of the algebraic system of OS that such a construction can be carried out successfully, and chapters 12, 13 are concerned with recursion theory on pairs of spaces $\mathscr{S}, \mathscr{S}'$.

Proposition 12.1. Take $\mathscr{F}' = \{\varphi' / \varphi' : \mathscr{F} \to \mathscr{F} \; \& \; \varphi' \text{ is monotonic}\}$, $\varphi' \leq \psi'$ iff $\forall \theta (\varphi'(\theta) \leq \psi'(\theta))$, $\varphi'\psi' = \lambda\theta . \varphi'(\psi'(\theta))$, $(\varphi', \psi') = \lambda\theta . (\varphi'(\theta), \; \psi'(\theta))$, $I' = \lambda\theta . \theta$, $L' = \lambda\theta . L\theta$ and $R' = \lambda\theta . R\theta$. Then $\mathscr{S}' = (\mathscr{F}', I', \Pi', L', R')$ is an OS.

Proof. We have

$$I'\varphi'(\theta) = I'(\varphi'(\theta)) = \varphi'(\theta),$$
$$\varphi' I'(\theta) = \varphi'(I'(\theta)) = \varphi'(\theta),$$
$$\varphi'(\psi'\chi')(\theta) = \varphi'(\psi'(\chi'(\theta))) = (\varphi'\psi')\chi'(\theta)$$

for all θ, hence $I'\varphi' = \varphi' I' = \varphi'$ and $\varphi'(\psi'\chi') = (\varphi'\psi')\chi'$.

If $\varphi' \leq \varphi'_1, \psi' \leq \psi'_1$, then

$$\varphi'\psi'(\theta) = \varphi'(\psi'(\theta)) \leq \varphi'(\psi'_1(\theta)) \leq \varphi'_1(\psi'_1(\theta)) = \varphi'_1\psi'_1(\theta)$$

for all θ, hence $\varphi'\psi' \leq \varphi'_1\psi'_1$. Therefore, $\mathscr{F}'$ is a partially ordered semigroup with unit I'.

If $\varphi' \leq \varphi'_1$, $\psi' \leq \psi'_1$, then

$$(\varphi', \psi')(\theta) = (\varphi'(\theta), \psi'(\theta)) \leq (\varphi'_1(\theta), \psi'_1(\theta)) = (\varphi'_1, \psi'_1)(\theta)$$

for all θ; hence $(\varphi', \psi') \leq (\varphi'_1, \psi'_1)$. It also follows that

$$(\varphi', \psi')\chi'(\theta) = (\varphi', \psi')(\chi'(\theta)) = (\varphi'(\chi'(\theta)), \; \psi'(\chi'(\theta)))$$
$$= (\varphi'\chi'(\theta), \psi'\chi'(\theta)) = (\varphi'\chi', \psi'\chi')(\theta)$$

for all θ; hence $(\varphi',\psi')\chi' = (\varphi'\chi',\psi'\chi')$. Finally,

$$L'(\varphi',\psi')(\theta) = L'((\varphi'(\theta),\psi'(\theta))) = L(\varphi'(\theta),\psi'(\theta)) = \varphi'(\theta)$$

implies $L'(\varphi',\psi') = \varphi'$. Similarly, $R'(\varphi',\psi') = \psi'$, which completes the proof.

Sometimes one may be interested in a smaller space $\mathscr{S}'$ composed of mappings which are not only monotonic but say, continuous. That is why we introduce the notion of consecutive spaces as follows.

Let $\mathscr{S} = (\mathscr{F}, I, \Pi, L, R)$ and $\mathscr{S}' = (\mathscr{F}', I', \Pi', L', R')$ be OS such that $\varphi' : \mathscr{F} \to \mathscr{F}$ for all $\varphi' \in \mathscr{F}'$, $\varphi' \leq \psi'$ iff $\forall\theta(\varphi'(\theta) \leq \psi'(\theta))$ and $(\varphi',\psi') = \lambda\theta.(\varphi'(\theta), \psi'(\theta))$. We write $\bar{\varphi} = \lambda\theta.\varphi\theta$, $\check{\varphi} = \lambda\theta.\varphi$ and $(\text{———})\check{}$ for $\lambda\theta.\text{———}$, provided θ does not occur in the expression ———, $Id = \check{I}$, $Ml = \lambda\theta.L\theta R\theta$, $\bar{\mathscr{B}} = \{\bar{\psi}/\psi \in \mathscr{B}\}$ and $\check{\mathscr{B}} = \{\check{\psi}/\psi \in \mathscr{B}\}$, where $\varphi \in \mathscr{F}$, $\mathscr{B} \subseteq \mathscr{F}$. Let also $\bar{\mathscr{F}} \subseteq \mathscr{F}'$ and $Id, Ml \in \mathscr{F}'$. Then $\mathscr{S}, \mathscr{S}'$ are said to be *consecutive* OS. (For such spaces appear consecutively in the hierarchies of OS studied in chapter 15.)

The OS $\mathscr{S}, \mathscr{S}'$ of 12.1 are consecutive since the mappings $\bar{\varphi}$, Id, Ml are monotonic, and hence in $\mathscr{F}'$. Similar constructions yielding consecutive spaces will be given in chapter 19, while particular consecutive spaces will be studied in chapters 28–30.

Several corollaries to the above definition follow, dashed letters standing for members of $\mathscr{F}'$.

Proposition 12.2. $I' = \bar{I}$. Therefore, $I' = \lambda\theta.\theta$.

Proof.

$$I' = \lambda\theta.I'(\theta) = \lambda\theta.I'(\bar{I}(\theta)) = I'\bar{I} = \bar{I}.$$

Proposition 12.3. $\bar{\varphi}Id = \check{\varphi}$.

Proof.

$$\bar{\varphi}Id = \lambda\theta.\bar{\varphi}(Id(\theta)) = \lambda\theta.\bar{\varphi}(I) = \lambda\theta.\varphi = \check{\varphi}.$$

Notice that 12.3 implies $\check{\mathscr{F}} \subseteq \mathscr{F}'$.

Proposition 12.4. $\check{\varphi}\psi' = \check{\varphi}$ for all φ, ψ'. In particular, $Id\varphi' = Id$ for all φ'.

Proof.

$$\check{\varphi}\psi' = \lambda\theta.\check{\varphi}(\psi'(\theta)) = \lambda\theta.\varphi = \check{\varphi}.$$

Proposition 12.5. $Id \notin \bar{\mathscr{F}}$.

Proof. Suppose that $Id = \bar{\varphi}$. Then $\varphi = \bar{\varphi}(I) = Id(I) = I$, hence $Id = \bar{I} = I'$. Therefore,

$$L' = I'L' = IdL' = Id = I',$$

which is not the case.

Proposition 12.6. $\check{\varphi} \leq \check{\psi}$ iff $\varphi \leq \psi$.

Proposition 12.7. The members of $\mathscr{F}'$ are monotonic mappings.

Proof. Let $\varphi' \in \mathscr{F}'$ and $\theta \leq \tau$. Then $\check{\theta} \leq \check{\tau}$; hence $\varphi'\check{\theta} \leq \varphi'\check{\tau}$, which implies $\varphi'(\theta) \leq \varphi'(\tau)$.

Proposition 12.8. $\bar{\varphi} \leq \bar{\psi}$ iff $\varphi \leq \psi$.

Proof. If $\varphi \leq \psi$, then $\bar{\varphi}(\theta) = \varphi\theta \leq \psi\theta = \bar{\psi}(\theta)$ for all θ, hence $\bar{\varphi} \leq \bar{\psi}$. If $\bar{\varphi} \leq \bar{\psi}$, then $\varphi = \bar{\varphi}(I) \leq \bar{\psi}(I) = \psi$.

Proposition 12.9. $\bar{\varphi}\bar{\psi} = \overline{\varphi\psi}$.

Proof.

$$\bar{\varphi}\bar{\psi} = \lambda\theta.\bar{\varphi}(\bar{\psi}(\theta)) = \lambda\theta.\varphi\psi\theta = \overline{\varphi\psi}.$$

Proposition 12.10. $(\bar{\varphi}, \bar{\psi}) = \overline{(\varphi, \psi)}$.

Proof.

$$(\bar{\varphi}, \bar{\psi}) = \lambda\theta.(\bar{\varphi}(\theta), \bar{\psi}(\theta)) = \lambda\theta.(\varphi\theta, \psi\theta) = \lambda\theta.(\varphi, \psi)\theta = \overline{(\varphi, \psi)}.$$

Since $\bar{L}(\varphi', \psi') = \varphi'$ and $\bar{R}(\varphi', \psi') = \psi'$ for all φ', ψ', one may also assume that $L' = \bar{L}$ and $R' = \bar{R}$. Therefore, the following statement holds.

Proposition 12.11. The OS $\mathscr{S}$ is isomorphic with the proper subspace

$\bar{\mathscr{S}} = (\bar{\mathscr{F}}, \bar{I}, \Pi' \restriction \bar{\mathscr{F}}^2, \bar{L}, \bar{R})$ of $\mathscr{S}'$.

Some immediate statements follow.

Proposition 12.12. $\bar{\varphi}\psi' = \lambda\theta.\varphi\psi'(\theta)$.

Proposition 12.13. $\varphi'\bar{\psi} = \lambda\theta.\varphi'(\psi\theta)$.

Proposition 12.14. $\varphi' Id = (\varphi'(I))\check{}$.

Proposition 12.15. $\varphi'\check{\psi} = (\varphi'(\psi))\check{}$. In particular, $\bar{\varphi}\check{\psi} = (\varphi\psi)\check{}$.

Proposition 12.16. $Ml(\varphi', \psi') = \lambda\theta.\varphi'(\theta)\psi'(\theta)$.

Proof.

$$Ml(\varphi', \psi') = \lambda\theta.Ml((\varphi', \psi')(\theta)) = \lambda\theta.Ml((\varphi'(\theta), \psi'(\theta))) = \lambda\theta.\varphi'(\theta)\psi'(\theta).$$

Therefore, the element Ml represents in $\mathscr{F}'$ the semigroup multiplication of $\mathscr{F}$.

Proposition 12.17. $Ml \notin \bar{\mathscr{F}}$.

Proof. Suppose that $Ml = \bar{\varphi}$. Then $\varphi = \bar{\varphi}(I) = Ml(I) = LR$, hence $LR^2 = \overline{LR}(R) = Ml(R) = LR^3$, which is not the case.

Proposition 12.18. $Ml(\check{\varphi}, \psi') = \bar{\varphi}\psi'$.

This follows from 12.16.

Proposition 12.19. $Ml(\check{\varphi}, I') = \bar{\varphi}$.

This follows from 12.18.

Proposition 12.20. $Ml(\check{\varphi}, \check{\psi}) = (\varphi\psi)\check{}$.

This follows from 12.18, 12.15.

Proposition 12.21. $Ml(\varphi' Id, I') = \overline{\varphi'(I)}$. Whenever $\bar{\psi} Id = \varphi' Id$, then $\psi = \varphi'(I)$.
This follows from 12.14, 12.19.

Being especially interested in IOS, we assume from now on that both $\mathscr{S}$ and $\mathscr{S}'$ are iterative. The μ-axiom of $\mathscr{S}$ used will as usual be indicated by the corresponding number of asterisks. In order to ensure that $\bar{\mathscr{S}}$ is a subspace of $\mathscr{S}'$ isomorphic with $\mathscr{S}$ as an IOS, we assume finally that $\langle I' \rangle = \overline{\langle I \rangle}$.

Notice in connection with the last equality that $(L', \overline{\langle I \rangle} R') = \overline{(L, \langle I \rangle R)} = \overline{\langle I \rangle}$ implies $\langle I' \rangle \leq \overline{\langle I \rangle}$. The converse inequality can also be proved by μA_2: $R'I' = I'R'$, $(L', \langle I' \rangle R') = \langle I' \rangle$ imply $\overline{\langle I \rangle} \leq \langle I' \rangle$ by the suggested hint to exercise 12.1. Moreover, $\langle I' \rangle = \langle I' \rangle \overline{\langle I \rangle}$ by 6.22; hence $\langle I' \rangle = \overline{\langle I \rangle}$ whenever $\langle I' \rangle = I'$.

An immediate example of consecutive IOS is that of the IOS $\mathscr{S}$ of example 3.1 and the space $\mathscr{S}'$ obtained from it by 12.1. The latter is iterative since being isomorphic with the IOS of example 3.2, while the equality $\overline{\langle I \rangle} = \langle I' \rangle = I'$ also holds in this case.

Proposition 12.22. $\langle \bar{\varphi} \rangle = \overline{\langle \varphi \rangle}$.
Proof.

$$(\bar{\varphi} L', \overline{\langle \varphi \rangle} R') = \overline{(\varphi L, \langle \varphi \rangle R)} = \overline{\langle \varphi \rangle}$$

implies $\langle \bar{\varphi} \rangle = \langle I' \rangle \overline{\langle \varphi \rangle}$ by 6.22. Therefore,

$$\langle \bar{\varphi} \rangle = \overline{\langle I \rangle}\, \overline{\langle \varphi \rangle} = \overline{\langle I \rangle \langle \varphi \rangle} = \overline{\langle \varphi \rangle}.$$

Proposition 12.23. $[\bar{\varphi}] = \overline{[\varphi]}$.
Proof. We have

$$(I', \bar{\varphi} \overline{[\varphi]}) = \overline{(I, \varphi[\varphi])} = \overline{[\varphi]},$$

hence $[\bar{\varphi}] \leq \overline{[\varphi]}$. On the other hand,

$$(\theta, \varphi[\bar{\varphi}](\theta)) = (I', \bar{\varphi}[\bar{\varphi}])(\theta) = [\bar{\varphi}](\theta)$$

implies $[\varphi]\theta \leq [\bar{\varphi}](\theta)$ by (££). Therefore, $\overline{[\varphi]}(\theta) \leq [\bar{\varphi}](\theta)$ for all θ, hence $\overline{[\varphi]} \leq [\bar{\varphi}]$. The proof is complete.

Proposition 12.24 (Imbedding Theorem). $\bar{\mathscr{S}}$ is an iterative subspace of $\mathscr{S}'$ isomorphic with the IOS $\mathscr{S}$.
This follows from 12.11, 12.22, 12.23.

Proposition 12.25. $[\varphi'] = \lambda\theta . \mu\theta_1 . (\theta, \varphi'(\theta_1))$.
Proof. We have

$$[\varphi'](\theta) = (I', \varphi'[\varphi'])(\theta) = (\theta, \varphi'([\varphi'](\theta))).$$

Suppose that $(\theta, \varphi'(\tau)) \leq \tau$. Then $(\check{\theta}, \varphi' \check{\tau}) \leq \check{\tau}$, hence $[\varphi']\check{\theta} \leq \check{\tau}$ by (££), i.e. $[\varphi'](\theta) \leq \tau$. Therefore, $[\varphi'](\theta) = \mu\theta_1 . (\theta, \varphi'(\theta_1))$ for all θ, which completes the proof. (Compare with the proof of 3.4.)

Proposition 12.26. $R'[\varphi'] = \lambda\theta.\mu\theta_1.\varphi'((\theta,\theta_1))$.
The proof follows that of 12.25, making use of 6.11 instead of (££).

Proposition 12.27. $R'[\varphi' R'] = (\mu\theta.\varphi'(\theta))^{\check{}}$.
This follows from 12.26.

Proposition 12.28. Let $Tr = \langle\ \ \rangle$. Then Tr is a member of $\mathscr{F}'$ prime recursive in *Id, Ml.*
Proof. Take $\sigma' = (Ml(L', \check{L}), Ml(R', \check{R}))$. Then $\sigma'((\theta,\theta_1)) = (\theta L, \theta_1 R)$ for all θ, θ_1, hence

$$Tr = \lambda\theta.\mu\theta_1.(\theta L, \theta_1 R) = R'[\sigma']$$

by 12.26.

Proposition 12.29. Let $It = [\ \ ]$. Then It is a member of $\mathscr{F}'$ prime recursive in *Id, Ml.*
Proof. Taking $\sigma' = (Id, Ml)$, we get, using 12.26

$$It = \lambda\theta.\mu\theta_1.(I, \theta\theta_1) = R'[\sigma'].$$

Proposition 12.30. $Tr,\ It \notin \bar{\mathscr{F}}$.
Proof. Suppose $Tr = \bar{\varphi}$. Then $\varphi = \bar{\varphi}(I) = Tr(I) = \langle I \rangle$, hence $\langle I \rangle L = \bar{\varphi}(L) = \langle L \rangle$, which is not the case. Iteration is treated similarly.

Proposition 12.31. If φ' is recursive, then it is prime recursive in *Id, Ml.*
Proof. All the recursive members of $\mathscr{F}'$ are in $\bar{\mathscr{F}}$. An easy induction on the construction of such an element $\bar{\varphi}$ shows that it is prime recursive in *Id, Ml*: we use of the equalities $\langle \bar{\varphi} \rangle = Ml(Tr\bar{\varphi}Id, I')$, $[\bar{\varphi}] = Ml(It\bar{\varphi}Id, I')$ which follow from 12.21 and take 12.28, 12.29 into account.

Proposition 12.32. $\langle \check{\varphi} \rangle = [I']\check{\varphi}$. In particular, $\langle Id \rangle = [I']Id$
Proof. We have

$$(\check{\varphi}L', [I']\check{\varphi}R') = (\check{\varphi}, [I']\check{\varphi}) = [I']\check{\varphi};$$

hence $\langle \check{\varphi} \rangle \leq [I']\check{\varphi}$. On the other hand,

$$(\check{\varphi}, \langle \check{\varphi} \rangle O') = (\check{\varphi}L', \langle \check{\varphi} \rangle R')O' = \langle \check{\varphi} \rangle O'$$

implies $[I']\check{\varphi} \leq \langle \check{\varphi} \rangle O'$ by (££); hence $[I']\check{\varphi} \leq \langle \check{\varphi} \rangle I' = \langle \check{\varphi} \rangle$. The proof is complete.

Proposition 12.33. $\langle Ml \rangle = \langle L' \rangle D' Ml(TrMl(I', \check{R}), I')$.
Proof. Writing ρ' for $\langle L' \rangle D' Ml(TrMl(I', \check{R}), I')$, we have

$$\begin{aligned}\rho'(\theta) &= \langle L \rangle D \langle \theta R \rangle \theta = (L^2, \langle L \rangle R) D \langle \theta R \rangle \theta \\ &= (L^3 \langle \theta R \rangle \theta, \langle L \rangle D \langle R \rangle R \langle \theta R \rangle \theta) \\ &= (L^2 \theta R L \theta, \langle L \rangle D \langle R\theta R \rangle R\theta) = (Ml(L\theta), \rho'(R\theta))\end{aligned}$$

for all θ, i.e. $\rho' = (MlL', \rho' R')$, which implies $\langle Ml \rangle = \langle I' \rangle \rho' = \rho'$ by 6.22. The proof is complete.

The following Pull Back Theorem generalizes 12.31.

Proposition 12.34. Let $\mathscr{B} \subseteq \mathscr{F}$, $\mathscr{B}' \subseteq \mathscr{F}'$. Then the following are equivalent.

(1) φ' is recursive in $\{Id, Ml\} \cup \breve{\mathscr{B}} \cup \mathscr{B}'$.

(2) φ' is prime recursive in $\{Id, Ml\} \cup \breve{\mathscr{B}} \cup \langle \mathscr{B}' \rangle$.

(3) $\varphi' \in cl(\{L', A', Id, Ml\} \cup \breve{\mathscr{B}} \cup \langle \mathscr{B}' \rangle / \circ, [\ \ ])$.

If all the members of $\langle \mathscr{B}' \rangle$ are prime recursive in $\{Id, Ml\} \cup \breve{\mathscr{B}} \cup \mathscr{B}'$, then $\mathscr{B}'$ can be substituted for $\langle \mathscr{B}' \rangle$ in (2), (3).

Proof. The implication (1)$\Rightarrow$(2) follows from 7.11, 12.31–12.33 and 12.28, while the other implications are immediate.

It should be mentioned that $\{Id, Ml\} \cup \breve{\mathscr{B}}$ can be replaced by $\{Id, Ml\} \cup \bar{\mathscr{B}}$ above, in view of 12.3, 12.19. In future this fact will be employed without further mention.

The notion of inductive mapping over $\mathscr{F}$ can be relativized by substituting $\{I, L, R\} \cup \mathscr{B}$ for $\{I, L, R\}$ in the first clause of the definition given in chapter 5. The following statement characterizes this notion of relative inductiveness in the terms of $\mathscr{S}'$.

Proposition 12.35. Let Γ be a $n+1$-ary mapping over $\mathscr{F}$ inductive in $\mathscr{B}$ and let Γ^* correspond to Γ by 7.20. Then Γ^* is a member of $\mathscr{F}'$ prime recursive in $\{Id, Ml\} \cup \breve{\mathscr{B}}$. Moreover, $\mu\theta.\Gamma(\theta_1, \ldots, \theta_n, \theta)$ exists for all $\theta_1, \ldots, \theta_n$.

Proof. By induction on the construction of Γ.

Let $\Gamma = \lambda\theta_1 \ldots \theta_n.\theta_i$, $1 \leq i \leq n$. If $i < n$, then $\Gamma^* = \lambda\theta.\bar{i}\theta = L'R'^i$. If $i = n$, then $\Gamma^* = \lambda\theta.R^{n-1}\theta = R'^{n-1}$.

If $\Gamma = \lambda\theta_1 \ldots \theta_n.\psi$, $\psi \in \{I, L, R\} \cup \mathscr{B}$, then $\Gamma^* = \lambda\theta.\psi = \breve{\psi}$.

Let Γ_1, Γ_2 be n-ary and Γ_1^*, $\Gamma_2^* \in \mathscr{F}'$ correspond to Γ_1, Γ_2.

If $\Gamma = \lambda\theta_1 \ldots \theta_n.\Gamma_1(\theta_1, \ldots, \theta_n)\Gamma_2(\theta_1, \ldots, \theta_n)$, then $\Gamma^* = \lambda\theta.\Gamma_1^*(\theta)\Gamma_2^*(\theta) = Ml(\Gamma_1^*, \Gamma_2^*)$.

If $\Gamma = \lambda\theta_1 \ldots \theta_n.(\Gamma_1(\theta_1, \ldots, \theta_n), \Gamma_2(\theta_1, \ldots, \theta_n))$, then $\Gamma^* = \lambda\theta.(\Gamma_1^*(\theta), \Gamma_2^*(\theta)) = (\Gamma_1^*, \Gamma_2^*)$.

Let Γ_1 be $n+1$-ary, $n > 0$, and $\Gamma_1^* \in \mathscr{F}'$ correspond to Γ_1. Take

$$\Gamma^* = R'[\Gamma_1^*(L'^2, \ldots, L'R'^{n-2}L', R'^{n-1}L', R')].$$

Then it follows that $\Gamma^* = \lambda\theta.\mu\theta_1.\Gamma_1^*((\bar{0}\theta, \ldots, \overline{n-2}\theta, R^{n-1}\theta, \theta_1))$ by 12.26, hence Γ^* corresponds to the maping $\Gamma = \lambda\theta_1 \ldots \theta_n.\mu\theta.\Gamma_1(\theta_1, \ldots, \theta_n, \theta)$. This completes the proof.

Proposition 12.36. Whenever $\varphi' \in \mathscr{F}'$ is prime recursive in $\{Id, Ml\} \cup \breve{\mathscr{B}}$, then φ' is a mapping inductive in $\mathscr{B}$.

This follows from an easy induction on the construction of φ'. In particular, if φ' is inductive in $\mathscr{B}$, then so is $[\varphi']$ since $[\varphi'] = \lambda\theta.\mu\theta_1.(\theta, \varphi'(\theta_1))$ by 12.25.

Proposition 12.37. Let Γ be a unary mapping over $\mathscr{F}$. Then the following are equivalent.

(1) Γ is a mapping inductive in $\mathscr{B}$.
(2) Γ is a member of $\mathscr{F}'$ prime recursive in $\{Id, Ml\} \cup \check{\mathscr{B}}$.
(3) Γ is a member of $\mathscr{F}'$ recursive in $\{Id\ Ml\} \cup \check{\mathscr{B}}$.

This follows from 12.34–12.36.

The following Conservativeness Theorem is the central result of this chapter.

Proposition 12.38*. Let Γ be a unary mapping over $\mathscr{F}$. Then the following are equivalent.

(1) Γ is a mapping recursive in $\mathscr{B}$.
(2) Γ is a member of $\mathscr{F}'$ prime recursive in $\{Id, Ml\} \cup \check{\mathscr{B}}$.
(3) Γ is a member of $\mathscr{F}'$ recursive in $\{Id, Ml\} \cup \check{\mathscr{B}}$.

This follows from 12.37, 9.9 and 9.15*. Notice that $\check{\mathscr{B}}$ can be replaced by $\bar{\mathscr{B}}$.

Therefore, the relative recursiveness of $\mathscr{S}$ is conservative in the next space $\mathscr{S}'$. In other words, the initial operations of $\mathscr{S}'$ preserve it. This is immediate for $\circ$, Π and quite apparent for [ ] since the iteration of $\mathscr{S}'$ is in essence the μ-operation over $\mathscr{F}$ by 12.25. However, the fact that the operation translation of $\mathscr{S}'$ preserves the relative recursiveness of $\mathscr{S}$ is curious and worth stating separately.

Proposition 12.39. If $\varphi' \in \mathscr{F}'$ is recursive in $\mathscr{B}$ as a mapping over $\mathscr{F}$, then so is $\langle \varphi' \rangle$.

Proof. Take Γ^* to correspond to φ' by the Transition Theorem. Then Γ^* is recursive in $\mathscr{B}$ and $\Gamma^*(\theta) = (\varphi'(L\theta), \Gamma^*(R\theta))$ for all θ. It follows from 12.35 that $\Gamma^* \in \mathscr{F}'$, hence $\Gamma^* = (\varphi' L', \Gamma^* R')$, which implies $\langle \varphi' \rangle = \langle I' \rangle \Gamma^*$ by 6.22. However, $\langle I \rangle \Gamma^*(\theta) = \Gamma^*(\theta)$ for all θ, hence $\langle I' \rangle \Gamma^* = \overline{\langle I \rangle \Gamma^*} = \Gamma^*$ and $\langle \varphi' \rangle = \Gamma^*$. The proof is complete.

Exercise 7.6 and the above proof imply that the operation translation of $\mathscr{S}'$ also preserves the relative primitive recursiveness of $\mathscr{S}$. Therefore, propositions 7.21, 12.39, exercise 7.6 and proposition 12.25 throw some more light on the nature of the operations translation and iteration.

Assume now that $\langle\!\langle\ \ \rangle\!\rangle$ is a t-operation over $\mathscr{F}$ with a corresponding set of functional elements $\mathscr{B}_0$ such that $To = \langle\!\langle\ \ \rangle\!\rangle \in \mathscr{F}'$. Then one may establish analogues to the above statements by adding $\check{\mathscr{B}}_0$ and To to the initial elements Id, Ml of $\mathscr{F}'$. Thus let us write $\mathscr{B}'_0$ for $\{Id, Ml, To\} \cup \check{\mathscr{B}}_0$.

Proposition 12.40. The element $\langle To \rangle$ is prime recursive in $\mathscr{B}'_0$.

Proof. Take the mappings $\langle\!\langle\ \ \rangle\!\rangle^*$, Γ_1^*, Γ_3 considered in the proof of 10.7. The mapping Γ_1^*, Γ_3 are recursive in $\mathscr{B}_0$, hence they are members of $\mathscr{F}'$ prime recursive in $\{Id, Ml\} \cup \check{\mathscr{B}}_0$ by 12.35. Therefore, $\langle\!\langle\ \ \rangle\!\rangle^*$ is a member of $\mathscr{F}'$ prime recursive in $\mathscr{B}'_0$ since $\langle\!\langle\ \ \rangle\!\rangle^* = \Gamma_1^* \Gamma_3 To$. It follows that $\langle To \rangle = \langle\!\langle\ \ \rangle\!\rangle^*$ by 6.22. The proof is complete.

10.7 guarantees that the operation $\langle\ \ \rangle$ of $\mathscr{S}'$ preserves the relative t-recursiveness of $\mathscr{S}$.

Making use of 12.40 and the proof of 12.34, one gets the following analogue to the latter statement.

Proposition 12.41. Let $\mathscr{B} \subseteq \mathscr{F}$, $\mathscr{B}' \subseteq \mathscr{F}'$. Then the following are equivalent.

(1) φ' is recursive in $\mathscr{B}'_0 \cup \check{\mathscr{B}} \cup \mathscr{B}'$.

(2) φ' is prime recursive in $\mathscr{B}'_0 \cup \check{\mathscr{B}} \cup \langle \mathscr{B}' \rangle$.

(3) $\varphi' \in cl(\{A'\} \cup \mathscr{B}'_0 \cup \check{\mathscr{B}} \cup \langle \mathscr{B}' \rangle / \circ, [\ \ ])$.

If all the member of $\langle \mathscr{B}' \rangle$ are prime recursive in $\mathscr{B}'_0 \cup \check{\mathscr{B}} \cup \mathscr{B}'$, then $\mathscr{B}'$ can be substituted for $\langle \mathscr{B}' \rangle$ in (2), (3).

Finally, the Parametrized First t-Recursion Theorem and an analogue to 12.37 give the following t-Conservativeness Theorem.

Proposition 12.42*. Let Γ be a unary mapping over $\mathscr{F}$. Then the following are equivalent.

(1) Γ is a mapping t-recursive in $\mathscr{B}$.

(2) Γ is a member of $\mathscr{F}'$ prime recursive in $\mathscr{B}'_0 \cup \check{\mathscr{B}}$.

(3) Γ is a member of $\mathscr{F}'$ recursive in $\mathscr{B}'_0 \cup \check{\mathscr{B}}$.

EXERCISES TO CHAPTER 12

Exercise 12.1. Let $\mathscr{S}$, $\mathscr{S}'$ be consecutive OS, and suppose that $\mathscr{S}$ is μA_2-iterative and the equality $(\varphi' L', \theta' R') = \theta'$ has a solution in $\mathscr{F}'$ for all φ'. Show that $\mathscr{S}'$ satisfies (£).

Hint. It suffices by 5.11 to find an element $\langle I' \rangle$ satisfying (£). Take $\langle I' \rangle = \overline{\langle I \rangle}$ and show that $(L', \langle I' \rangle R') = \langle I' \rangle$. Supposing $R'\psi' \leq \psi'\psi'_1$ and $(L'\psi', \tau'\psi'_1) \leq \tau'$, show that for all $\rho \in \mathscr{F}$ the normal segment $\mathscr{E} = \{\theta / \forall n(\theta\psi'\psi_1'^n(\rho) \leq \tau'\psi_1'^n(\rho))\}$ is closed under $\lambda\theta.(L, \theta R)$.

Exercise 12.2. Let $\mathscr{S}$, $\mathscr{S}'$ be consecutive OS and suppose that $\mathscr{F}'$ is closed under μ-operation over $\mathscr{F}$, i.e. whenever $\varphi' \in \mathscr{F}'$, then $\mu\theta_1.(\theta, \varphi'(\theta_1))$ exists for all θ and $\lambda\theta.\mu\theta_1.(\theta, \varphi'(\theta_1)) \in \mathscr{F}'$. Show that $\mathscr{S}'$ meets (££).

Hint. Take $[\varphi'] = \lambda\theta.\mu\theta_1.(\theta, \varphi'(\theta_1))$ and show that $(I', \varphi'[\varphi']) = [\varphi']$. If $(\psi', \varphi'\tau') \leq \tau'$, then $(\psi'(\theta), \varphi'(\tau'(\theta))) \leq \tau'(\theta)$; hence $[\varphi'](\psi'(\theta)) \leq \tau'(\theta)$ for all θ, i.e. $[\varphi']\psi' \leq \tau'$.

By 12.25, the requirement of the last exercise are necessary and sufficient for (££).

Assume now that $\mathscr{S}$, $\mathscr{S}'$ are consecutive IOS. Let $\Gamma_1, \ldots, \Gamma_m$ be unary mappings over $\mathscr{F}$ recursive in $\mathscr{B}$ and $\mathscr{V}(\theta, \Gamma)$ be the expression considered in exercise 9.8. Then $\Gamma_1, \ldots, \Gamma_m$ are members of $\mathscr{F}'$ recursive in $\{Id, Ml\} \cup \check{\mathscr{B}}$ by 12.37 and it follows that there is a unary mapping Γ' over $\mathscr{F}'$ polynomial in $\Gamma_1, \ldots, \Gamma_m$ such that $\Gamma'(\Gamma) = \lambda\theta.\mathscr{V}(\theta, \Gamma)$ for all $\Gamma \in \mathscr{F}'$. The First Recursion Theorem for $\mathscr{S}'$ implies that Γ' has a least fixed point Γ recursive in $\Gamma_1, \ldots, \Gamma_m$, hence recursive in $\{Id, Ml\} \cup \check{\mathscr{B}}$. (Proposition 9.11 may also serve, provided $c(\Gamma') \leq 1$.) Therefore, Γ is recursive in $\mathscr{B}$ by 12.38*,

(1) $$\mathscr{V}(\theta, \Gamma) = \Gamma(\theta), \text{ all } \theta,$$

and $\Gamma \leq \Gamma^*$ whenever $\Gamma^* \in \mathscr{F}'$ meets (1). However, (1) may have solutions which are nonmonotonic mappings and hence not in $\mathscr{F}'$. The next two exercises will show that Γ is least among all solutions of (1).

Take $\mathscr{F}'' = \{\varphi''/\varphi'':\mathscr{F} \to \mathscr{F}\}$ and introduce $\leq, \circ, \Pi''$ as in 12.1. Then $\mathscr{S}'' = (\mathscr{F}'', I', \Pi'', L', R')$ is almost an OS except for the nonmonotonicity of $\circ$ on its second argument. Notice that the OS $\mathscr{S}'$ is a 'subspace' of $\mathscr{S}''$ and whenever $\psi'' \leq \psi_1''$, then $\varphi'\psi'' \leq \varphi'\psi_1''$ for all $\varphi' \in \mathscr{F}'$.

Exercise 12.3. Let $\mathscr{S}'$ be μA_2-iterative and Γ' a unary mapping over $\mathscr{F}'$ polynomial in $\mathscr{F}'$. Show that $\mu\theta'.\Gamma'(\theta') = \mu\theta''.\Gamma'(\theta'')$ where the domain of Γ' is naturally extended to $\mathscr{F}''$.

Hint. Show first by induction on the construction of Γ' that $\theta' \leq \tau''$ implies $\Gamma'(\theta') \leq \Gamma'(\tau'')$. Assuming $\Gamma'(\tau'') \leq \tau''$, show that the normal segment

$$\mathscr{E}' = \{\theta'/\theta' \leq \tau''\} = \{\theta'/\forall\theta(\theta'\breve{\theta} \leq (\tau''(\theta))\breve{\ })\}$$

is closed under Γ'; hence $\mu\theta'.\Gamma'(\theta') \in \mathscr{E}'$ by μA_2.

While the above set of inequalities defining $\mathscr{E}'$ is most probably uncountable, its role could be played by countable one reflecting the construction of Γ'. A simple example outlines the idea. Let $\Gamma' = \lambda\theta'.\varphi'(I', \theta'\psi'\theta')$ and $\Gamma'(\tau'') \leq \tau''$. Then fix $\theta \in \mathscr{F}$ and show that the normal segment

$$\mathscr{E}' = \{\theta'/\forall n(\theta'(\psi'\tau'')^n\breve{\theta} \leq \tau''(\psi'\tau'')^n\breve{\theta})\}$$

is closed under Γ'.

The following exercise establishes a Generalized First Recursion Theorem for $\mathscr{S}$.

Exercise 12.4*. Let $\mathscr{S}'$ be μA_2-iterative, let $\mathscr{V}(\theta, \Gamma)$ be the expression considered in exercise 9.8 and suppose that $\Gamma_1, \ldots, \Gamma_m$ are recursive in $\mathscr{B}$. Show that there is a unary mapping Γ recursive in $\mathscr{B}$ such that $\mathscr{V}(\theta, \Gamma) = \Gamma(\theta)$ for all θ and whenever $\Gamma^*:\mathscr{F} \to \mathscr{F}$ and $\mathscr{V}(\theta, \Gamma^*) \leq \Gamma^*(\theta)$ for all θ, then $\Gamma(\theta) \leq \Gamma^*(\theta)$ for all θ.

Hint. Use 9.13* for $\mathscr{S}'$, exercise 12.3 and 12.38*.

The following exercise establishes a Generalized First t-Recursion Theorem.

Exercise 12.5*. Let $\mathscr{S}'$ be μA_2-iterative, let $\langle\ \ \rangle \in \mathscr{F}'$ be a t-operation over $\mathscr{F}$, let $\mathscr{V}(\theta, \Gamma)$ be the expression of exercise 9.8 and suppose that $\Gamma_1, \ldots, \Gamma_m$ are t-recursive in $\mathscr{B}$. Show that there is a unary mapping Γ t-recursive in $\mathscr{B}$ such that $\mathscr{V}(\theta, \Gamma) = \Gamma(\theta)$ for all θ and, if $\Gamma^*:\mathscr{F} \to \mathscr{F}$ and $\mathscr{V}(\theta, \Gamma^*) \leq \Gamma^*(\theta)$ for all θ, then $\Gamma(\theta) \leq \Gamma^*(\theta)$ for all θ.

Hint. Follow the hint to the previous exercise, using 12.42* instead of 12.38*.

As was the case for exercise 9.8, exercises 12.4*, 12.5* can be restated for n-ary mappings as well.

Exercise 12.6. Show that φ' is recursive in $\{\breve{\psi}_1, \ldots, \breve{\psi}_m\} \cup \mathscr{B}'$ iff there is a χ recursive in $\mathscr{B}'$ such that $\varphi' = \chi'(\breve{\psi}_1, \ldots, \breve{\psi}_m, I')$.

Hint. Make use of 12.4, 12.32, 6.11.

CHAPTER 13

$\mathscr{B}'$-Recursiveness

Let us recall that the concept of consecutive spaces was introduced in chapter 12 because we wanted to study monotonic operations other than the initial IOS-operations. So take consecutive IOS $\mathscr{S}$, $\mathscr{S}'$ and fix a subset $\mathscr{B}'$ of $\mathscr{F}'$. We are interested in elements of $\mathscr{F}$ and mappings over $\mathscr{F}$ which can be constructed by making use of operations from $\mathscr{B}'$. This leads to the following definitions.

A unary mapping Γ and $\mathscr{F}$ is said to be *$\mathscr{B}'$-recursive in* $\mathscr{B} \subseteq \mathscr{F}$ iff Γ is a member of $\mathscr{F}'$ recursive in $\{Id, Ml\} \cup \check{\mathscr{B}} \cup \mathscr{B}'$. An element φ is *$\mathscr{B}'$-recursive in* $\mathscr{B}$ iff $\check{\varphi}$ is a mapping $\mathscr{B}'$-recursive in $\mathscr{B}$. Therefore, the class $\mathscr{R}(\mathscr{B}, \mathscr{B}')$ of unary mappings $\mathscr{B}'$-recursive in $\mathscr{B}$ equals

$$cl(\check{\mathscr{B}} \cup \{L', R', Id, Ml\} \cup \mathscr{B}'/\circ, \Pi, \langle\ \ \rangle, [\ \ ]).$$

While the initial operations of $\mathscr{S}'$ are allowed by the above definition, 12.16, 12.28, 12.29 imply that those of $\mathscr{S}$ are also available. In some favourable cases the operations $\langle\ \ \rangle$, $[\ \ ]$ of $\mathscr{S}'$ can be eliminated and one may consider the class

$$\mathscr{R}_0(\mathscr{B}, \mathscr{B}') = cl(\check{\mathscr{B}} \cup \{L', R', Id, Ml, Tr, It\} \cup \mathscr{B}'/\circ, \Pi)$$

instead. For instance, $\mathscr{R}_0(\mathscr{B}, \varnothing) = \mathscr{R}(\mathscr{B}, \varnothing)$ by 12.38*.

It is obvious that $\mathscr{R}_0(\mathscr{B}, \mathscr{B}') \subseteq \mathscr{R}(\mathscr{B}, \mathscr{B}')$. In order to get equality, one should prove first that all the mappings in $\mathscr{R}_0(\mathscr{B}, \mathscr{B}')$ satisfy the transition property, thus ensuring that $\mathscr{R}_0(\mathscr{B}, \mathscr{B}')$ is closed under the operation $\langle\ \ \rangle$ of $\mathscr{S}'$. This is not a major obstacle since translation can be pulled back to the members of $\mathscr{B}'$ by 12.34 anyway. Secondly, one should prove a Parametrized First Recursion Theorem to ensure that $\mathscr{R}_0(\mathscr{B}, \mathscr{B}')$ is closed under μ-operation over $\mathscr{F}$, hence closed under the operation $[\ \ ]$ of $\mathscr{S}'$. This cannot be expected to be the case for arbitrary subsets $\mathscr{B}'$ of $\mathscr{F}'$ and as a result $\mathscr{R}_0(\mathscr{B}, \mathscr{B}')$ may differ from $\mathscr{R}(\mathscr{B}, \mathscr{B}')$.

Now if $\mathscr{B}'$ is *finite*, then there is by 9.18 a σ' recursive in $\{Id, Ml\} \cup \mathscr{B}'$ and universal for all the members of $\mathscr{F}'$ recursive in $\{Id, Ml\} \cup \mathscr{B}'$. Therefore, $\mathscr{R}(\mathscr{B}, \mathscr{B}') = cl(\check{\mathscr{B}} \cup \{I', L', R', \sigma'\}/\circ, \Pi)$ by exercise 12.6, hence $\mathscr{R}_0(\mathscr{B}, \{\sigma'\}) = \mathscr{R}(\mathscr{B}, \{\sigma'\})$ for all $\mathscr{B}$. However, while the replacement of $\mathscr{B}'$ by $\langle \mathscr{B}' \rangle$ can still be tolerated, the replacement of $\mathscr{B}'$ by σ' cannot. The mapping σ' can hardly be accepted as an initial operation, being far too sophisticated.

The definition of $\mathscr{B}'$-recursiveness assumes a First Recursion Theorem at the very beginning; it would be preferable to prove such a property rather than include it in the definition. Therefore, our ability to treat arbitrary monotonic operations over $\mathscr{F}$ is achieved only at a price.

Notions of mappings and elements *prime $\mathscr{B}'$-recursive* (respectively, *primitive $\mathscr{B}'$-recursive, $\mathscr{B}'$-primitive, $\mathscr{B}'$-polynomial) in $\mathscr{B}$* are introduced in the same way. For example, a unary mapping Γ is prime $\mathscr{B}$-recursive in $\mathscr{B}$ iff it is a member of $\mathscr{F}'$ prime recursive in $\{Id, Ml\} \cup \breve{\mathscr{B}} \cup \mathscr{B}'$, an element φ is prime $\mathscr{B}'$-recursive in $\mathscr{B}$ iff so is the mapping $\breve{\varphi}$. Notice that $\bar{\varphi}$, $\bar{\mathscr{B}}$ can be substituted for $\breve{\varphi}$, $\breve{\mathscr{B}}$ by 12.3, 12.19. Further on $\bar{n}$ will stand for LR^n or $L'R'^n$, depending on the context.

The notions thus introduced have certain ordinary properties similar to those stated at the beginning of chapter 7. Some of them are listed below, mostly concerning mappings. Obvious proofs are omitted.

Proposition 13.1. If $\varphi \in \mathscr{B}$, then φ is $\mathscr{B}'$-polynomial in $\mathscr{B}$. If $\Gamma \in \mathscr{B}'$, then Γ is $\mathscr{B}'$-polynomial in $\mathscr{B}$.

Proposition 13.2. If Γ is $\mathscr{B}'$-polynomial in $\mathscr{B}$, then it is both $\mathscr{B}'$-primitive in $\mathscr{B}$ and prime $\mathscr{B}'$-recursive in $\mathscr{B}$. If Γ is $\mathscr{B}'$-primitive in $\mathscr{B}$, then it is primitive $\mathscr{B}'$-recursive in $\mathscr{B}$. If Γ is prime or primitive $\mathscr{B}'$-recursive in $\mathscr{B}$, then it is $\mathscr{B}'$-recursive in $\mathscr{B}$.

Proposition 13.3. If $\mathscr{B} \subseteq \mathscr{B}_1$, $\mathscr{B}' \subseteq \mathscr{B}'_1$ and Γ is $\mathscr{B}'$-recursive (prime $\mathscr{B}'$-recursive etc.) in $\mathscr{B}$, then Γ is $\mathscr{B}'_1$-recursive (prime $\mathscr{B}'_1$-recursive etc.) in $\mathscr{B}_1$.

Proposition 13.4. If Γ is $\mathscr{B}'$-recursive (prime $\mathscr{B}'$-recursive etc.) in $\mathscr{B}$ and all the members of $\mathscr{B}$ are $\mathscr{B}'$-recursive (prime $\mathscr{B}'$-recursive etc.) in $\mathscr{B}_1$, then so is Γ.

Proposition 13.5. If Γ is $\mathscr{B}'$-recursive (prime $\mathscr{B}'$-recursive etc.) in $\mathscr{B}$ and all the members of $\mathscr{B}'$ are $\mathscr{B}'_1$-recursive (prime $\mathscr{B}'_1$-recursive etc.) in $\mathscr{B}$, then so is Γ.

Proposition 13.6. If Γ is $\mathscr{B}'$-recursive (prime $\mathscr{B}'$-recursive etc.) in $\mathscr{B}$, then there are finite subsets $\mathscr{B}_1$, $\mathscr{B}'_1$ of $\mathscr{B}$, $\mathscr{B}'$ respectively, such that Γ is $\mathscr{B}'_1$-recursive (prime $\mathscr{B}'_1$-recursive etc.) in $\mathscr{B}_1$.

Proposition 13.7. A mapping Γ is $\mathscr{B}'$-recursive in $\psi_1, \ldots, \psi_m$ iff there is a $\mathscr{B}'$-recursive mapping Γ^* such that $\Gamma = \lambda\theta.\Gamma^*((\psi_1, \ldots, \psi_m, \theta))$.

This follows from exercise 12.6.

Proposition 13.8. Γ is $\mathscr{B}'$-recursive in $\mathscr{B}$ iff Γ is prime $\langle \mathscr{B}' \rangle$-recursive in $\mathscr{B}$.

This follows from 12.34.

Proposition 13.9*. Γ is $\varnothing$-recursive in $\mathscr{B}$ iff Γ is recursive in $\mathscr{B}$.

This follows from 12.38*.

It is also the case that Γ is $\varnothing$-polynomial in $\mathscr{B}$ iff Γ is polynomial in $\mathscr{B}$. However, such an equivalence may fail for the other notions. Indeed, the mappings prime $\emptyset$-recursive in $\mathscr{B}$ are those recursive in $\mathscr{B}$ by 12.38*, while the mappings prime recursive in $\mathscr{B}$ are those $\{It\}$-polynomial in $\mathscr{B}$.

We next prove several Normal Form Theorems for prime $\mathscr{B}'$-recursiveness and $\mathscr{B}'$-recursiveness.

Proposition 13.10. If Γ is prime $\mathscr{B}'$-recursive in $\mathscr{B}$, then $\Gamma = \lambda\theta.L\mu\theta_1.\Gamma_0((\theta,\theta_1))$ for a certain Γ_0 strictly $\mathscr{B}'$-polynomial in $\mathscr{B}$. (That is, Γ_0 is a member of $\mathscr{F}'$ strictly polynomial in $\breve{\mathscr{B}} \cup \{Id, Ml\} \cup \mathscr{B}'$.)

Proof. It follows from 9.2 that $\Gamma = \bar{1}[\Gamma_0]$ with Γ_0 strictly $\mathscr{B}'$-polynomial in $\mathscr{B}$. Therefore, Γ has the desired normal form by 12.26.

Proposition 13.11. If φ is prime $\mathscr{B}'$-recursive in $\mathscr{B}$, then $\varphi = L\mu\theta.\Gamma(\theta)$ for a certain Γ strictly $\mathscr{B}'$-polynomial in $\mathscr{B}$.

Proof. This proof needs some care since a direct application of 13.10 to $\breve{\varphi}$ would only give a normal form with Γ $\mathscr{B}'$-polynomial in $\mathscr{B}$.

Proposition 9.2 implies that $\breve{\varphi} = \bar{1}[\Gamma_0]$ with Γ_0 strictly $\mathscr{B}'$-polynomial in $\mathscr{B}$. Multiplying by Id and using 6.13, 6.14, we get

$$\breve{\varphi} = \bar{2}[(Id, \Gamma_0 R')] = \bar{1}[(\bar{4}, \bar{0}, Id, \Gamma_0 R'^3)],$$

hence $\breve{\varphi} = \lambda\theta.L\mu\theta_1.(\bar{3}\theta_1, \theta, I, \Gamma_0 R'^2(\theta_1))$ by 12.26. The mapping $\Gamma = (\bar{3}, Id, Id, \Gamma_0 R'^2)$ is strictly $\mathscr{B}'$-polynomial in $\mathscr{B}$ and $\varphi = \breve{\varphi}(I) = L\mu\theta.\Gamma(\theta)$. This completes the proof.

Proposition 13.12. If Γ is $\mathscr{B}'$-recursive in $\mathscr{B}$, then $\Gamma = \lambda\theta.L\mu\theta_1.\Gamma_0((\theta,\theta_1))$ for a certain Γ_0 strictly $\langle\mathscr{B}'\rangle$-polynomial in $\mathscr{B}$.

This follows from 13.8, 13.10. Of course, Γ_0 is $\mathscr{B}'$-primitive in $\mathscr{B}$.

Proposition 13.13. If φ is $\mathscr{B}'$-recursive in $\mathscr{B}$, then $\varphi = L\mu\theta.\Gamma(\theta)$ with Γ strictly $\langle\mathscr{B}'\rangle$-polynomial in $\mathscr{B}$.

This follows from 13.8, 13.11.

Further normal form results are provided by the Normal Form Theorem 9.5 for $\mathscr{S}'$. Take $\mathscr{B}_0$ as in chapter 9. Its role in $\mathscr{S}'$ can be played by $\bar{\mathscr{B}}_0$, so the strictly primitive members of $\mathscr{F}'$ will be those of the form $\bar{\varphi}$ with φ a strictly primitive member of $\mathscr{F}$.

Proposition 13.14. If Γ is $\mathscr{B}'$-recursive in $\mathscr{B}$, then

$$\Gamma = \lambda\theta.L\mu\theta_1.\psi(\psi_0,\ldots,\psi_m,\theta,\theta_1,\langle\Gamma_0\rangle(\theta_1),\ldots,\langle\Gamma_n\rangle(\theta_1))$$

for certain $\psi_0,\ldots,\psi_m \in \{I\} \cup \mathscr{B}, \Gamma_0,\ldots,\Gamma_n \in \{Ml\} \cup \mathscr{B}'$ and a strictly primitive ψ.

Proof. Proposition 9.5 implies that

$$\Gamma = \bar{1}[\bar{\chi}(I', \langle\breve{\psi}_0\rangle,\ldots,\langle\breve{\psi}_m\rangle,\langle\Gamma_0\rangle,\ldots,\langle\Gamma_n\rangle)]$$

with $\psi_0,\ldots,\psi_m \in \{I\} \cup \mathscr{B}$, $\Gamma_0,\ldots,\Gamma_n \in \{Ml\} \cup \mathscr{B}'$ and a recursive χ. Substituting $\overline{[I]}\breve{\psi}_i$ for $\langle\breve{\psi}_i\rangle$ by 12.32, we get

$$\Gamma = \bar{1}[\bar{\chi}_1(I', \breve{\psi}_0,\ldots,\breve{\psi}_m,\langle\Gamma_0\rangle,\ldots,\langle\Gamma_n\rangle)]$$

for a recursive χ_1. Repeating the last part of the proof of 9.4 and making use of 12.4, we get

$$\Gamma = \bar{1}[\breve{\psi}(\breve{\psi}_0,\ldots,\breve{\psi}_m, L', R', \langle\Gamma_0\rangle R',\ldots,\langle\Gamma_n\rangle R')]$$

with a strictly primitive ψ, which completes the proof by 12.26.

Proposition 13.15. If φ' is $\mathscr{B}'$-recursive in $\mathscr{B}$, then

$$\varphi = \bar{1}\mu\theta.\psi(\psi_0,\ldots,\psi_m,\theta,\langle\Gamma_0\rangle(\theta),\ldots,\langle\Gamma_n\rangle(\theta))$$

for certain $\psi_0,\ldots,\psi_m \in \{I\}\cup\mathscr{B}, \Gamma_0,\ldots,\Gamma_n \in \{Ml\}\cup\mathscr{B}'$ and a strictly primitive ψ.

Proof. Following the previous proof, we get

$$\breve{\varphi} = \bar{1}[\bar{\chi}(\breve{\psi}_0,\ldots,\breve{\psi}_m, I', \langle\Gamma_0\rangle,\ldots,\langle\Gamma_n\rangle)]$$

for a strictly primitive χ. Multiplying by $\breve{\psi}_0$ and taking $\psi = (L, \chi)$, we get by 6.13

$$\breve{\varphi} = \bar{2}[\breve{\psi}(\breve{\psi}_0,\ldots,\breve{\psi}_m, R', \langle\Gamma_0\rangle R',\ldots,\langle\Gamma_n\rangle R')],$$

which completes the proof by 12.27.

One may assume without any loss of generality that $\psi_0 = I$, $\Gamma_0 = Ml$ and $\psi_i \neq \psi_j$, $\Gamma_i \neq \Gamma_j$ for $i \neq j$. It should also be mentioned that the above normal form results do not imply those of chapter 9 by taking $\mathscr{B}' = \varnothing$.

The following two statements are called respectively First $\mathscr{B}'$-Recursion Theorem and Parametrized First $\mathscr{B}'$-Recursion Theorem. In contrast to the corresponding theorems of chapter 9 their proofs are quite trivial and need no μ-axiom stronger then (££).

Proposition 13.16. If Γ is a mapping $\mathscr{B}'$-recursive in $\mathscr{B}$, then $\mu\theta.\Gamma(\theta)$ exists and is an element $\mathscr{B}'$-recursive in $\mathscr{B}$.

This follows from 12.27.

Proposition 13.17. If Γ is a mapping $\mathscr{B}'$-recursive in $\mathscr{B}$, then $\lambda\theta.\mu\theta_1.(\theta,\Gamma(\theta_1))$ exists and is a mapping $\mathscr{B}'$-recursive in $\mathscr{B}$.

This follows from 12.25.

The following statements are called respectively Parametrized $\mathscr{B}'$-Enumeration Theorem and $\mathscr{B}'$-Enumeration Theorem.

Proposition 13.18. Let $\mathscr{B}$, $\mathscr{B}'$ be finite and $\mathscr{M} = \mathscr{R}(\mathscr{B},\mathscr{B}')$. Then there is a mapping $\Sigma \in \mathscr{M}$ universal for $\mathscr{M}$.

Proof. $\mathscr{M}$ is the set of all the members of $\mathscr{F}'$ recursive in $\breve{\mathscr{B}} \cup \{Id, Ml\} \cup \mathscr{B}'$, hence there is a $\Sigma \in \mathscr{M}$ universal for $\mathscr{M}$ by 9.18. If $\Gamma \in \mathscr{M}$, then there is a n such that $\Gamma = \bar{n}\Sigma$ in $\mathscr{S}'$, hence $\Gamma(\theta) = \bar{n}\Sigma(\theta)$ for all θ. The proof is complete.

By 13.7, it follows that whenever $\mathscr{B}'$ is finite, then there is a $\mathscr{B}'$-recursive mapping Σ such that $\Gamma = \lambda\theta.\bar{n}\Sigma((\psi_1,\ldots,\psi_m,\theta))$ for a certain n, provided that Γ is $\mathscr{B}'$-recursive in $\psi_1,\ldots,\psi_m$.

Proposition 13.19. Let $\mathscr{B}, \mathscr{B}'$ be finite and $\mathscr{U} = \{\varphi/\breve{\varphi} \in \mathscr{M}\}$. Then there is a $\sigma \in \mathscr{U}$ which is universal for $\mathscr{U}$.

Proof. Take $\sigma = \Sigma(I)$, where Σ is the mapping considered in the previous proof. It follows that $\sigma \in \mathscr{U}$ since $\check{\sigma} = \Sigma Id$. If $\varphi \in \mathscr{U}$, then $\check{\varphi} \in \mathscr{M}$, hence $\check{\varphi} = \bar{n}\Sigma$ for a certain n. Therefore, $\check{\varphi} = \bar{n}\Sigma Id = \bar{n}\check{\sigma}$ which implies $\varphi = \bar{n}\sigma$. The proof is complete.

The notion of element (respectively mapping) principal universal for $\mathscr{U}$ (for $\mathscr{M}$) is introduced exactly as in chapter 9. All the statements of chapter 9 concerning such elements and mappings remain valid with '$\mathscr{B}'$-recursive in $\mathscr{B}$' substituted for 'recursive in $\mathscr{B}$'. It is worth mentioning that these considerations on universal elements and mappings make use of the operation $\langle\ \ \rangle$ of $\mathscr{S}'$ since so do the proofs of 13.18, 13.19, while propositions 13.16, 13.17 hold for prime $\mathscr{B}'$-recursiveness as well.

If $\langle\!\langle\ \ \rangle\!\rangle = To \in \mathscr{F}'$ is a t-operation over $\mathscr{F}$ with a corresponding set of functional elements $\mathscr{B}_0$, then the following analogue to 13.9* shows that t-recursiveness is a particular instance of $\mathscr{B}'$-recursiveness.

Proposition 13.20*. A unary mapping Γ over $\mathscr{F}$ is t-recursive in $\mathscr{B}$ iff Γ is $\check{\mathscr{B}}_0 \cup \{To\}$-recursive in $\mathscr{B}$.

This follows from 12.42*.

Accordingly, the equality

$$\mathscr{R}_0(\mathscr{B}, \check{\mathscr{B}}_0 \cup \{To\}) = \mathscr{R}(\mathscr{B}, \check{\mathscr{B}}_0 \cup \{To\})$$

holds for all $\mathscr{B}$.

There is another possible situation of some interest, viz. consecutive spaces $\mathscr{S}$, $\mathscr{S}'$ with a t-operation $\langle\!\langle\ \ \rangle\!\rangle$ over $\mathscr{F}'$. In particular, such spaces and operations appear in the next chapter (the spaces $\mathscr{S}'$, $\mathscr{S}''$ and the operation Tf), as well as in chapter 28 (the spaces $\mathscr{S}$, $\mathscr{S}_1$ and the operation St). We shall not adduce the relevant general considerations, which are in the spirit of the last two chapters.

A possible direction for further investigations in recursion theory on consecutive spaces is to study $\mathscr{B}'$-recursiveness for certain specific $\mathscr{B}'$. We introduce some monotonic mappings over $\mathscr{F}$ involving quantification, which may be taken to be in $\mathscr{B}'$, provided they are in $\mathscr{F}'$.

Let $\sigma \in \mathscr{F} \setminus \{O\}$ and Q be a *monotonic quantifier* over the set $\mathscr{A}_\sigma = \{\psi / O < \psi \leq \sigma\}$, i.e. $\varnothing \subset Q \subset 2^{\mathscr{A}_\sigma}$ and, if $\mathscr{A} \in Q$ and $\mathscr{A} \subseteq \mathscr{B} \subseteq \mathscr{A}_\sigma$, then $\mathscr{B} \in Q$. Let $Q^\cup$ be the quantifier dual to Q, i.e. $Q^\cup = \{\mathscr{A} / \mathscr{A} \subseteq \mathscr{A}_\sigma \& \mathscr{A}_\sigma \setminus \mathscr{A} \in Q\}$. Consider the mapping $Q_\sigma : \mathscr{F} \to \mathscr{F}$ such that

$$Q_\sigma(\theta) = \begin{cases} L, & \text{if } Q\psi \in \mathscr{A}_\sigma(\psi \leq \theta), \\ R, & \text{if } \forall \tau \geq \theta Q^\cup \psi \in \mathscr{A}_\sigma(\psi \not\leq \tau), \\ O, & \text{otherwise}, \end{cases}$$

where $Q\psi \in \mathscr{A}_\sigma(\text{———})$ stands for $\{\psi / \psi \in \mathscr{A}_\sigma\ \&\ \text{———}\} \in Q$ as usual.

Proposition 13.21. Q_σ is monotonic.

Proof. Suppose that $\theta \leq \theta_1$. If $Q_\sigma(\theta) = O$, then $Q_\sigma(\theta) \leq Q_\sigma(\theta_1)$ trivially. If $Q_\sigma(\theta) = L$, then $Q\psi \in \mathscr{A}_\sigma(\psi \leq \theta)$; hence $Q\psi \in \mathscr{A}_\sigma(\psi \leq \theta_1)$ by the monotonicity of Q; and so $Q_\sigma(\theta_1) = L$ If $Q_\sigma(\theta) = R$, then $\forall \tau \geq \theta Q^\cup \psi \in \mathscr{A}_\sigma(\psi \not\leq \tau)$; hence $\forall \tau \geq$

$\theta_1 Q^{\cup}\psi\in\mathscr{A}_\sigma(\psi\not\leq\tau)$, which implies $Q_\sigma(\theta_1)=R$. The proof is complete.

Taking $Q=\exists,\forall$ in particular, one gets

$$E_\sigma(\theta)=\begin{cases} L, & \text{if } \exists\psi(O<\psi\leq\sigma,\theta),\\ R, & \text{if } \forall\tau\geq\theta\forall\psi\neg(O<\psi\leq\sigma,\tau),\\ O, & \text{oherwise}\end{cases}$$

$$A_\sigma(\theta)=\begin{cases} L, & \text{if } \sigma\leq\theta,\\ R, & \text{if } \forall\tau\geq\theta(\sigma\not\leq\tau),\\ O, & \text{otherwise.}\end{cases}$$

The above mappings fit the case when the elements L, R have no upper bound in $\mathscr{F}$. If L, R have an upper bound U, then each two members of $\mathscr{F}$ have an upper bound (see the hint to exercise 7.7); hence the mapping Q_σ becomes

$$Q_\sigma(\theta)=\begin{cases} L, & \text{if } Q\psi\in\mathscr{A}_\sigma(\psi\leq\theta),\\ O, & \text{otherwise.}\end{cases}$$

However, appropriate mappings embodying quantification can be introduced in this instance as well. Take a fixed subset $\mathscr{C}$ of $\mathscr{F}$, $\sigma\in\mathscr{F}\setminus\{O\}$ and a monotonic quantifer Q over $\mathscr{A}_\sigma$. Writing $\mathscr{A}_Q$, $\mathscr{C}_\exists$ respectively for $\{\theta/Q\psi\in\mathscr{A}_\sigma(\psi\leq\theta)\}$ $\{\theta/\exists\psi\in\mathscr{C}(\psi\leq\theta)\}$, the following mapping is monotonic

$$Q_{\sigma,\mathscr{C}}(\theta)=\begin{cases} L, & \text{if } \theta\in\mathscr{A}_Q\,\&\,\forall\tau\leq\theta(\tau\notin\mathscr{C}_\exists\setminus\mathscr{A}_Q),\\ R, & \text{if } \theta\in\mathscr{C}_\exists\setminus\mathscr{A}_Q,\\ U, & \text{if } \theta\in\mathscr{A}_Q\,\&\,\exists\tau\leq\theta(\tau\in\mathscr{C}_\exists\setminus\mathscr{A}_Q),\\ O, & \text{if } \theta\notin\mathscr{C}_\exists\cup\mathscr{A}_Q.\end{cases}$$

Recursion in Q_σ and $Q_{\sigma,\mathscr{C}}$ has been intensively studied in some particular instances for, as will be shown in the exercises below, these mappings generalize the well known functionals $F^{\#}_Q$ and E of Kechris and Moschovakis [1977] and Moschovakis [1969] respectively. In the general theory however, the usefulness of Q_σ and $Q_{\sigma,\mathscr{C}}$ is yet to be tested.

EXERCISES TO CHAPTER 13

Exercise 13.1. Let $\mathscr{S}$ be the IOS of example 4.7 and let $\mathscr{S}'$ be obtained from it by 12.1. Let 0 be a fixed member of M, $\sigma=\lambda s.0$, let Q_0 be a monotonous quantifier over M and let

$$Q=\{\mathscr{A}/\mathscr{A}\subseteq\mathscr{A}_\sigma\,\&\cup\{\operatorname{Dom}\psi/\psi\in\mathscr{A}\}\in Q_0\}.$$

Show that $\mathscr{S}$, $\mathscr{S}'$ are consecutive IOS, $Q_\sigma\in\mathscr{F}'$ and

$$Q_\sigma(\theta)=\begin{cases} L, & \text{if } Q_0 s(\theta(s)=0),\\ R, & \text{if } Q_0^{\cup}s(\theta(s)\downarrow\,\&\,\theta(s)\neq 0),\\ O, & \text{otherwise.}\end{cases}$$

Hint. Use exercises 12.1, 12.2, 5.3 to show that $\mathscr{S}'$ is iterative. Employ the equality

$$Q_0=\{X/X\subseteq M\,\&\,\{\psi/O<\psi\leq\sigma\upharpoonright X\}\in Q\}.$$

Exercise 13.1 implies that the mapping Q_σ is nothing else than the functional $F^{\#}_{Q_0}$ of Kechris and Moschovakis [1977],

$$F^{\#}_{Q_0}(f)=\begin{cases}0, & \text{if } Q_0 s(f(s)=0),\\ 1, & \text{if } Q_0^{\cup} s(f(s)\downarrow \& f(s)\neq 0),\\ \uparrow, & \text{otherwise,}\end{cases}$$

where f ranges over the subset $\{f/f: M \longrightarrow \omega\}$ of $\mathscr{F}$, assuming $\omega \subseteq M$. Indeed, each of Q_σ, $F^{\#}_{Q_0}$ can be expressed by the other. In particular, E_σ corresponds to $F^{\#}_{\exists}=E^{\#}$.

Notice that Q_σ and $Q_\sigma^{\cup}$ can be reduced to each other in the above example.

Exercise 13.2. Let $\mathscr{S}$ be the IOS of example 4.8, let $\mathscr{S}'$ be obtained from it by 12.1, let σ, Q_0, Q be the same as in exercise 13.1, let $U=\lambda s.\{L(s),R(s)\}$ and $\mathscr{C}=\{\varphi/\varphi: M\to M\}$. Prove that $\mathscr{S}$, $\mathscr{S}'$ are consecutive IOS, $Q_{\sigma,\mathscr{C}}\in\mathscr{F}'$ and

$$E_{\sigma,\mathscr{C}}(\theta)=\begin{cases}L, & \text{if } \exists s(0\in\theta(s))\,\&\,\exists s(\theta(s)\uparrow \vee \theta(s)=0),\\ R, & \text{if } \forall s\exists t\neq 0(t\in\theta(s)\,\&\,0\notin\theta(s)),\\ U, & \text{if } \exists s(0\in\theta(s))\,\&\,\forall s\exists t\neq 0\ (t\in\theta(s)),\\ O, & \text{otherwise.}\end{cases}$$

Therefore, $E_{\sigma,\mathscr{C}}$ is in essence the functional $E:\mathscr{F}\longrightarrow\{0,1\}$ of Moschovakis [1969], where

$$0\in E(\theta),\quad \text{if } \exists s(0\in\theta(s))\,\&\,\forall s(\theta(s)\downarrow),$$
$$1\in E(\theta),\quad \text{if } \forall s\exists t\neq 0\ (t\in\theta(s)).$$

(To be more precise, $E_{\sigma,\mathscr{C}}$ gives an extension $E^{\#}$ of E such that $0\in E(\theta)$, if $\exists s(0\in\theta(s))$, without requiring $\forall s(\theta(s)\downarrow)$.)

If $\mathscr{S}$, $\mathscr{S}'$ are consecutive spaces, then $\mathscr{S}'$ can play the role of $\mathscr{S}$ and monotonic mappings $Q_{\sigma'}$, $Q_{\sigma',\mathscr{C}'}$ over $\mathscr{F}'$ can be considered. Take $\sigma'\in\mathscr{F}'\backslash\{O'\}$. We recall that

$$E_{\sigma'}(\theta')=\begin{cases}L' & \text{if } \exists\psi'(O'<\psi'\leq\sigma',\theta'),\\ R', & \text{if } \forall\tau'\geq\theta'\forall\psi'\neg(O'<\psi'\leq\sigma',\tau'),\\ O', & \text{otherwise,}\end{cases}$$

$$A_{\sigma'}(\theta')=\begin{cases}L', & \text{if } \sigma'\leq\theta',\\ R', & \text{if } \forall\tau'\geq\theta'(\sigma'\not\leq\tau'),\\ O', & \text{otherwise.}\end{cases}$$

Exercise13.3. Let $A_\sigma\in\mathscr{F}'$ for all σ. Show that $E_{\sigma'}(\theta')=L'$ iff $\exists\theta(E_{\sigma'(\theta)}(\theta'(\theta))=L)$. In particular, $E_{\dot\sigma}(\theta')=L'$ iff $\exists\theta(E_\sigma(\theta'(\theta))=L)$.

Hint. It suffices to show that $\exists\psi'(O'<\psi'\leq\sigma',\theta')$ iff $\exists\theta\exists\psi(O<\psi\leq\sigma'(\theta),\theta'(\theta))$. Assuming $O<\psi\leq\sigma'(\theta),\theta'(\theta)$, take $\psi'=Ml(A_\theta,\dot\psi,O')$.

Exercise 13.4. Prove that if $E_{\sigma'(\theta)}(\theta'(\theta))=R$ whenever $\sigma'(\theta)\neq O$, then $E_{\sigma'}(\theta')=R'$. In particular, $\forall\theta(E_\sigma(\theta'(\theta))=R)$ implies $E_{\dot\sigma}(\theta')=R'$.

Hint. Show that $\exists\tau' > \theta'\exists\psi'(O' < \psi' < \sigma', \tau')$ implies $\exists\theta\exists\tau > \theta'(\theta)\exists\psi\ (O < \psi \leq \sigma'(\theta), \tau)$.

Exercise 13.5. Show that $A_{\sigma'}(\theta') = L'$ iff $A_{\sigma'(\theta)}\ (\theta'(\theta)) = L$ for all θ such that $\sigma'(\theta) \neq O$. In particular, $A_{\check{\sigma}}(\theta') = L'$ iff $\forall\theta(A_{\sigma}(\theta'(\theta)) = L)$.

Exercise 13.6. Show that $\exists\theta(A_{\sigma'(\theta)}(\theta'(\theta)) = R)$ implies $A_{\sigma'}(\theta') = R'$. In particular, $\exists\theta(A_{\sigma}(\theta'(\theta)) = R)$ implies $A_{\check{\sigma}}(\theta') = R'$.

Hint. Suppose that $\tau' \geq \theta'$ and $\sigma' \leq \tau'$. Then $\sigma'(\theta), \theta'(\theta) \leq \tau'(\theta)$ for all θ; hence $A_{\sigma'(\theta)}(\theta'(\theta)) \neq R$ for all θ such that $\sigma'(\theta) \neq O$, which is not the case.

The last four exercises show that $E_{\sigma'}$, $A_{\sigma'}$ embody $\exists$, $\forall$-quantification over $\mathscr{F}$.

CHAPTER 14

Transfer operation

The present chapter studies a specific t-operation called *transfer*, which is connected with the operator nature of the elements in consecutive spaces.

Suppose that $\mathscr{S}, \mathscr{S}', \mathscr{S}''$ are consecutive IOS (i.e. both $\mathscr{S}, \mathscr{S}'$ and $\mathscr{S}', \mathscr{S}''$ are consecutive) and $Tf: \mathscr{F}'' \to \mathscr{F}''$ satisfies

$$Tf(\varphi'')(\theta')(\theta) = \varphi''(\theta'(L'\check{\theta}, I'))(R\theta)$$

for all $\varphi'' \in \mathscr{F}''$, $\theta' \in \mathscr{F}'$ and $\theta \in \mathscr{F}$. Consecutive spaces to be constructed in chapter 19 admit such an operation.

In order to show that Tf is a storing operation in the sense of chapter 10, take $\mathscr{L}'' = \tilde{\bar{\mathscr{F}}} = \{\tilde{\rho}/\rho \in \mathscr{F}\}$, where $\tilde{\rho} = \lambda\theta'.\theta'(\check{\rho}, I') = Ml'(I'', \rho^{\check{}\check{}}, Id') = Ml'(I'', \bar{\bar{\rho}} Id^{\check{}}, Id')$. We recall that Id, $Ml \in \mathscr{F}'$ and Id', $Ml' \in \mathscr{F}''$, where $Id = \lambda\theta.I$, $Ml = \lambda\theta.L\theta R\theta$, $Id' = \lambda\theta'.I'$ and $Ml' = \lambda\theta'.L'\theta' R'\theta'$.

Proposition 14.1. $\tilde{\rho}\overline{\varphi'} = \overline{\varphi'}\,\tilde{\rho}$.

Proof.

$$\tilde{\rho}\overline{\varphi'}(\theta') = \tilde{\rho}(\overline{\varphi'}(\theta')) = \varphi'\theta'(\check{\rho}, I') = \overline{\varphi'}\tilde{\rho}(\theta')$$

for all θ'; hence $\tilde{\rho}\overline{\varphi'} = \overline{\varphi'}\,\tilde{\rho}$

Proposition 14.2. $\tilde{\rho}(\varphi'', \psi'') = (\tilde{\rho}\varphi'', \tilde{\rho}\psi'')$.

Proof. Using 14.1 and the equalities $L'' = \overline{L'}$, $R'' = \overline{R'}$, $(L'', R'') = \overline{(L', R')}$, we get

$$\begin{aligned}\tilde{\rho}(\varphi'', \psi'') &= \tilde{\rho}(L'', R'')(\varphi'', \psi'') = (L'', R'')\tilde{\rho}(\varphi'', \psi'') = (L''\tilde{\rho}, R''\tilde{\rho})(\varphi'', \psi'') \\ &= (\tilde{\rho}L'', \tilde{\rho}R'')(\varphi'', \psi'') = (\tilde{\rho}\varphi'', \tilde{\rho}\psi'').\end{aligned}$$

Proposition 14.3. $Tf(\overline{\varphi'}) = \overline{\varphi'}\,Ml'(I'', \check{L}', \check{R}') = Ml'(\overline{\varphi'}, \check{L}', \check{R}')$. Therefore, $Tf(I'') = Ml'(I'', \check{L}', \check{R}')$ and $Tf(\overline{\varphi'}) = \overline{\varphi'}\,Tf(I'') = Tf(I'')\overline{\varphi'}$.

Proof. We have

$$Tf(\overline{\varphi'})(\theta')(\theta) = \varphi'\theta'(L'\check{\theta}, I')(R\theta) = \varphi'\theta'(L', R')(\theta)$$

for all θ', θ; hence $Tf(\overline{\varphi'})(\theta') = \varphi'\theta'(L', R')$ for all θ', which completes the proof.

Proposition 14.4. $Tf(\check{\varphi}') = \overline{\varphi' \check{R}'}$.

Proof.

$$Tf(\check{\varphi}')(\theta')(\theta) = \varphi'(R\theta) = \varphi' R'(\theta)$$

for all θ', θ, hence $Tf(\check{\varphi}')(\theta') = \varphi' R'$ for all θ'.

Proposition 14.5. $\tilde{\rho} Tf(\varphi'') = \varphi'' \tilde{\rho}$.

Proof.

$$\tilde{\rho} Tf(\varphi'')(\theta')(\theta) = \tilde{\rho}(Tf(\varphi'')(\theta'))(\theta) = Tf(\varphi'')(\theta')((\rho, \theta)) = \varphi''(\theta'(\check{\rho}, I'))(\theta)$$

for all θ', θ; hence $\tilde{\rho} Tf(\varphi'')(\theta') = \varphi''(\theta'(\check{\rho}, I')) = \varphi'' \tilde{\rho}(\theta')$ for all θ'. The proof is complete.

Proposition 14.6. Whenever $\varphi'' \tilde{\rho} \psi'' \leq \chi'' \tilde{\rho} \tau''$ for all ρ, then $Tf(\varphi'')\psi'' \leq Tf(\chi'')\tau''$.

Proof. We have

$$\begin{aligned} T(\varphi'')\psi''(\theta')(\theta) &= \varphi''(\psi''(\theta')(L'\check{\theta}, I'))(R\theta) = \varphi''(L\theta)\tilde{\ }\psi''(\theta')(R\theta) \\ &\leq \chi''(L\theta)\tilde{\ }\tau''(\theta')(R\theta) = Tf(\chi'')\tau''(\theta')(\theta) \end{aligned}$$

for all θ', θ, which completes the proof.

Now take

$$K_0'' = Tf(I''),\ K_1'' = MI'(I'', (\check{L}', L''\check{R}'), R''\check{R}'),\ K_2'' = MI'(I'', L''\check{L}', R''\check{L}', \check{R}').$$

Proposition 14.7. $K_0''(\varphi'', \psi'') = (K_0''\varphi'', K_0''\psi'')$.

This follows from the proof of 14.2 since $K_0''\overline{\varphi'} = \overline{\varphi'} K_0''$ by 14.3.

Proposition 14.8. $\tilde{\rho} K_0''(L'', R'') = (\tilde{\rho} L'', \tilde{\rho} R'')$.

Proof. 14.5, 14.2 give

$$\tilde{\rho} K_0''(L'', R'') = \tilde{\rho}(L'', R'') = (\tilde{\rho} L'', \tilde{\rho} R'').$$

Proposition 14.9. $\tilde{\rho}\tilde{\sigma} K_1'' = (\sigma, \rho)\tilde{\ }$, $(\sigma, \rho)\tilde{\ } K_2'' = \tilde{\rho}\tilde{\sigma}$.

Proof. We have

$$\begin{aligned} \tilde{\rho}\tilde{\sigma} K_1''(\theta') &= \tilde{\sigma} K_1''(\theta')(\check{\rho}, I') = K_1''(\theta')(\check{\sigma}, I')(\check{\rho}, I') = \theta'((L', L'R'), R'^2)(\check{\sigma}, \check{\rho}, I') \\ &= \theta'((\check{\sigma}, \check{\rho}), I') = \theta'((\sigma, \rho)\check{\ }, I') = (\sigma, \rho)\tilde{\ }(\theta') \end{aligned}$$

for all θ'. On the other hand,

$$\begin{aligned} (\sigma, \rho)\tilde{\ } K_2''(\theta') &= K_2''(\theta')((\sigma, \rho)\check{\ }, I') = \theta'(L'^2, R'L', R')((\check{\sigma}, \check{\rho}), I') \\ &= \theta'(\check{\sigma}, \check{\rho}, I') = \tilde{\rho}\tilde{\sigma}(\theta') \end{aligned}$$

for all θ'. The proof is complete.

Notice that $Tf(I'')K_0'' = K_0''$ by 14.6, while $Tf(Tf(I''))K_1'' = K_1''$ is valid as well.

Proposition 14.10. Tf is a t-operation with functional elements $K_0'' - K_2''$, $K_3'' = \tilde{I}, K_4'' = MI'(I'', \check{R}')$.

Proof. Transfer is obviously monotonic and 14.8, 14.9, 14.5, 14.6 imply that it is a storing operation. The equality $\tilde{\rho}K_4''(\theta') = \theta' R'(\check{\rho}, I') = \theta'$ gives $\tilde{\rho}K_4'' = I''$, which completes the proof by 10.18.

Proposition 14.11. $\overline{\varphi'}\,Tf(\psi'') = Tf(\overline{\varphi'}\psi'')$, $Tf(\psi'')\overline{\varphi'} = Tf(\psi''\overline{\varphi'})$.
This follows from 14.3, 10.12.

Proposition 14.12. $Tf((\varphi'', \psi'')) = (Tf(\varphi''), Tf(\psi''))$, $Tf(\Delta(\varphi'', \psi'')) = \Delta(Tf(\varphi''), Tf(\psi''))$, $Tf([\varphi'']) = [Tf(\varphi'')]Tf(I'')$. In particular, $Tf(\langle \varphi'' \rangle) = \Delta(Tf(\varphi'' L''), Tf(R''))$.

Proof. The first equality follows from 10.13, 14.7, 10.12. Using 6.34, one gets $\Delta(Tf(\varphi''), Tf(\psi'')) = \langle I'' \rangle Tf(\Delta(\varphi'', \psi''))$ which implies the second equality by 14.11. The third equality follows by the first one and the proof of 10.15.

Proposition 14.13. $\varphi'' = Ml'(I'', Id', Id')Tf(\varphi'')K_4''$, hence K_3'' can be replaced by $Ml'(I'', Id', Id')$.

Proof. We have

$$Ml'(I'', Id', Id')Tf(\varphi'')K_4''(\theta')(\theta) = Tf(\varphi'')(\theta' R')(I', I')(\theta)$$
$$= Tf(\varphi'')(\theta' R')((\theta, \theta)) = \varphi''(\theta')(\theta)$$

for all θ', θ. The proof is complete.

Proposition 14.14. $Tf(Id') = R'' Id'$.
This follows from 14.4.

Proposition 14.15. $Tf(Ml') = Ml'(L'', \check{L}', Ml'(R'', \check{L}', \check{R}'))$.

Proof.

$$Tf(Ml')(\theta')(\theta) = Ml'(\theta'(L'\check{\theta}, I'))(R\theta) = L'\theta'(L'\check{\theta}, I')R'\theta'(L'\check{\theta}, I')(R\theta)$$
$$= L'\theta'(L'\check{\theta}, R'\theta'(L'\check{\theta}, I'))(R\theta) = L'\theta'(\check{L}', R'\theta'(\check{L}', \check{R}'))(\theta)$$
$$= Ml'(L'', \check{L}', Ml'(R'', \check{L}', \check{R}'))(\theta')(\theta)$$

for all θ', θ, which completes the proof.

Proposition 14.10 can be sharpened as follows.

Proposition 14.16 (Transfer Operation Theorem). Tf is a t-operation with a set of functional elements $\mathscr{B}_0'' = \{L'', R'', Id', Ml'\}$ and mappings Σ_0'', Σ_1'', $\Sigma_3'' - \Sigma_5''$ prime recursive in $\mathscr{B}_0''$.

Proof. The elements I'', $Tf(I'')$, $Tf(L'')$, $Tf(R'')$, $Tf(Id')$, $Tf(Ml')$ may be removed from $\mathscr{B}_0''$ since they are prime recursive in Id', Ml' by 5.12, 14.3, 14.14, 14.15. The mappings Σ_0'', Σ_1'', $\Sigma_3'' - \Sigma_5''$ satisfying the equalities (0), (1), (3)–(5) of chapter 10 by 10.12, 14.12, 10.16, 14.13 are prime recursive in $\mathscr{B}_0''$. This completes the proof.

To indicate that notions of chapter 10 concern the operation Tf now, we write tf-recursive, tf-recursive, tf-simple segment, tf μ-axiom etc. Recall that φ'' is tf-recursive in $\mathscr{B}'' \subseteq \mathscr{F}''$ iff

$$\varphi'' \in cl(\mathscr{B}_0'' \cup \mathscr{B}''/\circ, \Pi, \langle\ \ \rangle, [\ \ ], Tf).$$

Proposition 14.16 allows to apply the theory developed in chapter 10 to transfer operation and tf-recursiveness. While the proof of the First tf-Recursion Theorem depends on tfμA, the verification of this axiom in particular instances is simplified by the fact that all tf-simple segments are normal by 10.18.

The spaces $\mathscr{S}'$, $\mathscr{S}''$ are consecutive, hence tf-recursiveness can be characterized by making use of 10.2, 14.4, 12.34 as follows.

Proposition 14.17. Let $\mathscr{B}' \subseteq \mathscr{F}'$, $\mathscr{B}'' \subseteq \mathscr{F}''$. Then the following are equivalent.

(1) φ'' is tf-recursive in $\check{\mathscr{B}}' \cup \mathscr{B}''$.

(2) φ'' is recursive in $\{Id', Ml'\} \cup \check{\mathscr{B}}' \cup Tf(\mathscr{B}'')$.

(3) φ'' is prime recursive in $\{Id', Ml'\} \cup \check{\mathscr{B}}' \cup \langle Tf(\mathscr{B}'')\rangle$.

In particular, φ'' is tf-recursive in $\check{\mathscr{B}}'$ iff it is prime recursive in $\{Id', Ml'\} \cup \check{\mathscr{B}}'$.

Now assume that another IOS $\mathscr{S}'''$ is also given, such that $\mathscr{S}''$, $\mathscr{S}'''$ are consecutive and $Tf \in \mathscr{F}'''$. Then propositions 12.40–12.42* hold with $\mathscr{S}''$, $\mathscr{S}'''$, Tf playing the role of $\mathscr{S}$, $\mathscr{S}'$, $\langle\ \ \rangle$. Assume finally that $Tf' : \mathscr{F}''' \to \mathscr{F}'''$ is a transfer operation, i.e.

$$Tf'(\varphi''')(\theta'')(\theta') = \varphi'''(\theta''(L'\check{\theta}', I''))(R'\theta')$$

for all $\varphi''' \in \mathscr{F}'''$, $\theta'' \in \mathscr{F}''$, $\theta' \in \mathscr{F}'$. It is of interest to see how do the operations Tf, Tf' interact. Notice that unlike the basic IOS-operations, the isomorphism of $\mathscr{S}''$, $\overline{\mathscr{S}''}$ does not extent to ensure $\overline{Tf(\varphi'')} = Tf'(\overline{\varphi''})$ for all φ'': take $\varphi'' = Id'$ for example.

Proposition 14.18. $Tf \notin \overline{\mathscr{F}''}$.

Proof. Suppose that $Tf = \overline{\varphi''}$. Then $\varphi'' = \overline{\varphi''}(I'') = Tf(I'')$. Using 14.14, 14.3, we get

$$\check{R}' = Tf(Id') = \overline{\varphi''}\,(Id') = \varphi'' Id' = Tf(I'')Id' = (\check{L}', \check{R}'),$$

which is not the case.

Proposition 14.19. $Tf'(Tf) = Ml''(Tf, \overline{Ml'}(\check{L}'', R'''I\check{d}'), \check{R}'')$.

Proof. We have

$$\begin{aligned} Tf'(Tf)(\varphi'')(\theta')(\theta) &= Tf(\varphi''(L''\check{\theta}', I''))(R'\theta')(\theta) = \varphi''(L''\check{\theta}', I'')(R'\theta'(L'\check{\theta}, I'))(R\theta) \\ &= \varphi''((L'\theta', R'\theta'(L'\check{\theta}, I')))(R\theta) \\ &= \varphi''((L'\theta' R', R'\theta')(L'\check{\theta}, I'))(R\theta) \\ &= Tf(\varphi'')((L'\theta' R', R'\theta'))(\theta) \end{aligned}$$

for all φ'', θ', θ; hence

$$Tf'(Tf)(\varphi'')(\theta') = Tf(\varphi'')((L'\theta' R', R'\theta')) = Tf(\varphi'')(Ml'(L'', \check{R}'), R'')(\theta')$$

for all φ'', θ'. Therefore

$$Tf'(Tf)(\varphi'') = Tf(\varphi'')(Ml'(L'', \check{R}'), R'') = Ml''(Tf, \overline{Ml'}(\check{L}'', R'''I\check{d}'), \check{R}'')(\varphi'')$$

for all φ'', which completes the proof.

Proposition 12.42* can be sharpened to give the following tf-Conservativeness Theorem.

Proposition 14.20*. Let $\mathscr{B}'' \subseteq \mathscr{F}''$ and Γ'' be a unary mapping over $\mathscr{F}''$. Then the following are equivalent.

(1) Γ'' is a mapping tf-recursive in $\mathscr{B}''$.

(2) Γ'' is a member of $\mathscr{F}'''$ prime recursive in

$$\{\check{Id}', \check{Ml}', Id'', Ml'', Tf\} \cup \check{\mathscr{B}}''.$$

(3) Γ'' is a member of $\mathscr{F}'''$ tf'-recursive in

$$\{\check{Id}', \check{Ml}', Tf\} \cup \check{\mathscr{B}}''.$$

Proof. Propositions 14.19, 12.40 imply that the element $\langle Tf'(Tf) \rangle$ is prime recursive in $\check{Id}'$, $\check{Ml}'$, Id'', Ml'', Tf, hence (3) implies (2) by 14.17. The implication (2)$\Rightarrow$(1) follows by 12.42*, while (1)$\Rightarrow$(3) is immediate. This completes the proof.

Using 12.3, 12.19, $\overline{Id'}$, $\overline{Ml'}$, $\overline{\mathscr{B}''}$ can be substituted respectively for $\check{Id}'$, $\check{Ml}'$, $\check{\mathscr{B}}''$ in 14.20* since Id'', Ml'' are among the initial elements.

EXERCISES TO CHAPTER 14

Exercise 14.1. Let $\Pi^* : \mathscr{F}''^2 \to \mathscr{F}''$ be such that

$$\Pi^*(\varphi'', \psi'') = \lambda\theta'.\lambda\theta.\varphi''(\lambda\rho.\psi''(\lambda\sigma.\theta'((\rho,\sigma)))(\theta))(\theta).$$

Show that Tf can be expressed in terms of Π^*, Ml', Id'.

Hint.

$$Tf(\varphi'') = \Pi^*(Ml'(I'', \check{L}'), Ml'(\varphi'', \check{R}')).$$

Exercise 14.2. Show that the operation Π^* of the previous exercise can be expressed in terms of Tf, Ml', Id'.

Hint.

$$\Pi^*(\varphi'', \psi'') = Ml'(Tf(\varphi'')Ml'(Tf(\psi''), \check{R}', \check{L}'), Id', Id').$$

Exercise 14.3.** Show that the t-operation Tf satisfies the assumptions of exercises 10.2–10.5, provided the elements $\check{Id}$, $\check{Ml}$ are added to $\mathscr{B}''_0$.

Hint. Take $\rho = \Delta((L, L^2R, R^2), \ (L, RLR, R^2))$, $\psi''_0 = \langle Ml' \rangle \bar{\bar{\rho}}(I'', \check{Id}, Id')$ and $\psi''_1 = Tf(Ml')(\check{Ml}, Ml'(I'', \check{R}'))$. These elements are prime recursive in $\check{Id}$, $\check{Ml}$, Id', Ml' by 12.31, 12.33, 14.15. Show that $\bar{n}\psi''_0 = \bar{\bar{n}}$, $\bar{\bar{n}}\psi''_1 = \bar{n}$, $\bar{n}$ standing for LR^n or $L''R''^n$. Use exercise 10.8**.

Exercise 14.4.** Let $\mathscr{B}' \subseteq \mathscr{F}'$, $\mathscr{B}'' \subseteq \mathscr{F}''$. Show that the following are equivalent.

(1) φ'' is tf-recursive in $\{\check{Id}, \check{Ml}\} \cup \check{\mathscr{B}}' \cup \mathscr{B}''$.

(2) φ'' is prime recursive in $\{\check{Id}, \check{Ml}, Id', Ml'\} \cup \check{\mathscr{B}}' \cup Tf(\mathscr{B}'')$.

Hint. Use 14.17, exercises 14.3**, 10.2 and proposition 10.16.

Exercise 14.5. Determine elements K''_4, K''_5, K''_6 to satisfy the assumptions of exercise 10.9 with $\tilde{\bar{\mathscr{F}}}$ playing the role of $\mathscr{L}$.

Hint. Take $K''_4 = Ml'(I'', \check{R}')$, $K''_5 = Ml'(I'', \check{L}, \check{L}, \check{R}')$, $K''_6 = Ml'(I'', L'\check{R}', \check{L}, R''\check{R}')$.

Exercise 14.6. Let $Co: \mathscr{F}^{\omega} \to \mathscr{F}$, $Co': \mathscr{F}'^{\omega} \to \mathscr{F}'$ and $Co'': \mathscr{F}''^{\omega} \to \mathscr{F}''$ be collection operations such that $Co''\{\varphi''_n\}(\theta') = Co'\{\varphi''_n(\theta')\}$ for all $\theta' \in \mathscr{F}'$ and all sequences $\{\varphi''_n\}$ in $\mathscr{F}''$. Determine an element K''_7 to satisfy the assumptions of exercise 11.7 with $\mathscr{L}''_0 = \mathscr{L}'' = \tilde{\bar{\mathscr{F}}}$.

Hint. Take the element ψ''_1 of exercise 14.3*, then take $K''_7 = [I'']Co''\{Ml'(K''_2\psi''_1, \bar{n}\check{L}', \check{R}')\}$ and assign $Co\{(\overline{k_n}, \rho_n)\}^{\sim}\bar{0}$ to $\{\tilde{\rho}_n \overline{k_n}\}$.

Exercise 14.7.** Let $\mathscr{B}'' \subseteq \mathscr{F}''$, $\mathscr{B}''' \subseteq \mathscr{F}'''$ and $\varphi''' \in \mathscr{F}'''$. Show that the following are equivalent.

(1) φ''' is tf'-recursive in $\{I\check{d}', M\check{l}', Tf\} \cup \check{\mathscr{B}}'' \cup \mathscr{B}'''$.

(2) φ''' is prime recursive in

$$\{I\check{d}', M\check{l}', Id'', Ml'', Tf\} \cup \check{\mathscr{B}}'' \cup Tf'(\mathscr{B}''').$$

Hint. Use exercise 14.4** and 14.19 to get (1)⇒(2).

Exercise 14.8. Let Tf^* be an operation over $\mathscr{F}'''$ such that

$$Tf^*(\varphi''') = \lambda\theta''.\lambda\theta'.\lambda\theta.\varphi'''(\lambda\tau'.\theta''(\tau')(L'\check{\theta}, I'))(\theta')(R\theta).$$

Show that Tf^* is a storing operation and $Tf^*(\overline{\varphi''}) = \overline{Tf(\varphi'')}$ for all φ''.

Hint. Take $\mathscr{L}^* = \tilde{\bar{\bar{\mathscr{F}}}}$ and $K^*_i = \bar{K}''_i$, $i = 0, 1, 2$.

Exercise 14.9. Show that the operation Tf^* of the previous exercise can be expressed in terms of Tf', Tf.

Hint. Take $\sigma''' = \overline{Ml'}(Ml''(I''', \check{R}''), \check{L}'')$ and show that $Tf^*(\varphi''') = Ml''(Tf\,Tf'(\varphi''')\sigma''', I\check{d}', \overline{Ml'}(Id'', R'^{\check{\sim}}))$.

Applying the construction of 12.1 to example 3.1 thrice, one gets consecutive IOS $\mathscr{S}$, $\mathscr{S}'$, $\mathscr{S}''$, $\mathscr{S}'''$ which admit transfer operations Tf, Tf' and collection operations Co, Co', Co'' (by exercise 11.1) satisfying all the assumptions made in this chapter.

CHAPTER 15

Hierarchies of operative spaces

The study of consecutive spaces naturally develops into the study of hierarchies of spaces. For, given an IOS $\mathscr{S}_0$, one is tempted to construct by 12.1 a sequence $\{\mathscr{S}_n\}$ of consecutive spaces, then to try to bring them all together into a single space $\mathscr{S}_\omega$ and iterate, thereby constructing a transfinite sequence of spaces. We feel that such hierarchies of spaces provide the appropriate domains for an intrinsic generalized recursion theory. In this chapter we aim only to establish some basic properties.

Let $\mathscr{S}$ be an IOS. A *hierarchy of* IOS *based on* $\mathscr{S}$ is a transfinite sequence $\{\mathscr{S}_\xi\}$ of IOS such that $\mathscr{S}_0 = \mathscr{S}$ and whenever $\xi < \eta$, then $\mathscr{S}_\xi$ is (isomorphic to) a proper subspace of $\mathscr{S}_\eta$. Hierarchies $\{\mathscr{S}_\xi\}_{\xi<\xi_0}$ of length ξ_0 could also be of interest, allowing us to work either in the single space $\mathscr{S}_{\xi_0}$, if ξ_0 is a limit, or in $\mathscr{S}_\xi$, if $\xi_0 = \xi_1 + 1$.

A hierarchy is called *monotonic* if for all ξ the IOS $\mathscr{S}_\xi, \mathscr{S}_{\xi+1}$ are consecutive and $\mathscr{F}_{\xi_1} = \bigcup_{\eta<\xi}\mathscr{F}_\eta$ whenever $\xi > 0$ is limit, identifying (this is crucial!) the subspace $\bar{\mathscr{S}}_\xi$ of $\mathscr{S}_{\xi+1}$ with $\mathscr{S}_\xi$ for all ξ. Two constructions yielding monotonic hierarchies will be given in chapter 20. They really exploit the idea outlined above, except that beyond ω one takes in $\mathscr{F}_{\xi+1}$ only *some* of the monotonic unary mappings over $\mathscr{F}_\xi$ as opposed to *all* of them.

Now assume that a monotonic hierarchy $\{\mathscr{S}_\xi\}$ is given. The proper class $\bigcup_\xi \mathscr{F}_\xi$ is denoted by $\mathscr{MH}$ (for 'monotonic hierarchy'). The operations of multiplication, pairing, translation and iteration are well defined on the whole class $\mathscr{MH}$, so $\mathscr{MH}$ could be regarded as the carrier of a single 'universal' IOS. Notice that if $\mathscr{B} \subseteq \mathscr{MH}$ is a set, then cardinality considerations imply that $\mathscr{B} \subseteq \mathscr{F}_\xi$ for a certain ξ. For all $\varphi \in \mathscr{MH}$ the ordinal $r(\varphi) = \min\{\xi/\varphi \in \mathscr{F}_\xi\}$ (which is zero or a successor) is called the *rank of* φ. Accordingly, $\varphi \in \mathscr{F}_\xi$ iff $r(\varphi) \leq \xi$.

If $\varphi \in \mathscr{F}_{\xi+1}$, $\psi \in \mathscr{F}_\xi$ and φ_1 is the member of $\mathscr{F}_{\xi+1}$ identified with φ, we write $\varphi(\psi)_\xi$ for $\varphi_1(\psi)$. It follows that $\varphi(\psi)_\xi = \varphi(\psi)$, provided $r(\varphi) = \xi + 1$. If $r(\varphi) < \xi + 1$, then a certain member φ_2 of $\mathscr{F}_\xi$ is identified with φ so that $\varphi_1 = \overline{\varphi_2}$; hence $\varphi(\psi)_\xi = \overline{\varphi_2}(\psi) = \varphi_2\psi$, which gives $\varphi(\psi)_\xi = \varphi\psi$. In particular, if $\varphi \in \mathscr{F}_\xi$, then $\varphi = \varphi(I)_\xi$; this will often be used in the sequel. Moreover, $\varphi, \psi \in \mathscr{F}_{\xi+1}$, $\chi \in \mathscr{F}_\xi$ and $\varphi \leq \psi$ imply $\varphi(\chi)_\xi \leq \psi(\chi)_\xi$.

Another convenient notation is $\lambda_\xi\theta.$——— for the mapping which assigns to each $\theta \in \mathscr{F}_\xi$ the expression———.

Let Id_ξ be the mapping Id of chapter 12 corresponding to $\mathscr{S}_\xi$. Then

$Id_\xi = \lambda_\xi \theta . I$ and $Id_\xi \in \mathscr{F}_{\xi+1}$. On the other hand, 12.5 implies that $Id_\xi \notin \mathscr{F}_\xi$; hence $r(Id_\xi) = \xi + 1$.

Proposition 15.1. $Id_\xi \varphi = Id_\xi$ for all $\varphi \in \mathscr{M}\mathscr{H}$.

Proof. By transfinite induction on $r(\varphi)$.

If $r(\varphi) \leq \xi + 1$, then $\varphi \in \mathscr{F}_{\xi+1}$ and $Id_\xi \varphi = Id_\xi$ follows from 12.4.

Suppose that $r(\varphi) = \eta + 1 > \xi + 1$ and $Id_\xi \psi = Id_\xi$ whenever $r(\psi) < \eta + 1$. Let $\theta \in \mathscr{F}_\eta$. Then $Id_\xi \varphi \in \mathscr{F}_{\eta+1}$, $\varphi(\theta)_\eta \in \mathscr{F}_\eta$, hence $Id_\xi \varphi(\theta)_\eta = Id_\xi(\varphi(\theta)_\eta)_\eta = Id_\xi \circ (\varphi(\theta)_\eta)$ since $r(Id_\xi) < \eta + 1$. The induction assumption gives $Id_\xi \circ (\varphi(\theta)_\eta) = Id_\xi$; hence

$$Id_\xi \varphi(\theta)_\eta = Id_\xi = Id_\xi \theta = Id_\xi(\theta)_\eta .$$

This holds for all $\theta \in \mathscr{F}_\eta$; hence $Id_\xi \varphi = Id_\xi$. This completes the proof.

Proposition 15.2. If $\varphi \in \mathscr{F}_\xi$, then $\varphi Id_\xi = \lambda_\xi \theta . \varphi$. If $\varphi \in \mathscr{F}_{\xi+1}$, then $\varphi Id_\xi = \varphi(I)_\xi Id_\xi$. If $\varphi \in \mathscr{F}_{\eta+1}$, $\eta > \xi$, then $\varphi Id_\xi = \varphi(Id_\xi)_\eta Id_\eta$.

Proof. The first two assertions follows from 12.3 and 12.14.

Whenever $\varphi \in \mathscr{F}_{\eta+1}$, $\eta > \xi$, then $\varphi Id_\xi \in \mathscr{F}_{\eta+1}$, hence

$$\begin{aligned} \varphi Id_\xi &= \lambda_\eta \theta . (\varphi Id_\xi)(\theta)_\eta = \lambda_\eta \theta . \varphi(Id_\xi(\theta)_\eta)_\eta = \lambda_\eta \theta . \varphi(Id_\xi \theta)_\eta \\ &= \lambda_\eta \theta . \varphi(Id_\xi)_\eta = \varphi(Id_\xi)_\eta Id_\eta \end{aligned}$$

using 15.1. The proof is complete.

Let Ml_ξ be the mapping Ml of chapter 12 corresponding to $\mathscr{S}_\xi$. Then $Ml_\xi = \lambda_\xi \theta . L\theta R\theta$ and $r(Ml_\xi) = \xi + 1$ by 12.17.

Proposition 15.3. If φ, $\psi \in \mathscr{F}_{\xi+1}$, then $Ml_\xi(\varphi, \psi) = \lambda_\xi \theta . \varphi(\theta)_\xi \psi(\theta)_\xi$. In particular, if φ, $\psi \in \mathscr{F}_\xi$, then $Ml_\xi(\varphi Id_\xi, \psi Id_\xi) = \varphi \psi Id_\xi$.

This follows from 12.16, 12.20.

Proposition 15.4. If $\varphi \in \mathscr{F}_{\xi+1}$, $\psi \in \mathscr{F}_\xi$, then $\varphi(\psi)_\xi = Ml_\xi(\varphi \psi Id_\xi, I)$.

Proof. Using 12.21, we get

$$Ml_\xi(\varphi \psi Id_\xi, I) = (\varphi \psi)(I)_\xi = \varphi(\psi(I)_\xi)_\xi = \varphi(\psi)_\xi .$$

Let Tr_ξ, It_ξ be the mappings Tr, It of chapter 12 corresponding to $\mathscr{S}_\xi$. Then $Tr_\xi = \lambda_\xi \theta . \langle \theta \rangle$, $It_\xi = \lambda_\xi \theta . [\theta]$ and $r(Tr_\xi) = r(It_\xi) = \xi + 1$ by 12.30.

We are now going to specify abstract concepts of effective computability within the hierarchy, bearing in mind those introduced in chapter 7.

Let $\varphi \in \mathscr{M}\mathscr{H}$ and $\mathscr{B} \subseteq \mathscr{M}\mathscr{H}$. Then the notions of relative recursiveness, prime recursiveness, primitive recursiveness, primitiveness and polynomiality reappear as follows. We say that φ is *recursive* (*prime recursive* etc.) *in* $\mathscr{B}$ iff $\varphi \in \mathscr{F}_\xi$ and φ is recursive (prime recursive etc.) in $\mathscr{B} \cap \mathscr{F}_\xi$ in the sense of $\mathscr{S}_\xi$ for a certain ξ. This is a sound definition since if $\xi \leq \eta$, then $\mathscr{S}_\xi$ is a subspace of $\mathscr{S}_\eta$ and $\mathscr{B} \cap \mathscr{F}_\xi \subseteq \mathscr{F}_\xi$; hence $\varphi \in \mathscr{F}_\xi$ is recursive (prime recursive etc.) in $\mathscr{B} \cap \mathscr{F}_\xi$ in the sense of $\mathscr{S}_\xi$ iff φ is recursive (prime recursive etc.) in $\mathscr{B} \cap \mathscr{F}_\xi$ in the sense of $\mathscr{S}_\eta$.

The above notions satisfy the ordinary properties of chapter 7 since those properties hold for $\mathscr{S}_\xi$ for all ξ. In particular it is worth mentioning that whenever φ is recursive (prime recursive etc.) in $\mathscr{B}$, then φ is recursive (prime recursive etc.) in a finite subclass of $\mathscr{B}$.

Proposition 15.5. The elements Tr_ξ, It_ξ are prime recursive in Id_ξ, Ml_ξ.
This follows from 12.28, 12.29.

Proposition 15.6. Let $\varphi \in \mathscr{F}_\xi$ and $\mathscr{B} \subseteq \mathscr{F}_\xi$. Then the following are equivalent.
(1) φ is recursive in $\mathscr{B}$.
(2) $\varphi \in cl(\mathscr{B} \cup \{L, R, Id_\xi, Ml_\xi, Tr_\xi, It_\xi\}/\circ, \Pi)$.
(3) $\varphi \in cl(\mathscr{B} \cup \{L, A, Id_\xi, Ml_\xi, Tr_\xi, It_\xi\}/\circ)$.
Proof. The implications (1)$\Rightarrow$(2)$\Rightarrow$(3) are immediate. Supposing (3), we get that φ is a mapping over $\mathscr{F}_\xi$ recursive in $\mathscr{B}$, hence $\varphi = \varphi(I)_\xi$ is a member of $\mathscr{F}_\xi$ recursive in $\mathscr{B}$. The proof is complete.

A version of 15.6 for prime recursiveness can be given with Tr_ξ omitted, and similarly for the other notions considered. Further on we shall concentrate on recursiveness, taking the element A as initial and correspondingly omitting R and the pairing operation.

The following Hierarchy Conservativeness Theorem is an extension of 12.38*.

Proposition 15.7*. Let $\varphi \in \mathscr{F}_\xi$, $\mathscr{B} \subseteq \mathscr{F}_\xi$ and φ be recursive in $\mathscr{B} \cup \{Id_\eta, Ml_\eta/\eta \geq \xi\}$. Then φ is recursive in $\mathscr{B}$.

Proof. Writing $\mathscr{B}_\eta$ for $\mathscr{B} \cup \{Id_\zeta, Ml_\zeta/\xi \leq \zeta \leq \eta\}$, there is a η_0 such that φ is recursive in $\mathscr{B}_{\eta_0}$. Suppose that $\eta > \xi$. Applying 12.38* to the consecutive spaces $\mathscr{S}_{\eta_0}, \mathscr{S}_{\eta_0+1}$, we see that φ is a mapping over $\mathscr{F}_{\eta_0}$ recursive in $\mathscr{B} \cup \{Id_\zeta, Ml_\zeta/\xi \leq \zeta < \eta_0\}$; hence $\varphi = \varphi(I)_{\eta_0}$ is an element recursive in $\mathscr{B} \cup \{Id_\zeta, Ml_\zeta/\xi \leq \zeta < \eta_0\}$. Therefore, φ is recursive in $\mathscr{B}_{\eta_1}$ for a certain $\eta_1 < \eta_0$. Since any decreasing sequence of ordinals is finite, we obtain in finitely many steps an $\eta_n \leq \xi$ such that φ is recursive in $\mathscr{B}_{\eta_n}$. Therefore φ is recursive in $\mathscr{B} \cup \{Id_\xi, Ml_\xi\}$; hence $\varphi = \varphi(I)_\xi$ is recursive in $\mathscr{B}$ by 12.38*. The proof is complete.

By 15.5, $\{Id_\eta, Ml_\eta/\eta \geq \xi\}$ can be replaced by $\{Id_\eta, Ml_\eta, Tr_\eta, It_\eta/\eta \geq \xi\}$.

The following conservativeness result called the Hierarchy First Recursion Theorem is in a sense analogous to the First Recursion Theorem 5.3.2 of Platek [1966].

Proposition 15.8*. Let $\varphi \in \mathscr{F}_{\xi+1}$ and $\mathscr{B} \subseteq \mathscr{F}_\xi$. Then the following are equivalent.
(1) φ is (prime) recursive in $\mathscr{B} \cup \{Id_\xi, Ml_\xi\}$.
(2) $\varphi \in cl(\mathscr{B} \cup \{L, A, Id_\xi, Ml_\xi, Tr_\xi, It_\xi\}/\circ)$.
(3) $\varphi \in cl(\mathscr{B} \cup \{L, A, Id_\eta, Ml_\eta, Tr_\eta, It_\eta/\eta \geq \xi\}/\circ)$.
If $\varphi \in \mathscr{F}_\xi$, then (1) can be replaced by 'φ is recursive in $\mathscr{B}$'.

Proof. The sentence (1) means that φ is (prime) recursive in $\mathscr{B} \cup \{Id_\xi, Ml_\xi\}$ within $\mathscr{S}_{\xi+1}$. Therefore, (1) and (2) are equivalent by the Conservativeness Theorem 12.38*. It is immediate that (2) implies (3), while (3) implies (2) by 15.7*.

If $\varphi \in \mathscr{F}_\xi$, then (1) implies that φ is recursive in $\mathscr{B}$ by 12.38*. The proof is complete.

Now assume that the hierarchy $\{\mathscr{S}_\xi\}$ admits the operation Tf of chapter 14, i.e. for all ξ the mapping.

$$Tf_{\xi+2} = \lambda_{\xi+2}\varphi . \lambda_{\xi+1}\rho . \lambda_\xi \theta . \varphi(\rho(L\theta Id_\xi, I))_{\xi+1}(R\theta)_\xi$$

is a member of $\mathscr{F}_{\xi+3}$. Of course, $r(Tf_{\xi+2}) = \xi + 3$ by 14.18. The hierarchies to be constructed in chapter 20 admit transfers.

Let $\varphi \in \mathscr{M}\mathscr{H}$ and $\mathscr{B} \subseteq \mathscr{M}\mathscr{H}$. Then we say that φ is *$tf_{\xi+2}$-recursive in $\mathscr{B}$* iff $\varphi \in \mathscr{F}_{\xi+2}$ and φ is tf-recursive in $\mathscr{B} \cap \mathscr{F}_{\xi+2}$ in the sense of chapter 14. If φ is $tf_{\xi+2}$-recursive in $\mathscr{B}$ and $\xi < \eta$, then it can not be claimed that φ is also $tf_{\eta+2}$-recursive in $\mathscr{B}$. Instead, φ is (prime) recursive in $\mathscr{B} \cup \{Id_{\xi+1}, Ml_{\xi+1}, Tf_{\xi+2}\}$ since the set $\mathscr{B}_0$ of functional elements corresponding to the t-operation $Tf_{\xi+2}$ is $\{L, R, Id_{\xi+1}, Ml_{\xi+1}\}$.

Proposition 14.20* may be extended to the following two Hierarchy tf-Conservativeness Theorems.

Proposition 15.9*. Let $\varphi \in \mathscr{F}_{\xi+2}$, $\mathscr{B} \subseteq \mathscr{F}_{\xi+2}$ and φ be recursive in $\mathscr{B} \cup \{Id_\eta, Ml_\eta / \eta > \xi\} \cup \{Tf_{\eta+2} / \eta \geq \xi\}$. Then φ is $tf_{\xi+2}$-recursive in $\mathscr{B}$.

Proof. Writing $\mathscr{B}_\eta$, $\mathscr{B}_{<\eta}$ respectively for

$$\mathscr{B} \cup \{Id_\zeta, Ml_\zeta / \xi < \zeta \leq \eta\} \cup \{Tf_{\zeta+2} / \xi + 2 \leq \zeta + 2 \leq \eta\},$$
$$\mathscr{B} \cup \{Id_\zeta, Ml_\zeta / \xi < \zeta < \eta\} \cup \{Tf_{\zeta+2} / \xi + 2 \leq \zeta + 2 < \eta\},$$

φ is recursive in $\mathscr{B}_{\eta_0}$ for a certain η_0. Assuming that $\eta_0 > \xi + 2$, there are two possible cases.

If η_0 is not of the form $\zeta + 2$, then 12.38* implies that φ is a mapping over $\mathscr{F}_{\eta_0}$ recursive in $\mathscr{B}_{<\eta_0}$, hence $\varphi = \varphi(I)_{\eta_0}$ is an element recursive in $\mathscr{B}_{<\eta_0}$.

If $\eta_0 = \zeta + 2$, then an application of 14.20* to the consecutive spaces $\mathscr{S}_{\eta_0}$, $\mathscr{S}_{\eta_0+1}$ implies that φ is a mapping over $\mathscr{F}_{\eta_0}$ tf_{η_0}-recursive in $\mathscr{B}_{<\eta_0}$; hence $\varphi = \varphi(I)_{\eta_0}$ is an element tf_{η_0}-recursive in $\mathscr{B}_{<\eta_0}$. The implication (3)$\Rightarrow$(2) of 14.20* implies that φ is recursive in $\mathscr{B}_{<\eta_0}$ in this case as well.

Therefore, there is an $\eta_1 < \eta_0$ such that φ is recursive in $\mathscr{B}_{\eta_1}$. We get in finitely many steps an $\eta_n \leq \xi + 2$ such that φ is recursive in $\mathscr{B}_{\eta_n}$; hence φ is recursive in

$$\mathscr{B} \cup \{Id_{\xi+1}, Ml_{\xi+1}, Id_{\xi+2}, Ml_{\xi+2}, Tf_{\xi+2}\}.$$

The implication (2)$\Rightarrow$(1) of 14.20* ensures that φ is a mapping over $\mathscr{F}_{\xi+2}$ $tf_{\xi+2}$-recursive in $\mathscr{B}$, hence $\varphi = \varphi(I)_{\xi+2}$ is an element $tf_{\xi+2}$-recursive in $\mathscr{B}$. The proof is complete.

Proposition 15.10*. Let ξ be a limit ordinal or a successor to a limit ordinal, $\varphi \in \mathscr{F}_\xi$, $\mathscr{B} \subseteq \mathscr{F}_\xi$ and φ be recursive in

$$\mathscr{B} \cup \{Id_\eta, Ml_\eta / \eta \geq \xi\} \cup \{Tf_{\eta+2} / \eta + 2 > \xi\}.$$

Then φ is recursive in $\mathscr{B}$.

Proof. Writing $\mathscr{B}_\eta$ for

$$\mathscr{B} \cup \{Id_\zeta, Ml_\zeta / \xi \leq \zeta \leq \eta\} \cup \{Tf_{\zeta+2} / \xi < \zeta + 2 \leq \eta\}$$

and following the proof of 15.9*, we get an $\eta_n \leq \xi$ such that φ is recursive in $\mathscr{B}_{\eta_n} \subseteq \mathscr{B} \cup \{Id_\xi, Ml_\xi\}$; hence φ is recursive in $\mathscr{B}$ by 12.38*. This completes the proof.

There are two statements to extend 14.20* since we have no transfer operations over $\mathscr{F}_\xi, \mathscr{F}_{\xi+1}$ for limit ξ. The corresponding two Hierarchy First tf-Recursion Theorems below seem closer to the original theorem of Platek.

Proposition 15.11*. Let $\varphi \in \mathscr{F}_{\xi+3}$ and $\mathscr{B} \subseteq \mathscr{F}_{\xi+2}$. Then the following are equivalent.

(1) φ is (prime) recursive in

$$\mathscr{B} \cup \{Id_{\xi+1}, Ml_{\xi+1}, Id_{\xi+2}, Ml_{\xi+2}, Tf_{\xi+2}\}.$$

(2) $\varphi \in cl(\mathscr{B} \cup \{L, A, Id_{\xi+1}, Ml_{\xi+1}, Id_{\xi+2}, Ml_{\xi+2}, Tr_{\xi+2}, It_{\xi+2}, Tf_{\xi+2}\}/\circ)$.

(3) $\varphi \in cl(\mathscr{B} \cup \{L, A\} \cup \{Id_\eta, Ml_\eta / \eta > \xi\}$
$\cup \{Tr_\eta, It_\eta / \eta > \xi + 1\} \cup \{Tf_{\eta+2} / \eta \geq \xi\}/\circ)$.

If $\varphi \in \mathscr{F}_{\xi+2}$, then (1) can be replaced by 'φ is $tf_{\xi+2}$-recursive in $\mathscr{B}$'. Exercise 14.3** allows us to drop $Tr_{\xi+2}$ in (2).

Proof. The first two assertions are equivalent by 14.20*. The implication (1)$\Rightarrow$(3) is immediate. Assuming (3), we obtain that φ is $tf_{\xi+3}$-recursive in $\mathscr{B} \cup \{Id_{\xi+1}, Ml_{\xi+1}, Tf_{\xi+2}\}$ by 15.9*, which gives (1) by the implication (3)$\Rightarrow$(2) of 14.20*.

If $\varphi \in \mathscr{F}_{\xi+2}$, then (1) implies that φ is $tf_{\xi+2}$-recursive in $\mathscr{B}$ by 14.20*. The proof is complete.

Proposition 15.12*. Let $\varphi \in \mathscr{F}_{\xi+1}$, $\mathscr{B} \subseteq \mathscr{F}_\xi$ and suppose that ξ is not of the form $\xi_1 + 2$. Then the following are equivalent.

(1) φ is (prime) recursive in $\mathscr{B} \cup \{Id_\xi, Ml_\xi\}$.

(2) $\varphi \in cl(\mathscr{B} \cup \{L, A, Id_\xi, Ml_\xi, Tr_\xi, It_\xi\}/\circ)$.

(3) $\varphi \in cl(\mathscr{B} \cup \{L, A\} \cup \{Id_\eta, Ml_\eta, Tr_\eta, It_\eta / \eta \geq \xi\} \cup \{Tf_{\eta+2} / \eta + 2 > \xi\}/\circ)$.

If $\varphi \in \mathscr{F}_\xi$, then (1) can be replaced by 'φ is recursive in $\mathscr{B}$'.

Proof. The equivalence (1)$\Leftrightarrow$(2) was established in 15.8*, while (1)$\Rightarrow$(3) is immediate. Finally, (3) implies (1) by 15.10*.

If $\varphi \in \mathscr{F}_\xi$, then (1) implies that φ is recursive in $\mathscr{B}$ by 12.38*. The proof is complete.

Some concluding remarks.

The carrier of a monotonic hierarchy with transfers seems to form a model of a typed $\lambda\mu$-calculus with combinators I, L, R, Ml_0 and $\Pi_0 = \Pi \upharpoonright \mathscr{F}_0^2$; the μ-operation concerned is introduced in the exercises below. Instead of $\mathscr{MH}$ one may consider initial segments $\mathscr{F}_\xi$, ξ a limit ordinal.

Transfers in a hierarchy are most probably independent, i.e. $Tf_{\xi+2}$ is not recursive in $\{Id_\eta, Ml_\eta / \eta \geq 0\} \cup \{Tf_{\eta+2} / \eta \neq \xi\}$ for all ξ. (Though we have not even proved that $Tf_{\xi+2}$ is not recursive in $\{Id_\eta, Ml_\eta / \eta \geq 0\}$.) Therefore, the transfer operation is necessary for the abovementioned combinatorial completeness.

It may be of some interest to study recursion in certain specific elements, e.g. elements embodying quantifiers. Among the possible candidates are the mappings Q_σ, $Q_{\sigma,\mathscr{C}}$ of chapter 13.

The work of Platek referred to above studies what can be described in IOS-terms as the ω-segment of a monotonic hierarchy based on example 4.7. Therefore, the considerations of this chapter embody Platek's work in Generalized Recursion Theory in a more general setting.

EXERCISES TO CHAPTER 15

Exercise 15.1. Let $\varphi, \psi \in \mathscr{F}_\xi$ and $\varphi, \psi \leq \chi$. Show that there is a $\chi_1 \in \mathscr{F}_\xi$ such that $\varphi, \psi \leq \chi_1$.

Hint. Use transfinite induction on $r(\chi)$.

Exercise 15.2. Show that $\varphi \in \mathscr{F}_\xi$ iff $\varphi = Ml_\xi(\varphi Id_\xi, I)$. Therefore, if $\varphi \in \mathscr{F}_\xi$, then φ is polynomial in φId_ξ, Ml_ξ.

Hint. $\varphi \in \mathscr{F}_\xi$ iff $\varphi = \varphi(I)_\xi$.

Exercise 15.3. Show that whenever $\xi < \eta$, then

$$Tr_\xi = RMl_\eta(It_\eta(Ml_\xi(LId_\eta, LId_\xi), Ml_\xi(RId_\eta, \mathrm{RId}_\xi)), I).$$

Hint. Observing that $\mathscr{S}_{\xi+1}$ is a subspace of $\mathscr{S}_\eta$, get $Tr_\xi = RIt_\eta((Ml_\xi(L, LId_\xi), Ml_\xi(R, RId_\xi)))_\eta$ by the proof of 12.28, then use 15.4.

Exercise 15.4. Show that if $\xi < \eta$, then

$$It_\xi = RMl_\eta(It_\eta(Id_\xi, Ml_\xi Id_\eta), I).$$

Hint. As for the previous exercise, with 12.28 replaced by 12.29.

Write $\mu_\xi\theta.$ ——— for the least θ in $\mathscr{F}_\xi$ such that——— $\leq \theta$, provided it exists.

Exercise 15.5. Let $\varphi \in \mathscr{F}_{\xi+1}$. Show that $\mu_\xi\theta.\varphi(\theta)_\xi$ exists and equals $Ml_\xi(R[\varphi R], I)$. Show that $\mu_\eta\theta.\varphi(\theta)_\eta = R[\varphi R]$, provided $\xi < \eta$.

Hint. 12.27 implies that $\theta_0 = \mu_\xi\theta.\varphi(\theta)_\xi$ exists and $\theta_0 Id_\xi = R[\varphi R]$. Make use of exercise 15.2.

Exercise 15.6. Prove that there is no collection operation over $\mathscr{F}_\xi$, $\xi \geq \omega$.

Hint. Suppose that $Co{:}\mathscr{F}_\xi^\omega \to \mathscr{F}_\xi$ satisfies the axiom cA_3 of chapter 11. Take $\varphi = Co\{\varphi_n\}$, where $r(\varphi_n) = n$ for all n. Construct a sequence $\xi_1 > \cdots > \xi_n \geq \omega$ such that $\varphi(I)_{\xi_1} \ldots (I)_{\xi_i} \in \mathscr{F}_{\xi_{i+1}+1}$ for all $1 \leq i < n$, and $\underline{\psi = \varphi(I)_{\xi_1}} \ldots (I)_{\xi_n} \in \mathscr{F}_\omega$. Assuming $r(\psi) = m$, get $\varphi_{m+1} = \varphi_{m+1}(I)_{\xi_1} \ldots (I)_{\xi_n} = m + 1\psi \in \mathscr{F}_m$, which is not the case.

PART D

Constructing operative spaces

CHAPTER 16

Pairing spaces

In this chapter we introduce certain algebraic systems, 'semi-manufactures' from which one can construct OS. Roughly speaking, these systems are OS without multiplication. The necessity of considering such simpler algebraic systems is suggested by 12.1, where an OS $\mathscr{S}'$ was constructed from a given space $\mathscr{S}$ assumed (somewhat excessively) to be an OS.

Some preliminary notions. Let $\mathscr{F}$ be a nonempty partially ordered set, let $\mathscr{T}$ be a nonempty *right directed set* (i.e., a partially ordered set each two members of which have an upper bound) and let $t:\mathscr{F}\to\mathscr{T}$. Letters $\varphi,\psi,\chi,\theta,\tau$ will stand for members of $\mathscr{F}$ and a,b,c for members of $\mathscr{T}$. We call $t(\varphi)$ the *type of* φ and write $\mathscr{F}_a$ for $\{\varphi/\varphi\in\mathscr{F}\,\&\,t(\varphi)\le a\}$. A mapping $\Gamma:\mathscr{F}^n\to\mathscr{F}$ *preserves types* iff $\Gamma(\mathscr{F}_a^n)\subseteq\mathscr{F}_a$ for all a, while Γ is *normal* if $\exists a\forall b\ge a$ $(\Gamma(\mathscr{F}_b^n)\subseteq\mathscr{F}_b)$. Clearly, any type preserving mapping is normal. An n-ary mapping Γ is $\mathscr{T}$*-continuous* iff whenever for all i, $1\le i\le n$, $\{\varphi_{i,m}\}_m$ is a chain (i.e., increasing countable sequence) in $\mathscr{F}_a$, $\varphi_i\in\mathscr{F}$ and $\varphi_i=\sup_m\varphi_{i,m}$, then $\sup_m\Gamma(\varphi_{1,m},\ldots,\varphi_{n,m})$ exists and equals $\Gamma(\varphi_1,\ldots,\varphi_n)$. Notice that all $\mathscr{T}$-continuous mappings are monotonic. If $\mathscr{T}$ is a singleton, then the $\mathscr{T}$-continuous mappings are exactly the continuous ones and all the mappings over $\mathscr{F}$ preserve types. ('Continuous' means countable continuous here.)

Let $\Pi:\mathscr{F}^2\to\mathscr{F}$ and $L',R':\mathscr{F}\to\mathscr{F}$. As usual, (φ,ψ) will stand for $\Pi(\varphi,\psi)$. The quadruple $\mathscr{S}=(\mathscr{F},\Pi,L',R')$ is a *pairing space* iff the following hold.

1. Π,L',R' are monotonic.
2. $L'((\varphi,\psi))=\varphi$, $R'((\varphi,\psi))=\psi$.

In other words, Π is a *pairing operation* and L',R' are its inverses. To exclude trivialities assume that $\mathscr{F}$ is not a singleton.

A pairing space $\mathscr{S}$ is *continuous* iff so are Π,L',R'.

The following two statements are immediate.

Proposition 16.1. Let $\mathscr{F}$ be a set with at least two distinct members, let $\Pi:\mathscr{F}^2\to\mathscr{F}$ be injective and let $L',R':\mathscr{F}\to\mathscr{F}$ be its inverses. Take $\varphi\le\psi$ iff $\varphi=\psi$. Then $\mathscr{S}=(\mathscr{F},\Pi,L',R')$ is a continuous pairing space.

Proposition 16.2. Whenever $(\mathscr{F},I,\Pi,L,R)$ is an OS, then $(\mathscr{F},\Pi,\lambda\theta.L\theta,\lambda\theta.R\theta)$ is a pairing space.

We now give several constructions which yield pairing spaces.

Proposition 16.3. Let M be a nonempty set, let $f_1, f_2 : M \to M$ be injective and suppose that $f_1(M) \cap f_2(M) = \emptyset$. Let E be a partially ordered set with at least two distinct members and let e be a fixed member of it. Take $\mathscr{F} = \{\varphi/\varphi: M \to E\}$, $\varphi \leq \psi$ iff $\forall s(\varphi(s) \leq \psi(s))$, $(\varphi,\psi)(f_1(s)) = \varphi(s)$, $(\varphi,\psi)(f_2(s)) = \psi(s)$ and $(\varphi,\psi)(s) = e$ otherwise, $L'(\varphi) = \lambda s.\varphi(f_1(s))$ and $R'(\varphi) = \lambda s.\varphi(f_2(s))$. Then $\mathscr{S} = (\mathscr{F}, \Pi, L', R')$ is a continuous pairing space.

Proof. We have

$$L'((\varphi,\psi)) = \lambda s.(\varphi,\psi)(f_1(s)) = \lambda s.\varphi(s) = \varphi$$

and similarly $R'((\varphi,\psi)) = \psi$.

We are going to show that L', R' are continuous with respect to least upper bounds of arbitrary subsets of $\mathscr{F}$, while Π is continuous with respect to least upper bounds of nonempty subsets of $\mathscr{F}$. But let us show first that whenever $\varphi \in \mathscr{F}$ and $\mathscr{H} \subseteq \mathscr{F}$, then $\varphi = \sup \mathscr{H}$ iff $\varphi(s) = \sup\{\theta(s)/\theta \in \mathscr{H}\}$ for all s.

Suppose that $\varphi = \sup \mathscr{H}$. Then $\theta(s) \leq \varphi(s)$ for all s and $\theta \in \mathscr{H}$. Let $s \in M$, $d \in E$ and $\theta(s) \leq d$ for all $\theta \in \mathscr{H}$. Taking $\tau(s) = d$ and $\tau(t) = \varphi(t)$ otherwise, we get $\theta \leq \tau$ for all $\theta \in \mathscr{H}$, hence $\varphi \leq \tau$, which implies $\varphi(s) \leq d$. Therefore, $\varphi(s) = \sup\{\theta(s)/\theta \in \mathscr{H}\}$.

Conversely, suppose that $\varphi(s) = \sup\{\theta(s)/\theta \in \mathscr{H}\}$ for all s. Then $\theta \leq \varphi$ for all $\theta \in \mathscr{H}$. If $\theta \leq \tau$ for all $\theta \in \mathscr{H}$, then $\theta(s) \leq \tau(s)$ for all $\theta \in \mathscr{H}$ and all s, hence $\varphi(s) \leq \tau(s)$ for all s. Therefore, $\varphi \leq \tau$, which gives $\varphi = \sup \mathscr{H}$.

Now let $\varphi = \sup \mathscr{H}$. Then we have

$$L'(\varphi)(s) = \varphi(f_1(s)) = \sup\{\theta(f_1(s))/\theta \in \mathscr{H}\} = \sup\{L'(\theta)(s)/\theta \in \mathscr{H}\}$$

for all s, hence $L'(\varphi) = \sup L'(\mathscr{H})$ and similarly $R'(\varphi) = \sup R'(\mathscr{H})$.

Let $\mathscr{H}_1, \mathscr{H}_2 \neq \emptyset$, $\varphi = \sup \mathscr{H}_1$ and $\psi = \sup \mathscr{H}_2$. Then

$$(\varphi,\psi)(f_1(s)) = \varphi(s) = \sup\{\theta(s)/\theta \in \mathscr{H}_1\} = \sup\{(\theta,\tau)(f_1(s))/\theta \in \mathscr{H}_1 \& \tau \in \mathscr{H}_2\}.$$

Similarly, $(\varphi,\psi)(f_2(s)) = \sup\{(\theta,\tau)(f_2(s))/\theta \in \mathscr{H}_1 \& \tau \in \mathscr{H}_2\}$, while $(\varphi,\psi)(s) = e = \sup\{(\theta,\tau)(s)/\theta \in \mathscr{H}_1 \& \tau \in \mathscr{H}_2\}$ otherwise. Therefore, $(\varphi,\psi)(s) = \sup\{(\theta,\tau)(s)/\theta \in \mathscr{H}_1 \& \tau \in \mathscr{H}_2\}$ for all s, hence $(\varphi,\psi) = \sup \Pi(\mathscr{H}_1, \mathscr{H}_2)$. The proof is complete.

Some remarks are in order here.

If M, f_1, f_2 satisfy the assumptions of 16.3, then f_1, f_2 is said to be a *splitting scheme* for M. A slightly more general notion of splitting scheme is considered in Ivanov [1980, 1980b]. Multiple-valued splitting could also be of some interest.

Splitting schemes have already been used to construct OS in examples 4.7, 4.8. As mentioned there, all infinite sets admit splitting schemes. However, this general assertion depends on the axiom of choice, which is not the case in some particular instances. On the other hand, no set augmented with a splitting scheme can be finite since the members of the sequence $\{f_2^n(f_1(s))\}$ are pairwisely distinct for all s.

If $M = \omega$, then splitting schemes for M can be naturally specified. For instance, one may take $f_1 = \lambda s.2s$, $f_2 = \lambda s.2s + 1$ as in example 3.1, or $f_1 = \lambda s.3s$, $f_2 = \lambda s.3s + 1$ as in example 4.3 etc.

The set $M = \omega^* = \bigcup_n \omega^n$ can also be augmented with a natural splitting scheme. Writing Λ for the empty sequence and omitting brackets and commas, one may take $f_1 = \lambda x.0x$, $f_2(\Lambda) = \Lambda$ and $f_2(nx) = n + 1x$.

If ξ_0 is a limit ordinal and $M = \{\xi / \xi < \xi_0\}$, then one may take $f_1 = \lambda\xi.2\xi$, $f_2 = \lambda\xi.2\xi + 1$.

An arbitrary nonempty set M_0 can be extended to a wider one which admits natural splitting schemes, e.g. $M = \omega \times M_0$ or $M = \omega^* \times M_0$. If M_0 is finite, then another possible extension is $M = M_0 \cup \omega$.

More generally, whenever f_1, f_2 is a splitting scheme for M and N is a nonempty set, then $\lambda sx.(f_1(s), x)$, $\lambda sx.(f_2(s), x)$ is a splitting scheme for $M \times N$. It follows in particular that 16.3 holds with $M \times N$ playing the role of M.

Proposition 16.4. Let M, f_1, f_2, E, e be the same as in 16.3 and N be an arbitrary nonempty set. Take $\mathscr{F} = \{\varphi / \varphi : M \times N \to E\}$, $\varphi \leq \psi$ iff $\forall sx(\varphi(s, x) \leq \psi(s, x))$, $(\varphi, \psi)(f_1(s), x) = \varphi(s, x)$, $(\varphi, \psi)(f_2(s), x) = \psi(s, x)$ and $(\varphi, \psi)(s, x) = e$ otherwise, $L'(\varphi) = \lambda sx.\varphi(f_1(s), x)$ and $R'(\varphi) = \lambda sx.\varphi(f_2(s), x)$. Then $\mathscr{S} = (\mathscr{F}, \Pi, L', R')$ is a continuous pairing space.

The existence of a splitting scheme is the standard assumption we regard as necessary for the development of a computability theory over the object domain M. This requirement is essentially weaker than the existence of a computable pairing function for M assumed in Platek [1966], Moschovakis [1969], Fenstad [1980], Skordev [1980] and other works. Indeed, given a pairing function J and two fixed distinct members s_1, s_2 of M, then $f_1 = \lambda s. J(s_1, s)$ and $f_2 = \lambda s.J(s_2, s)$ is a splitting scheme for M.

Let M be the open interval of reals (0, 1). Then $f_1 = \lambda s.s/2$, $f_2 = \lambda s.(s + 1)/2$ is a splitting scheme for M. Notice that f_1, f_2 (and f_1^{-1}, f_2^{-1}) are continuous functions in the ordinary sense, while a well known theorem of Weierstrass rules out the possibility of a continuous pairing function for (0, 1). This fact is not insignificant since an approximation of reals by means of rationals would only be justified if continuous functions are used in the computations.

Another advantage of splitting as compared with pairing is that, as mentioned above, whenever M has a splitting scheme and N is a nonempty set, then we obtain a natural splitting scheme for $M \times N$. (This property is shared by the so called restricted pairing to be introduced in chapter 21.)

In order to consider a 'boldface' computability with arbitrary members of M allowed as enumeration indices however, one has to have a pairing function for M. This is arranged in our general theory by means of operation *St*. (Cf. exercises 10.1, 10.9.)

Given an arbitrary nonempty set M, then the wider set $\omega \times M^* = \omega \times \bigcup_n M^n$ admits a natural pairing function. One may take

$$J((k, s_1, \ldots, s_m), (l, t_1, \ldots, t_n)) = (J_0(k, l, m), s_1, \ldots, s_m, t_1, \ldots, t_n),$$

where J_0 codes triples of natural numbers. Of course, this extension of M does not differ essentially from that of Moschovakis [1969]. A common disadvantage of the extensions of all kinds is that one is interested in functions (relations etc.) over the set originally given which have to be distinguished from the others.

The following general construction produces new pairing spaces from given ones.

Proposition 16.5. Let $\mathscr{S}=(\mathscr{F},\Pi,L',R')$ be a pairing space. Take $\mathscr{F}'=\{\varphi'/\varphi'\colon \mathscr{F}\to\mathscr{F}\}$, $\varphi'\le\psi'$ iff $\forall\theta(\varphi'(\theta)\le\psi'(\theta))$, $\Pi'(\varphi',\psi')=\lambda\theta.(\varphi'(\theta),\psi'(\theta))$, $L''(\varphi')=\lambda\theta.L'(\varphi'(\theta))$ and $R''(\varphi')=\lambda\theta.R'(\varphi'(\theta))$. Then $\mathscr{S}'=(\mathscr{F}',\Pi',L'',R'')$ is a pairing space. Writing $\check{\varphi}$ for $\lambda\theta.\varphi$, the structure $\check{\mathscr{S}}=(\check{\mathscr{F}},\Pi'\restriction\check{\mathscr{F}}^2,L''\restriction\check{\mathscr{F}},R''\restriction\check{\mathscr{F}})$ is a subspace of $\mathscr{S}'$ isomorphic to $\mathscr{S}$.

Proof. If $\varphi'\le\psi'$, then

$$L''(\varphi')(\theta)=L'(\varphi'(\theta))\le L'(\psi'(\theta))=L''(\psi')(\theta)$$

for all θ, hence $L''(\varphi')\le L''(\psi')$. Therefore, L'' is monotonic. The monotonicity of R'',Π' is verified similarly. It follows that

$$L''((\varphi',\psi'))(\theta)=L'((\varphi'(\theta),\psi'(\theta)))=\varphi'(\theta)$$

for all θ; hence $L''((\varphi',\psi'))=\varphi'$. Similarly, $R''((\varphi',\psi'))=\psi'$; hence $\mathscr{S}'$ is a pairing space. $\check{\mathscr{S}}$ and $\mathscr{S}$ are isomophic since $\check{\varphi}\le\check{\psi}$ iff $\varphi\le\psi$, $(\check{\varphi},\check{\psi})=(\varphi,\psi)\check{}$, $L''(\check{\varphi})=L'(\varphi)\check{}$ and $R''(\check{\varphi})=R'(\varphi)\check{}$. The proof is complete.

Proposition 16.6. Proposition 16.5 holds with 'continuous pairing space' substituted for 'pairing space'.

Proof. The argument adduced in the proof of 16.3 applies here to ensure that if $\varphi'\in\mathscr{F}'$, $\mathscr{H}'\subseteq\mathscr{F}'$, then $\varphi'=\sup\mathscr{H}'$ iff $\varphi'(\theta)=\sup\{\theta'(\theta)/\theta'\in\mathscr{H}'\}$ for all θ.

Let $\{\varphi'_n\}$ be a chain in $\mathscr{F}'$ and $\varphi'=\sup_n\varphi'_n$. Then $\varphi'(\theta)=\sup_n\varphi'_n(\theta)$ for all θ; hence

$$L''(\varphi')(\theta)=L'(\varphi'(\theta))=\sup_n L'(\varphi'_n(\theta))=\sup_n L''(\varphi'_n)(\theta)$$

for all θ by the continuity of L'. Therefore, $L''(\varphi')=\sup_n L''(\varphi'_n)$. The continuity of R'',Π' is verified similarly. The proof is complete.

Pairing spaces will be used to construct OS, but not all the pairing spaces give iterative OS. We introduce certain special kinds of pairing spaces which do.

A pairing space $\mathscr{S}=(\mathscr{F},\Pi,L',R')$ is said to be *$\mathscr{T}$-complete* ($\mathscr{T}$-CPS) iff the following hold.

1. Π,L',R' preserve types.
2. For any $a\in\mathscr{T}$ and any well ordered subset $\mathscr{H}$ of $\mathscr{F}_a$ there is $\varphi\in\mathscr{F}_a$ such that $\varphi=\sup\mathscr{H}$ in $\mathscr{F}$.

A $\mathscr{T}$-complete pairing space is *strongly $\mathscr{T}$-complete* ($\mathscr{T}$-SCPS) if whenever $a\in\mathscr{T}$, $\mathscr{H}$ is a well ordered subet of $\mathscr{F}_a$ and $\varphi=\sup\mathscr{H}$ in $\mathscr{F}$, then $L'(\varphi)=\sup L'(\mathscr{H})$, $R'(\varphi)=\sup R'(\mathscr{H})$.

A pairing space $\mathscr{S}$ is *$\mathscr{T}$-continuously complete* ($\mathscr{T}$-CCPS) if the following hold.

1. Π,L',R' are $\mathscr{T}$-continuous.
2. Π,L',R' preserve types.
3. For any a and any chain $\{\varphi_n\}$ in $\mathscr{F}_a$ there is $\varphi\in\mathscr{F}_a$ such that $\varphi=\sup_n\varphi_n$ in $\mathscr{F}$.

4. $\mathscr{F}$ has a least element O.

If $\mathscr{T}$ is a singleton, then we say *complete pairing space* (CPS), *strongly complete pairing space* (SCPS), *continuously complete pairing space* (CCPS) respectively for $\mathscr{T}$-CPS, $\mathscr{T}$-SCPS and $\mathscr{T}$-CCPS. Notice that all CCPS are also continuous pairing spaces.

Proposition 16.7. If $\mathscr{S}$ is a $\mathscr{T}$-CPS, then there is $O \in \mathscr{F}$ such that $O \le \varphi$ for all φ and $t(O) \le a$ for all a.

Proof. For all a the set $\varnothing$ is a well ordered subset of $\mathscr{F}_a$; hence there is an $O \in \mathscr{F}_a$ such that $O = \sup \varnothing$ in $\mathscr{F}$. Therefore, the element in question is one and the same in all $\mathscr{F}_a$; hence $O \le \varphi$ for all φ and $t(O) \le a$ for all a. The proof is complete.

We shall assume without loss of generality that the element O in the definition of $\mathscr{T}$-CCPS also satisfies the inequality $t(O) \le a$ for all a. Otherwise one can embed $\mathscr{T}$ into an upper semilattice $\mathscr{T}'$ by the construction given in 16.11 below and then take $\mathscr{T}_1 = \{t(O) \vee a / a \in \mathscr{T}'\}$ and $t_1(\varphi) = t(O) \vee t(\varphi)$, thereby ensuring that $\mathscr{S}$ is a $\mathscr{T}_1$-CCPS and $t_1(O) \le a$ for all $a \in \mathscr{T}_1$.

The pairing space of 16.1 is neither a $\mathscr{T}$-CPS nor a $\mathscr{T}$-CCPS since it lacks a least element O.

Proposition 16.8. Let $\mathscr{S}, E$ be the same as in 16.3, and suppose that E has a least member $\perp$ and all chains in E have least upper bounds. Then $\mathscr{S}$ is a CCPS.

Proof. The element $O = \lambda s.\perp$ is a least member of $\mathscr{F}$. By the proof of 16.3, it suffices to prove that all chains in $\mathscr{F}$ have least upper bounds.

Let $\{\varphi_n\}$ be a chain in $\mathscr{F}$. Take $\varphi = \lambda s.\sup_n \varphi_n(s)$. Then $\varphi = \sup_n \varphi_n$ by the proof of 16.3. The proof is complete.

Proposition 16.9. Let $\mathscr{S}, E$ be the same as in 16.3 and suppose that all well ordered subsets of E have least upper bounds. Then $\mathscr{S}$ is both a SCPS and a CCPS.

Proof. By the proof of 16.3, it suffices to show that all well ordered subsets of $\mathscr{F}$ have least upper bounds.

If $\mathscr{H}$ is a well ordered subset of $\mathscr{F}$, then $\{\theta(s)/\theta \in \mathscr{H}\}$ is a well ordered subset of E for all s. Taking $\varphi = \lambda s.\sup\{\theta(s)/\theta \in \mathscr{H}\}$, we get $\varphi = \sup \mathscr{H}$, which completes the proof.

Proposition 16.10. Let $\mathscr{S}, E$ be the same as in 16.4. If E has a least member and all chains in E have least upper bounds, then $\mathscr{S}$ is a CCPS. If all well ordered subsets of E have least upper bounds, then $\mathscr{S}$ is both a SCPS and a CCPS.

This follows from 16.8, 16.9.

The construction of 16.3 is quite general, providing a variety of examples. Multiplication can be directly introduced in some of them to give OS or IOS; such examples will be studied in detail in chapter 21. By way of illustration we sketch a few possibilities here.

Take $E = \{\perp, \top\}$, $\perp < \top$. Then the space $\mathscr{S}$ of 16.3 is both a SCPS and a CCPS by 16.9. The members of $\mathscr{F}$ may be regarded as subsets of M. Namely, interpreting $\varphi(s)=\bot$, $\varphi(s)=\top$ as $s\notin\varphi$, $s\in\varphi$, it follows that $\mathscr{F}=\{\varphi/\varphi\subseteq M\}$ and $\varphi\leq\psi$ iff $\varphi\subseteq\psi$. Otherwise, $\varphi(s)=\bot$ could be interpreted as $s\in\varphi$, so that $\varphi\leq\psi$ iff $\psi\subseteq\varphi$ in this case.

Take $N=M$ and E as above. Then the space $\mathscr{S}$ of 16.4 is both a SCPS and a CCPS by 16.10. The members of $\mathscr{F}$ may be regarded as subsets of M^2, i.e. as binary relations on M. This can again be done in two ways, with either $\varphi\leq\psi$ iff $\varphi\subseteq\psi$ or $\varphi\leq\psi$ iff $\psi\subseteq\varphi$. Moreover, the members of $\mathscr{F}$ may be regarded as partial multiple-valued functions over M in two ways, with $t\in\varphi(s)$ iff $(s,t)\in\varphi$ or $t\in\varphi(s)$ iff $(t,s)\in\varphi$ respectively.

One may substitute the interval $[0,1]$ for $\{\bot,\top\}$ above. Then the resulting spaces consist of relations which can be regarded either as fuzzy or probabilistic, depending on how a multiplication operation is introduced.

The following basic construction is an adaptation of 16.5 in the case of complete spaces.

Proposition 16.11. Let $\mathscr{S}=(\mathscr{F},\Pi,L',R')$ be a $\mathscr{F}$-CPS ($\mathscr{T}$-SCPS). Take $\mathscr{F}'=\{\varphi'/\varphi':\mathscr{F}\to\mathscr{F}\,\&\,\varphi'$ is normal$\}$, $\leq,\Pi',L'',R'',\check{\mathscr{S}}$ as in 16.5,

$$\mathscr{T}'=\{a'/\varnothing\subset a'\subseteq\mathscr{T}\,\&\,\forall ab(a\in a'\,\&\,a\leq b\Rightarrow b\in a')\},$$

$a'\leq b'$ iff $b'\subseteq a'$ and $t'(\varphi')=\{a/\forall b\geq a(\varphi'(\mathscr{F}_b)\subseteq\mathscr{F}_b)\}$. Then $\mathscr{S}'=(\mathscr{F}',\Pi',L'',R'')$ is a $\mathscr{T}'$-CPS (respectively, $\mathscr{T}'$-SCPS) and $\mathscr{S},\check{\mathscr{S}}$ are isomorphic.

Proof. We first consider $\mathscr{T}'$. Obviously, $\mathscr{T}'$ is a partially ordered set. Notice that $a'\leq b'$ iff $\forall b\in b'\exists a\in a'(a\leq b)$. Let $a',b'\in\mathscr{T}'$. If $a\in a'$, $b\in b'$ and $c\geq a,b$, then $c\in a'\cap b'$, hence $a'\cap b'\neq\varnothing$ and $a'\cap b'\in\mathscr{T}'$. Moreover, $a',b'\leq a'\cap b'$, hence $\mathscr{T}'$ is a right directed set. In fact, $\mathscr{T}'$ is an upper semilattice and $\mathscr{T}$ is embedded in $\mathscr{T}'$ by assigning to each $a\in\mathscr{T}$ the member $\{b/a\leq b\}$ of $\mathscr{T}'$.

Let $\varphi'\in\mathscr{F}'$. Then $t'(\varphi')\neq\varnothing$ and $a\in t'(\varphi')$, $a\leq b$ imply $b\in t'(\varphi')$; hence $t'(\varphi')\in\mathscr{T}'$.

Suppose that $\varphi'\in\mathscr{F}'$ and $a\in t'(\varphi')$. Let $b\geq a$ and $\theta\in\mathscr{F}_b$. The mapping φ' is normal, so that $\varphi'(\theta)\in\mathscr{F}_b$. The mapping L' preserves types; hence $L'(\varphi'(\theta))\in\mathscr{F}_b$, i.e. $L''(\varphi')(\theta)\in\mathscr{F}_b$. Therefore the mapping $L''(\varphi')$ is normal, hence $L''(\varphi')\in\mathscr{F}'$. The above argument gives also $t'(\varphi')\subseteq t'(L''(\varphi'))$, i.e. $t'(L''(\varphi'))\leq t'(\varphi')$; hence L'' preserves types. Similarly, $R'':\mathscr{F}'\to\mathscr{F}'$, $\Pi':\mathscr{F}'^2\to\mathscr{F}'$ and R'',Π' preserve types. The verification that $\mathscr{S}'$ is a pairing space follows the proof of 16.5.

Suppose now that $a'\in\mathscr{T}'$ and $\mathscr{H}'$ is a well ordered subset of $\mathscr{F}'_{a'}$. Let $\theta\in\mathscr{F}$, $a\in a'$ and $b\geq a,t(\theta)$. We have $a\in t'(\theta')$ and $\theta'(\theta)\in\mathscr{F}_b$ for all $\theta'\in\mathscr{H}'$ since $t'(\theta')\leq a'$. Therefore, $\{\theta'(\theta)/\theta'\in\mathscr{H}'\}$ is a well ordered subset of $\mathscr{F}_b$, so there is a $\tau_\theta\in\mathscr{F}_b$ such that $\tau_\theta=\sup\{\theta'(\theta)/\theta'\in\mathscr{H}'\}$ in $\mathscr{F}$. The mapping $\varphi'=\lambda\theta.\tau_\theta$ is normal and $a'\subseteq t'(\varphi')$; hence $\varphi'\in F'_{a'}$. It is immediate that $\varphi'=\sup\mathscr{H}'$ in $\mathscr{F}'$. Therefore, $\mathscr{S}'$ is a $\mathscr{T}'$-CPS.

Suppose that $\mathscr{S}$ is strongly $\mathscr{T}$-complete. If $\mathscr{H}'$ is a well ordered subset of $\mathscr{F}'_{a'}$ and φ' is the same as above, then

$$\begin{aligned}L''(\varphi')(\theta)=L'(\varphi'(\theta))=L'(\sup\{\theta'(\theta)/\theta'\in\mathscr{H}'\})=\sup\{L'(\theta'(\theta))/\theta'\in\mathscr{H}'\}\\=\sup\{L''(\theta')(\theta)/\theta'\in\mathscr{H}'\}\end{aligned}$$

for all θ, hence $L''(\varphi') = \sup L''(\mathscr{H}')$. Similarly, $R''(\varphi') = \sup R''(\mathscr{H}')$, hence $\mathscr{S}'$ is a $\mathscr{T}'$-SCPS.

The isomorphism of $\mathscr{S}, \check{\mathscr{S}}$ is established as in 16.5, noting that the mappings $\check{\varphi}$ are normal, so that $\check{\mathscr{F}} \subseteq \mathscr{F}'$. It also follows that $t'(\check{\varphi}) = \{a/t(\varphi) \le a\}$ and whenever $\mathscr{H} \subseteq \mathscr{F}_a$ and $\varphi = \sup \mathscr{H}$, then $\check{\varphi} = \sup \check{\mathscr{H}}$ in $\mathscr{F}'$. This completes the proof.

It is worth mentioning that $\check{\mathscr{S}}$ is a proper subspace of $\mathscr{S}'$ since L', $R' \in \mathscr{F}' \backslash \check{\mathscr{F}}$.

Proposition 16.12. Proposition 16.11 holds with '$\mathscr{T}$-CCPS', '$\mathscr{T}'$-CCPS' substituted respectively for '$\mathscr{T}$-CPS', '$\mathscr{T}'$-CPS'.

The proof follows that of 16.11. The mappings $\check{O} = \lambda\theta.O$ is a least member of $\mathscr{F}'$.

EXERCISES TO CHAPTER 16

Exercise 16.1. Let $\mathscr{S} = (\mathscr{F}, \Pi, L', R')$ be a pairing space with a least element O. Take $\mathscr{F}' = \{\varphi'/\varnothing \subset \varphi' \subseteq \mathscr{F}\}$, $\varphi' \le \psi'$ iff $\forall \varphi \in \varphi' \exists \psi \in \psi' (\varphi \le \psi)$, $\Pi'(\varphi', \psi') = \Pi(\varphi', \psi')$, $L''(\varphi') = L'(\varphi')$ and $R''(\varphi') = R'(\varphi')$. Show that $\mathscr{S}' = (\mathscr{F}', \Pi', L'', R'')$ is both a SCPS and a CCPS.

Remarks. The relation $\le$ is not antisymmetric, hence $\mathscr{F}'$ is a quasi-ordered rather than partially ordered set and should be factorized. One may construct $\mathscr{S}'$ in a slightly different manner, taking $\mathscr{F}' = \{\varphi'/\varnothing \subset \varphi' \subseteq \mathscr{F} \,\&\, \forall \varphi\psi(\varphi \le \psi \,\&\, \psi \in \varphi' \Rightarrow \varphi \in \varphi')\}$, $\varphi' \le \psi'$ iff $\varphi' \subseteq \psi'$; no factorization is needed then.

One advantage of the construction of exercise 16.1 is that it gives a complete space without requiring $\mathscr{S}$ to be complete. Take for instance the pairing space $\mathscr{S}_1$ obtained from the OS of example 3.1 by 16.2, then take its subspace $\mathscr{S}$ consisting of functions with finite domains. The pairing space $\mathscr{S}$ has a least element but is neither a CPS nor a CCPS.

On the other hand, the above construction has a drawback. While the given pairing space $\mathscr{S}$ is isomorphic with the subspace $\mathscr{S}_0$ of $\mathscr{S}'$ based on $\mathscr{F}_0 = \{\{\varphi\}/\varphi \in \mathscr{F}\}$, one can not expect that whenever $\mathscr{S}$ is a CPS (or CCPS), then $\mathscr{S}_0$ is a subspace of $\mathscr{S}'$ as a CPS (a CCPS). It may well happen that $\varphi = \sup_n \varphi_n$ and $\varphi \ne \varphi_n$ for all n, so that $\{\varphi\} \ne \sup_n \{\varphi_n\}$.

Exercise 16.2. Let N be a nonempty set and $\mathscr{S}_s$ be a $\mathscr{T}_s$-CPS (respectively, $\mathscr{T}_s$-SCPS, $\mathscr{T}_s$-CCPS) for all $s \in N$. Take $\mathscr{F} = \times_{s \in N} \mathscr{F}_s$, i.e.

$$\mathscr{F} = \{\varphi/\varphi : N \to \bigcup_{s \in N} \mathscr{F}_s \,\&\, \forall s(\varphi(s) \in \mathscr{F}_s)\},$$

$\varphi \le \psi$ iff $\forall s(\varphi(s) \le \psi(s))$, $(\varphi, \psi) = \lambda s.\Pi_s(\varphi(s), \psi(s))$, $L'(\varphi) = \lambda s.L'_s(\varphi(s))$, $R'(\varphi) = \lambda s.R'_s(\varphi(s))$, $\mathscr{T} = \times_{s \in N} \mathscr{T}_s$, $a \le b$ iff $\forall s(a(s) \le b(s))$, and $t(\varphi) = \lambda s.t_s(\varphi(s))$. Show that $\mathscr{S} = (\mathscr{F}, \Pi, L', R')$ is a $\mathscr{T}$-CPS (respectively, $\mathscr{T}$-SCPS, $\mathscr{T}$-CCPS). In particular, if $\mathscr{S}_s$ is a CPS (SCPS, CCPS) for all s, then so is $\mathscr{S}$.

Exercise 16.3. Let $\mathscr{S}$ be a pairing space and $(L'(\varphi), R'(\varphi)) \le \varphi$ for all φ. Show that Π is continuous with respect to least upper bounds of nonempty subsets of $\mathscr{F}$.

Hint. Supposing $\varnothing \subset \mathscr{H}_1, \mathscr{H}_2 \subseteq \mathscr{F}, \varphi_1 = \sup \mathscr{H}_1, \quad \varphi_2 = \sup \mathscr{H}_2$ and $(\theta, \tau) \leq \varphi$ for all $\theta \in \mathscr{H}_1, \tau \in \mathscr{H}_2$, show that $\varphi_1 \leq L'(\varphi), \varphi_2 \leq R'(\varphi)$, which implies $(\varphi_1, \varphi_2) \leq \varphi$.

Exercise 16.4. Let $\mathscr{S}$ be a pairing space and $(L'(\varphi), R'(\varphi)) \geq \varphi$ for all φ. Show that L', R' are continuous with respect to least upper bounds.

Hint. Supposing $\varphi = \sup \mathscr{H}$ and $L'(\theta) \leq \tau$ for all $\theta \in \mathscr{H}$, get $\varphi \leq (\tau, R'(\varphi))$, which implies $L'(\varphi) \leq \tau$.

As a corollary to the last two exercises, whenever $\Pi(\mathscr{F}, \mathscr{F}) = \mathscr{F}$ and $\mathscr{S}$ is $\mathscr{T}$-complete, then it is also strongly and continuously $\mathscr{T}$-complete.

Exercise 16.5. Let $\mathscr{S}$ be the pairing space of 16.3 or 16.4. Prove the following assertions. If $f_1(M) \cup f_2(M) \subset M$, then $\forall \varphi((L'(\varphi), R'(\varphi)) \leq \varphi)$ iff $e \leq d$ for all $d \in E$, while $\forall \varphi((L'(\varphi), R'(\varphi)) \geq \varphi)$ iff $\forall d(e \geq d)$. Finally, $\Pi(\mathscr{F}, \mathscr{F}) = \mathscr{F}$ iff $f_1(M) \cup f_2(M) = M$.

Exercise 16.6. Let $\mathscr{S}$ be a $\mathscr{T}$-CPS ($\mathscr{T}$-SCPS, $\mathscr{T}$-CCPS) and $a \in \mathscr{T}$. Prove that $\mathscr{S}_a = (\mathscr{F}_a, \Pi \restriction \mathscr{F}_a^2, L' \restriction \mathscr{F}_a, R' \restriction \mathscr{F}_a)$ is a CPS (respectively SCPS, CCPS).

Exercise 16.7. Take $\mathscr{F} = \{\varphi/\varphi \subseteq \omega^2\} = \{\varphi/\varphi : \omega \to 2^\omega\}$, $\varphi \leq \psi$ iff $\varphi \subseteq \psi$, $(\varphi, \psi)(3s) = \varphi(s)$, $(\varphi, \psi)(3s+1) = \psi(s)$, $(\varphi, \psi)(3s+2)\uparrow$, $L'(\varphi) = \lambda s.\varphi(3s)$ and $R'(\varphi) = \lambda s.\varphi(3s+1)$, if $\operatorname{Dom} \varphi \subset \omega$, and $L'(\varphi), R'(\varphi) = \omega^2$ otherwise. Show that $\mathscr{S} = (\mathscr{F}, \Pi, L', R')$ is a CPS but neither a SCPS nor a CCPS.

Hint. Take a chain $\{\varphi_n\}$ of single-valued functions such that $\operatorname{Dom} \varphi_n \subset \omega$ for all n and $\bigcup_n \operatorname{Dom} \varphi_n = \omega$. Then $L'(\sup_n \varphi_n) \neq \sup_n L'(\varphi_n)$.

Exercise 16.8. Let $\mathscr{S} = (\mathscr{F}, \Pi, L', R')$ be the pairing space of 16.4 with $N = M^\omega$ and $E = \{\bot, \top\}$, $\bot < \top$. Take $\mathscr{F}_1 = \{\varphi \in \mathscr{F} / \exists m \forall k(\varphi(\{s_n\}) = \top \Rightarrow s_{m+k} = s_m)\}$, $\mathscr{T} = \omega$ and

$$t(\varphi) = \min \{m / \forall k(\varphi(\{s_n\}) = \top \Rightarrow s_{m+k} = s_m)\}.$$

show that while $\mathscr{S}_1 = (\mathscr{F}_1, \Pi \restriction \mathscr{F}_1^2, L' \restriction \mathscr{F}_1, R' \restriction \mathscr{F}_1)$ is both a $\mathscr{T}$-SCPS and a $\mathscr{T}$-CCPS, it is neither CPS nor CCPS.

Hint. Notice that $\varphi \leq \psi$ implies $t(\varphi) \leq t(\psi)$, so any chain $\{\varphi_n\}$ such that $t(\varphi_n) = n$ for all n will have no upper bound in $\mathscr{F}_1$.

CHAPTER 17

Hierarchies of pairing spaces

Starting with a $\mathscr{T}$-complete (strongly $\mathscr{T}$-complete, continuously $\mathscr{T}$-complete) pairing space $\mathscr{S}$, we construct in this chapter a whole hierarchy $\{\mathscr{S}_\xi\}$ of $\mathscr{S}_\xi$-complete (respectively, strongly $\mathscr{T}_\xi$-complete, continuously $\mathscr{T}_\xi$-complete) pairing spaces, obtaining new examples of such spaces. The construction employed will also be useful later when building hierarchies of IOS.

To begin with, we construct an associated hierarchy of type sets $\{\mathscr{T}_\xi\}$.

Proposition 17.1. Given a nonempty right directed set $\mathscr{T}$, for all ξ right directed sets $\mathscr{T}_\xi$ can be constructed such that $\mathscr{T}_0 = \mathscr{T}$, the set $\mathscr{T}_\xi$ is obtained from $\mathscr{T}_{\xi_1}$ by 16.11, if $\xi = \xi_1 + 1$, while $\mathscr{T}_\xi = \{(\eta, a)/\eta < \xi \,\&\, a \in \mathscr{T}_\eta\}$, if $\xi > 0$ is a limit, and if $\xi \leq \eta$, then the members of $\mathscr{T}_\xi$ are *represented* in $\mathscr{T}_\eta$ by members of $\mathscr{T}_\eta$ (by themselves, if $\xi = \eta$) in such a way that the following condition is satisfied.

$$(\eta)\;\left\{\begin{array}{l}\text{If } \eta_1 \leq \eta_2 \leq \eta,\, a \in \mathscr{T}_{\eta_1}, b \in \mathscr{T}_{\eta_2}, a', a'' \text{ represent } a \text{ respectively in } \mathscr{T}_{\eta_2}, \mathscr{T}_\eta \\ \text{and } b' \text{ represents } b \text{ in } \mathscr{T}_\eta, \text{ then } a' \leq b \text{ iff } a'' \leq b'.\end{array}\right.$$

Taking $\eta_1 = \eta_2 = \xi$ in particular, it follows that $\mathscr{T}_\xi$ is isomorphic to a partially ordered subset of $\mathscr{T}_\eta$, provided $\xi \leq \eta$.

Proof. By transfinite induction on ξ.

Take $\mathscr{T}_0 = \mathscr{T}$.

Suppose that $\xi > 0$, right directed sets $\mathscr{T}_\eta$ have been constructed for all $\eta < \xi$ and if $\eta_1 \leq \eta_2 \leq \eta$, then a member of $\mathscr{T}_{\eta_2}$ is assigned to each member of $\mathscr{T}_{\eta_1}$ in such a way that (η) holds. This is the induction assumption.

We consider two separate cases, depending on whether ξ is a successor or a limit ordinal.

1. $\xi = \xi_1 + 1$.

We obtain $\mathscr{T}_\xi$ from $\mathscr{T}_{\xi_1}$ by 16.11. A member a of $\mathscr{T}_{\xi_1}$ is represented in $\mathscr{T}_\xi$ by $a' = \{b/b \in \mathscr{T}_{\xi_1} \,\&\, a \leq b\}$. If $\eta < \xi_1$, $a \in \mathscr{T}_\eta$, a' represents a in $\mathscr{T}_{\xi_1}$ by the induction assumption and a'' represents a' in $\mathscr{T}_\xi$, take a'' to represent a in $\mathscr{T}_\xi$.

Suppose that $\eta_1 \leq \eta_2 < \xi$, $a \in \mathscr{T}_{\eta_1}$, $b \in \mathscr{T}_{\eta_2}$ and a' represents a in $\mathscr{T}_{\eta_2}$. Then $\eta_1 \leq \eta_2 \leq \xi_1$ and the induction assumption provides (ξ_1), hence $a' \leq b$ iff $a'' \leq b'$, where a'', b' represent a, b in $\mathscr{T}_{\xi_1}$. If a''', b'' represent a, b in $\mathscr{T}_\xi$, then $a'' \leq b'$ iff $a''' \leq b''$ by the definition of a''', b'' and the construction of $\mathscr{T}_\xi$. Therefore, condition (ξ) is satisfied. (The case of $\eta_2 = \xi$ is obvious.)

2. ξ is a limit ordinal.

Take $\mathscr{T}_\xi = \{(\eta, a)/\eta < \xi \,\&\, a \in \mathscr{T}_\eta\}$ and $(\eta, a) \leq (\zeta, b)$ iff $\eta \leq \zeta \,\&\, a' \leq b$, where a' represents a in $\mathscr{T}_\zeta$. The relation $\leq$ is obviously reflexive and antisymmetric.

Suppose that $(\eta_1, a_1) \leq (\eta_2, a_2)$ and $(\eta_2, a_2) \leq (\eta_3, a_3)$. Let a_1', a_1'' represent a_1 respectively in $\mathscr{T}_{\eta_2}$, $\mathscr{T}_{\eta_3}$ and a_2' represent a_2 in $\mathscr{T}_{\eta_3}$. It follows that $\eta_1 \leq \eta_2 \leq \eta_3$ and $a_1' \leq a_2$, hence $a_1'' \leq a_2'$ by (η_3). Then $a_2' \leq a_3$ implies $a_1'' \leq a_3$, hence $(\eta_1, a_1) \leq (\eta_3, a_3)$. Therefore, $\leq$ is transitive and $\mathscr{T}_\xi$ is a partially ordered set.

Let $(\eta, a), (\zeta, b) \in \mathscr{T}_\xi$. Take $\xi_1 = \max\{\eta, \zeta\}$, a', b' to represent a, b in $\mathscr{T}_{\xi_1}$, then take $c \in \mathscr{T}_{\xi_1}$ such that $c \geq a', b'$. It follows that $(\xi_1, c) \in \mathscr{T}_\xi$ and (η, a), $(\zeta, b) \leq (\xi_1, c)$; hence $\mathscr{T}_\xi$ is right directed.

If $a \in \mathscr{T}_\eta$, $\eta < \xi$, then take (η, a) to represent a in $\mathscr{T}_\xi$. Let $\eta_1 \leq \eta_2 < \xi$, $a \in \mathscr{T}_{\eta_1}$, $b \in \mathscr{T}_{\eta_2}$ and a' represent a in $\mathscr{T}_{\eta_2}$. Then $a' \leq b$ iff $(\eta_1, a) \leq (\eta_2, b)$ by the definition of $\leq$ in $\mathscr{T}_\xi$. Therefore, condition (ξ) is satisfied. The proof is complete.

Proposition 17.2. Let $\mathscr{S}$ be a $\mathscr{T}$-CPS (respectively, $\mathscr{T}$-SCPS, $\mathscr{T}$-CCPS) and $\{\mathscr{T}_\xi\}$ be obtained from $\mathscr{T}$ by 17.1. Then for each ordinal ξ a $\mathscr{T}_\xi$-CPS ($\mathscr{T}_\xi$-SCPS, $\mathscr{T}_\xi$-CCPS) $\mathscr{S}_\xi$ can be constructed such that whenever $\xi < \eta$, $\mathscr{S}_\xi$ is a proper subspace of $\mathscr{S}_\eta$.

Proof. By transfinite induction on ξ.

Take $\mathscr{S}_0 = \mathscr{S}$ and $t_0 = t$.

Suppose that $\xi > 0$ and $\mathscr{T}_\eta$-CPS (respectively, $\mathscr{T}_\eta$-SCPS, $\mathscr{T}_\eta$-CCPS) $\mathscr{S}_\eta$ with type functions t_η have been constructed for all $\eta < \xi$. Whenever $\eta = \eta_1 + 1 < \xi$, let $\mathscr{S}_\eta$ be obtained from $\mathscr{S}_{\eta_1}$ by 16.11 and let the subspace $\check{\mathscr{S}}_{\eta_1}$ of $\mathscr{S}_\eta$ be identified with $\mathscr{S}_{\eta_1}$. Whenever $\eta < \xi$ is a limit ordinal, then let all $\mathscr{S}_\zeta, \zeta < \eta$ be subspaces of $\mathscr{S}_\eta$ and $\mathscr{F}_\eta = \bigcup_{\zeta < \eta} \mathscr{F}_\zeta$. If $\varphi \in \bigcup_{\eta < \xi} \mathscr{F}_\eta$, then the ordinal $r(\varphi) = \min\{\eta/\varphi \in \mathscr{F}_\eta\}$ (zero or successor) is called the *rank of* φ. Finally, whenever $\eta < \xi$ and $\zeta = r(\varphi) < \eta$, let $t_\zeta(\varphi)$ be represented in $\mathscr{T}_\eta$ by $t_\eta(\varphi)$. This is the induction assumption.

There are two possibilities.

1. $\xi = \xi_1 + 1$.

Obtain $\mathscr{S}_\xi$, t_ξ from $\mathscr{S}_{\xi_1}$, t_{ξ_1} by 16.11.

Let $\varphi \in \mathscr{F}_\xi$ and $\zeta = r(\varphi) < \xi$. Then $\zeta \leq \xi_1$, hence $t_{\xi_1}(\varphi)$ represents $t_\zeta(\varphi)$ in $\mathscr{T}_{\xi_1}$ by the induction assumption. The proof of 16.11 gives $t_\xi(\varphi) = \{a \in \mathscr{T}_{\xi_1} / t_{\xi_1}(\varphi) \leq a\}$, which is exactly the member of $\mathscr{T}_\xi$ representing $t_{\xi_1}(\varphi)$ in $\mathscr{T}_\xi$. Therefore, $t_\xi(\varphi)$ represents $t_\zeta(\varphi)$ in $\mathscr{T}_\xi$. The remaining assertions of the induction clause follow from the proof of 16.11 (16.12).

2. ξ is a limit ordinal.

Take $\mathscr{F}_\xi = \bigcup_{\eta < \xi} \mathscr{F}_\eta$ and $t_\xi(\varphi) = (r(\varphi), t_{r(\varphi)}(\varphi))$ for all $\varphi \in \mathscr{F}_\xi$. Take $\varphi \leq_\xi \psi$ iff $\varphi \leq_\zeta \psi$, $\Pi_\xi(\varphi, \psi) = \Pi_\zeta(\varphi, \psi)$, writing simply $(\varphi, \psi)_\xi = (\varphi, \psi)_\zeta$, where $\zeta = \max\{r(\varphi), r(\psi)\}$, and $L_\xi'(\varphi) = L_\zeta'(\varphi)$, $R_\xi'(\varphi) = R_\zeta'(\varphi)$, where $\zeta = r(\varphi)$. It follows that whenever $\zeta \leq \eta < \xi$, then $(\varphi, \psi)_\xi = (\varphi, \psi)_\eta$, $L_\xi'(\varphi) = L_\eta'(\varphi)$, $R_\xi'(\varphi) = R_\eta'(\varphi)$ and $\varphi \leq_\xi \psi$ iff $\varphi \leq_\eta \psi$. Therefore, the quadruple $\mathscr{S}_\xi = (\mathscr{F}_\xi, \Pi_\xi, L_\xi, R_\xi)$ is a pairing space since so are $\mathscr{S}_\eta$ for all $\eta < \xi$.

If $\zeta = r(\varphi) < \xi$, then $t_\xi(\varphi) = (\zeta, t_\zeta(\varphi))$ represents $t_\zeta(\varphi)$ by the proof of 17.1.

Let $t_\xi(\varphi) = (\eta, t_\eta(\varphi))$, $t_\xi(\psi) = (\zeta, t_\zeta(\psi))$, $(\xi_1, a) \in \mathscr{T}_\xi$ and $t_\xi(\varphi)$, $t_\xi(\psi) \leq (\xi_1, a)$. Then $\eta, \zeta \leq \xi_1$, $t_{\xi_1}(\varphi)$, $t_{\xi_1}(\psi)$ represent $t_\eta(\varphi)$, $t_\zeta(\psi)$ in $\mathscr{T}_{\xi_1}$ and $t_{\xi_1}(\varphi)$, $t_{\xi_1}(\psi) \leq a$. It follows that $r(\varphi)$, $r(\psi) \leq \xi_1$, hence $\varphi, \psi \in \mathscr{F}_{\xi_1}$, so $(\varphi, \psi)_\xi \in \mathscr{F}_{\xi_1}$ and

$t_{\xi_1}((\varphi, \psi)_\xi) \leq a$ since Π_{ξ_1} preserves types. Let $\xi_2 = r((\varphi, \psi)_\xi)$. Then $\xi_2 \leq \xi_1$, $t_\xi((\varphi, \psi)_\xi) = (\xi_2, t_{\xi_2}((\varphi, \psi)_\xi))$ and $t_{\xi_1}((\varphi, \psi)_\xi)$ represents $t_{\xi_2}((\varphi, \psi)_\xi)$ in $\mathscr{T}_{\xi_1}$, hence $t_\xi((\varphi, \psi)_\xi) \leq (\xi_1, a)$. (Observe that $(\varphi, \psi)_\xi = (\varphi, \psi)_{\xi_1} = (\varphi, \psi)_{\xi_2}$.) Therefore, Π_ξ preserves types. Similarly, L'_ξ and R'_ξ preserve types.

Let $a \in \mathscr{T}_\xi$ and $\mathscr{H}$ be a well ordered subset of $\mathscr{F}_{\xi,a} = \{\varphi \in \mathscr{F}_\xi / t_\xi(\varphi) \leq a\}$. Then $a = (\eta, b)$ for certain $\eta < \xi$, $b \in \mathscr{T}_\eta$ and $t_\xi(\theta) \leq a$ implies $\theta \in \mathscr{F}_\eta$, $t_\eta(\theta) \leq b$ for all $\theta \in \mathscr{H}$. Therefore, $\mathscr{H}$ is a well ordered subset of $\mathscr{F}_{\eta,b}$. There is by the induction assumption a $\varphi \in \mathscr{F}_{\eta,b}$ such that $\varphi = \sup \mathscr{H}$ in $\mathscr{F}_\eta$. We get $t_\xi(\varphi) \leq a$ by condition (ξ) of 17.1, hence $\varphi \in \mathscr{F}_{\xi,a}$. It follows that $\theta \leq_\xi \varphi$ for all $\theta \in \mathscr{H}$. Suppose that $\tau \in \mathscr{F}_\xi$ and $\theta \leq_\xi \tau$ for all $\theta \in \mathscr{H}$. Taking $\zeta = \max\{\eta, r(\tau)\}$, we get $\zeta < \xi$, $\mathscr{H} \subseteq \mathscr{F}_\zeta$, $\tau \in \mathscr{F}_\zeta$ and $\theta \leq_\zeta \tau$ for all $\theta \in \mathscr{H}$. However, $\varphi = \sup \mathscr{H}$ in $\mathscr{F}_\zeta$ since $\mathscr{S}_\eta$ is a subspace of $\mathscr{S}_\zeta$ by the induction assumption, hence $\varphi \leq_\zeta \tau$ which implies $\varphi \leq_\xi \tau$. Therefore, $\varphi = \sup \mathscr{H}$ in $\mathscr{S}_\xi$. This argument implies that $\mathscr{S}_\xi$ is a $\mathscr{T}_\xi$-CPS and whenever $\eta < \xi$, then $\mathscr{S}_\eta$ is a subspace of $\mathscr{S}_\xi$.

Assuming $\mathscr{S}$ a $\mathscr{T}$-SCPS, we show in addition that $L'_\xi(\varphi) = \sup L'_\xi(\mathscr{H})$ in $\mathscr{F}_\xi$ using the fact that $L'_\eta(\varphi) = \sup L'_\eta(\mathscr{H})$ in $\mathscr{F}_\eta$, where $\mathscr{H}, \eta, \varphi$ are as above. The case of R'_ξ is treated similarly.

Finally, assume that $\mathscr{S}$ is a $\mathscr{T}$-CCPS. Let $\{\varphi_n\}$, $\{\psi_n\}$ be chains in $\mathscr{F}_{\xi,a}$, $a = (\eta, b)$, $\eta < \xi$ and $b \in \mathscr{T}_\eta$. Then $\{\varphi_n\}$, $\{\psi_n\}$ are chains in $\mathscr{F}_{\eta,b}$. There are by the induction assumption $\varphi, \psi \in \mathscr{F}_{\eta,b}$ such that $\varphi = \sup_n \varphi_n$, $\psi = \sup_n \psi_n$, $(\varphi, \psi)_n = \sup_n (\varphi_n, \psi_n)_\eta$, $L'_\eta(\varphi) = \sup_n L'_\eta(\varphi_n)$ and $R'_\eta(\varphi) = \sup_n R'_\eta(\varphi_n)$ in $\mathscr{F}_\eta$. Proceeding as above, we get $\varphi, \psi \in \mathscr{F}_{\xi,a}$ and $\varphi = \sup_n \varphi_n$, $\psi = \sup_n \psi_n$, $(\varphi, \psi)_\xi = \sup_n (\varphi_n, \psi_n)_\xi$, $L_\xi(\varphi) = \sup_n L'_\xi(\varphi_n)$, $R'_\xi(\varphi) = \sup_n R'_\xi(\varphi_n)$ in $\mathscr{F}_\xi$. The element $O \in \mathscr{F}$ satisfies the inequality $O \leq_\xi \varphi$ for all $\varphi \in \mathscr{F}_\xi$ since $O \leq_\eta \varphi$ for all $\varphi \in \mathscr{F}_\eta$, $\eta < \xi$. Therefore, $\mathscr{S}_\xi$ is a $\mathscr{T}_\xi$-CCPS, which completes the induction and the proof.

EXERCISE TO CHAPTER 17

Let $\mathscr{S}$ be a CPS let $\{\mathscr{S}_\xi\}$ be a hierarchy based on it constructed by 17.2. The following exercises show how the types arise in initial segments of the hierarchy; the same happens if one starts with a SCPS or a CCPS.

Exercise 17.1. Show that $\mathscr{S}_n$ is a CPS for all n.

Exercise 17.2. Show that $\mathscr{S}_\omega$ is a ω-CPS with a type function $t_\omega = r$.

Exercise 17.3. Show that $\mathscr{S}_{\omega+n+1}$ is a ω-CPS with a type function $t_{\omega+n+1}(\varphi) = \min\{m / \forall k \geq m (\varphi(\mathscr{F}_{\omega+n,k}) \subseteq \mathscr{F}_{\omega+n,k})\}$ for all n.

CHAPTER 18

Conditions sufficient for iterativeness

When constructing IOS one has to verify that a structure satisfies certain axioms. While the axioms of OS are verified more or less directly, it is not convenient to do so with the μ-axiom. Fortunately, there are simple sufficient conditions which ensure iterativeness. Several such conditions are studied in this chapter, including the case of spaces with t-operations.

We start with two μ-Induction Theorems asserting the existence of least fixed points with nice properties. These theorems originate from a well known fixed point result of Tarski [1955] and its continuous version in Kleene [1952], and also generalize corresponding theorems of Skordev [1980] by involving types.

Assume that a partially ordered set $\mathscr{F}$, a type set $\mathscr{T}$ and a type function $t:\mathscr{F} \to \mathscr{T}$ are given.

Proposition 18.1. For all $a \in \mathscr{T}$ suppose that all the well ordered subsets of $\mathscr{F}_a$ have least upper bounds in $\mathscr{F}$ which are in $\mathscr{F}_a$. Let $\Gamma:\mathscr{F} \to \mathscr{F}$ be monotonic and normal. Then Γ has a fixed point which is a member of all subsets $\mathscr{E}$ of $\mathscr{F}$ which are closed under Γ and such that $\mathscr{E} \cap \mathscr{F}_a$ is closed under least upper bounds of well ordered subsets for all $a \in \mathscr{T}$. In particular, the fixed point concerned is exactly $\mu\theta.\Gamma(\theta)$.

Proof. There is an $a \in \mathscr{T}$ such that $\Gamma(\mathscr{F}_a) \subseteq \mathscr{F}_a$. We construct by transfinite induction an increasing sequence $\{\theta_\xi\}$ in $\mathscr{F}_a$ such that $\theta_\xi = \sup\{\Gamma(\theta_\eta)/\eta < \xi\}$ for all ξ.

Suppose that for all $\eta < \xi$ a $\theta_\eta \in \mathscr{F}_a$ has been constructed such that $\{\theta_\zeta\}_{\zeta \le \eta}$ is increasing. Then $\{\theta_\eta\}_{\eta < \xi}$ is an increasing sequence in $\mathscr{F}_a$, hence so is $\{\Gamma(\theta_\eta)/\eta < \xi\}$ since Γ is monotonic and $\Gamma(\mathscr{F}_a) \subseteq \mathscr{F}_a$. Therefore, $\theta_\xi = \sup\{\Gamma(\theta_\eta)/\eta < \xi\}$ exists and is in $\mathscr{F}_a$. Certainly, the sequence $\{\theta_\eta\}_{\eta \le \xi}$ is increasing, which completes the induction step.

Notice that if $\theta_\zeta = \theta_{\zeta+1}$, then $\theta_\xi = \theta_\zeta$ for all $\xi \ge \zeta$. And such a ζ actually exists since $\theta_\zeta = \theta_{\zeta+1}$ whenever $\mathrm{Card}(\zeta) > \mathrm{Card}(\mathscr{F}_a)$. The element θ_ζ is the required fixed point of Γ. In fact

$$\Gamma(\theta_\zeta) = \sup\{\Gamma(\theta_\eta)/\eta \le \zeta\} = \theta_{\zeta+1} = \theta_\zeta.$$

In order to show that $\theta_\zeta \in \mathscr{E}$, we prove by transfinite induction that $\theta_\xi \in \mathscr{E}$ for all ξ.

Suppose that $\theta_\eta \in \mathscr{E}$ for all $\eta < \xi$. Then $\Gamma(\theta_\eta) \in \mathscr{E}$ for all $\eta < \xi$ since $\mathscr{E}$ is closed under Γ, hence $\{\Gamma(\theta_\eta)/\eta < \xi\}$ is a well ordered subset of $\mathscr{E} \cap \mathscr{F}_a$, which implies $\theta_\xi \in \mathscr{E}$.

Finally, let $\Gamma(\tau) \leq \tau$. Then the set $\mathscr{E} = \{\theta/\theta \leq \tau\}$ is closed both under Γ and under least upper bounds, hence $\theta_\zeta \in \mathscr{E}$. Therefore, $\theta_\zeta = \mu\theta.\Gamma(\theta)$ and the proof is complete.

Instead of well ordered sets one may use in the formulation of 18.1 totally ordered sets, increasing transfinite sequences and the like.

Least fixed points θ_ζ are obtained at level $\zeta \leq \omega$ in the continuous case considered below.

Proposition 18.2. Let $\mathscr{F}$ have a least member O and suppose that for all $a \in \mathscr{T}$ all the chains in $\mathscr{F}_a$ have least upper bounds in $\mathscr{F}$ which are in $\mathscr{F}_a$. Let $\Gamma:\mathscr{F} \to \mathscr{F}$ be $\mathscr{T}$-continuous and normal. Then Γ has a fixed point which is a member of all subsets $\mathscr{E}$ of $\mathscr{F}$ closed under Γ and such that $O \in \mathscr{E}$ and $\mathscr{E} \cap \mathscr{F}_a$ is closed under least upper bounds of chains for all $a \in \mathscr{T}$. The least fixed point in question is exactly $\mu\theta.\Gamma(\theta)$.

Proof. There is an $a \in \mathscr{T}$ such that $O \in \mathscr{F}_a$ and $\Gamma(\mathscr{F}_a) \subseteq \mathscr{F}_a$. Take $\theta_n = \Gamma^n(O)$ for all n. It follows that $\theta_0 \leq \theta_1$, while $\theta_n \leq \theta_{n+1}$ implies

$$\theta_{n+1} = \Gamma(\theta_n) \leq \Gamma(\theta_{n+1}) = \theta_{n+2}$$

by the monotonicity of Γ; hence $\{\theta_n\}$ is a chain. Moreover, $\theta_0 \in \mathscr{F}_a$ and $\theta_n \in \mathscr{F}_a$ implies $\theta_{n+1} \in \mathscr{F}_a$. Therefore, $\{\theta_n\}$ is a chain in $\mathscr{F}_a$, hence $\sigma = \sup_n \theta_n$ exists and is in $\mathscr{F}_a$. The element σ is the required fixed point of Γ. Actually,

$$\Gamma(\sigma) = \Gamma\left(\sup_n \theta_n\right) = \sup_n \Gamma(\theta_n) = \sup_n \theta_{n+1} = \sigma$$

by the $\mathscr{T}$-continuity of Γ. Moreover, $\theta_n \in \mathscr{E}$ for all n since $\theta_0 = O \in \mathscr{E}$ and $\mathscr{E}$ is closed under Γ. Therefore, $\sigma = \sup_n \theta_n \in \mathscr{E}$.

Let $\Gamma(\tau) \leq \tau$. Taking $\mathscr{E} = \{\theta/\theta \leq \tau\}$, we get $\sigma \leq \tau$, hence $\sigma = \mu\theta.\Gamma(\theta)$. The proof is complete.

Assume that an OS $\mathscr{S} = (\mathscr{F}, I, \Pi, L, R)$ with a type set $\mathscr{T}$ and a type function $t:\mathscr{F} \to \mathscr{T}$ are given. Bearing in mind the above fixed point theorems and the μ-axioms of chapter 5, we introduce the following conditions intended to ensure iterativeness.

(*) The operations $\circ, \Pi$ preserve types. If $a \in \mathscr{T}$ and $\mathscr{H}$ is a well ordered subset of $\mathscr{F}_a$, then there is a $\varphi \in \mathscr{F}_a$ such that $\varphi\psi = \sup(\mathscr{H}\psi)$ for all ψ.

(**) The operations $\circ, \Pi$ preserve types. Whenever $a \in \mathscr{T}$ and $\mathscr{H}$ is a well ordered subset of $\mathscr{F}_a$, then there is a $\varphi \in \mathscr{F}_a$ such that $L\varphi = \sup L\mathscr{H}$, $R\varphi = \sup R\mathscr{H}$ and $\varphi\psi = \sup(\mathscr{H}\psi)$ for all ψ.

(***) The operations $\circ, \Pi$ preserve types. If $a \in \mathscr{T}$ and $\{\varphi_n\}$ is a chain in $\mathscr{F}_a$, then there is a $\varphi \in \mathscr{F}_a$ such that $\varphi\psi = \sup_n(\varphi_n\psi)$, $\psi\varphi = \sup_n \psi\varphi_n$ for all ψ and $(I, \varphi) = \sup_n(I, \varphi_n)$. There is an element O such that $O \leq \psi$ and $O\psi = O$ for all ψ.

In view of the remark following 16.7 we may assume without loss of generality that $t(I)$, $t(L)$, $t(R)$, $t(O) \leq a$ for all a.

The following frequently used sufficient conditions are particular instances of the above ones, with $\mathscr{T}$ a singleton.

$(*)_0$ If $\mathscr{H}$ is a well ordered subset of $\mathscr{F}$, then there is a φ such that $\varphi\psi = \sup(\mathscr{H}\psi)$ for all ψ.

$(**)_0$ If $\mathscr{H}$ is a well ordered subset of $\mathscr{F}$, then there is a φ such that $L\varphi = \sup L\mathscr{H}$, $R\varphi = \sup R\mathscr{H}$ and $\varphi\psi = \sup(\mathscr{H}\psi)$ for all ψ.

$(***)_0$ If $\{\varphi_n\}$ is a chain in $\mathscr{F}$, then there is a φ such that $\varphi\psi = \sup_n(\varphi_n\psi)$, $\psi\varphi = \sup_n \psi\varphi_n$ for all ψ and $(I, \varphi) = \sup_n(I, \varphi_n)$. There is an element O such that $O \leq \psi$ and $O\psi = O$ for all ψ.

If $\mathscr{S}$ satisfies condition $(*)$ for certain $\mathscr{T}, t$, we shall say also that $\mathscr{S}$ is $(*)$-*complete*, and similarly for $(**)$ etc. The spaces of examples 3.1, 4.3, 4.7, 4.8 are $(**)_0$, $(***)_0$-complete, while that of example 3.2 is $(**)_0$-complete but not $(***)_0$-complete.

Proposition 18.3. Let $\mathscr{S}$, $\mathscr{T}$ satisfy $(**)$, $a \in \mathscr{T}$ and let $\mathscr{H}$ be a well ordered subset of $\mathscr{F}_a$. Then there is a $\varphi \in \mathscr{F}_a$ such that $\alpha\varphi\psi = \sup(\alpha\mathscr{H}\psi)$ for all $\alpha \in \mathscr{D}$, $\psi \in \mathscr{F}$.

This is immediate.

Proposition 18.4. Let $\mathscr{S}$, $\mathscr{T}$ satisfy $(***)$. Then the operations $\circ, \Pi$ are $\mathscr{T}$-continuous.

Proof. Suppose that $a \in \mathscr{T}$, $\{\varphi_n\}$, $\{\psi_n\}$ are chains in $\mathscr{F}_a$ and $\varphi = \sup_n \varphi_n$, $\psi = \sup_n \psi_n$.

It follows that $\varphi_n\psi_n \leq \varphi\psi$ for all n. Let $\tau \in \mathscr{F}$ and $\varphi_n\psi_n \leq \tau$ for all n. Then $\varphi_n\psi_m \leq \varphi_m\psi_m \leq \tau$ for all $m \geq n$, hence $\varphi_n\psi \leq \tau$ for all n by $(***)$, which implies $\varphi\psi \leq \tau$ by $(***)$. Therefore, $\varphi\psi = \sup_n \varphi_n\psi_n$.

Since $\{\varphi_n L\}$, $\{(I, \varphi_n L)\}$ are chains in $\mathscr{F}_a$, condition $(***)$ implies

$$(\varphi L, R) = (R, RL)(I, \varphi L) = A\left(I, \sup_n(\varphi_n L)\right) = A \sup_n(I, \varphi_n L)$$
$$= \sup_n A(I, \varphi_n L) = \sup_n(\varphi_n L, R).$$

Also $\{(\varphi_n L, R)\}$, $\{(I, \psi_n)\}$ are chains in $\mathscr{F}_a$, hence

$$(\varphi, \psi) = (\varphi L, R)(I, \psi) = \sup_n(\varphi_n L, R)\sup_n(I, \psi_n) = \sup_n(\varphi_n L, R)(I, \psi_n)$$
$$= \sup_n(\varphi_n, \psi_n).$$

The proof is complete.

Proposition 18.5. Let $\mathscr{S}, \mathscr{T}$ satisfy condition $(*)$ (condition $(**)$). Then $\mathscr{S}_1 = (\mathscr{F}, \Pi, \lambda\theta.L\theta, \lambda\theta.R\theta)$ is a $\mathscr{T}$-CPS ($\mathscr{T}$-SCPS respectively).

Proof. $\mathscr{S}_1$ is a pairing space by 16.2. It is $\mathscr{T}$-complete (strongly $\mathscr{T}$-complete) by $(*)$ (respectively, by $(**)$).

Proposition 18.6. Let $\mathscr{S}$, $\mathscr{T}$ satisfy $(***)$ and let $\mathscr{S}_1$ be the same as in the previous statement. Then $\mathscr{S}_1$ is a $\mathscr{T}$-CCPS.

Proof. The pairing space $\mathscr{S}_1$ is continuously $\mathscr{T}$-complete by $(***)$ and 18.4.

Proposition 18.7. Let $\mathscr{S}$, $\mathscr{T}$ satisfy $(*)$ or $(***)$. Then all inductive mappings over $\mathscr{F}$ preserve types.

Proof. Assume that $\mathscr{S}, \mathscr{T}$ satisfy $(*)$.

The mappings $\lambda\theta_1\ldots\theta_n.\theta_i$, $1 \leq i \leq n$ and $\lambda\theta_1\ldots\theta_n.\psi$, $\psi \in \{I, L, R\}$ preserve types, of course.

If $\Gamma_1, \Gamma_2 : \mathscr{F}^n \to \mathscr{F}$ preserve types, then so do

$$\Gamma = \lambda\theta_1\ldots\theta_n.\Gamma_1(\theta_1,\ldots,\theta_n)\Gamma_2(\theta_1,\ldots,\theta_n)$$

and

$$\Gamma = \lambda\theta_1\ldots\theta_n.(\Gamma_1(\theta_1,\ldots,\theta_n), \Gamma_2(\theta_1,\ldots,\theta_n))$$

since $\circ, \Pi$ preserve types.

Suppose that $\Gamma_1 : \mathscr{F}^{n+1} \to \mathscr{F}$ preserves types and

$$\Gamma = \lambda\theta_1\ldots\theta_n.\mu\theta.\Gamma_1(\theta_1,\ldots,\theta_n,\theta).$$

Let $\theta_1,\ldots,\theta_n \in \mathscr{F}_a$ and $\Gamma^* = \lambda\theta.\Gamma_1(\theta_1,\ldots,\theta_n,\theta)$. Then Γ^* is monotonic and normal and $\Gamma^*(\mathscr{F}_a) \subseteq \mathscr{F}_a$. If $\{\theta_n\}$ is the sequence assigned to Γ^* in the proof of 18.1, then $\theta_\xi \in \mathscr{F}_a$ for all ξ, hence $\mu\theta.\Gamma^*(\theta) \in \mathscr{F}_a$. Therefore, Γ preserves types.

If $\mathscr{S}, \mathscr{T}$ satisfy $(***)$, then we make use of 18.2 instead of 18.1. The proof is complete.

Proposition 18.8. If $\mathscr{S}, \mathscr{T}$ satisfy $(***)$, then all inductive mappings are $\mathscr{T}$-continuous.

Proof. Taking 18.4 into account, it suffices to show that whenever $\Gamma_1 : \mathscr{F}^{n+1} \to \mathscr{F}$ is inductive and $\mathscr{T}$-continuous, then $\Gamma = \lambda\theta_1\ldots\theta_n.\mu\theta.\Gamma_1(\theta_1,\ldots,\theta_n,\theta)$ is $\mathscr{T}$-continuous.

Let $a \in \mathscr{T}$, $\{\varphi_{i,m}\}$ be a chain in $\mathscr{F}_a$ and $\varphi_i = \sup_m \varphi_{i,m}$ for all i, $1 \leq i \leq n$. Then the mapping Γ is monotonic and preserves types by 18.7, hence $\{\Gamma(\varphi_{1,m},\ldots,\varphi_{n,m})\}_m$ is a chain in $\mathscr{F}_a$. Therefore, $\sigma = \sup_m \Gamma(\varphi_{1,m},\ldots,\varphi_{n,m})$ exists by $(***)$. It follows that $\Gamma(\varphi_{1,m},\ldots,\varphi_{n,m}) \leq \Gamma(\varphi_1,\ldots,\varphi_n)$ for all m; hence $\sigma \leq \Gamma(\varphi_1,\ldots,\varphi_n)$. On the other hand, the $\mathscr{T}$-continuity of Γ_1 implies

$$\begin{aligned}
&\Gamma_1(\varphi_1,\ldots,\varphi_n,\sigma)\\
&\quad = \Gamma_1\left(\sup_m \varphi_{1,m},\ldots,\sup_m \varphi_{n,m}, \sup_m \Gamma(\varphi_{1,m},\ldots,\varphi_{n,m})\right)\\
&\quad = \sup_m \Gamma_1(\varphi_{1,m},\ldots,\varphi_{n,m}, \Gamma(\varphi_{1,m},\ldots,\varphi_{n,m}))\\
&\quad = \sup_m \Gamma(\varphi_{1,m},\ldots,\varphi_{n,m}) = \sigma,
\end{aligned}$$

hence $\Gamma(\varphi_1,\ldots,\varphi_n) \leq \sigma$. Therefore, $\Gamma(\varphi_1,\ldots,\varphi_n) = \sigma$, hence Γ is $\mathscr{T}$-continuous. The proof is complete.

Proposition 18.9. Let $\mathscr{S}, \mathscr{T}$ satisfy $(*)$, let $\mathscr{E}$ be a normal segment and let $a \in \mathscr{T}$. Then $\mathscr{E} \cap \mathscr{F}_a$ is closed under least upper bounds of well ordered subsets.

Proof. Suppose that $\mathscr{A} \subseteq \mathscr{F}^2$,

$$\mathscr{E} = \{\theta/\theta\psi \leq \tau \quad \text{for all } \langle \psi, \tau \rangle \in \mathscr{A}\}$$

and $\mathscr{H}$ is a well ordered subset of $\mathscr{E} \cap \mathscr{F}_a$. Then $\theta\psi \leq \tau$ for all $\theta \in \mathscr{H}$, $\langle \psi, \tau \rangle \in \mathscr{A}$, hence $(\sup \mathscr{A})\psi \leq \tau$ for all $\langle \psi, \tau \rangle \in \mathscr{A}$ by $(*)$, i.e. $\sup \mathscr{H} \in \mathscr{E}$.

Proposition 18.10. Let $\mathscr{S}, \mathscr{T}$ satisfy $(**)$, let $\mathscr{E}$ be a regular segment and let $a \in \mathscr{T}$. Then $\mathscr{E} \cap \mathscr{F}_a$ is closed under least upper bounds of well ordered subsets.

The proof follows that of 18.9, using 18.3.

Proposition 18.11. Let $\mathscr{S}, \mathscr{T}$ satisfy $(***)$, let $\mathscr{E}$ be a regular segment and let $a \in \mathscr{T}$. Then $O \in \mathscr{E} \cap \mathscr{F}_a$ and $\mathscr{E} \cap \mathscr{F}_a$ is closed under least upper bounds of chains.

Proof. Suppose that $\mathscr{A} \subseteq \mathscr{D} \times \mathscr{F}^2$,

$$\mathscr{E} = \{\theta/\alpha\theta\psi \leq \tau \quad \text{for all } \langle \alpha, \psi, \tau \rangle \in \mathscr{A}\}$$

and $\{\varphi_n\}$ is chain in $\mathscr{E} \cap \mathscr{F}_a$. Then $\alpha\varphi_n\psi \leq \tau$ for all n and all $\langle \alpha, \psi, \tau \rangle \in \mathscr{A}$, hence $\alpha(\sup_n \varphi_n)\psi \leq \tau$ for all $\langle \alpha, \psi, \tau \rangle \in \mathscr{A}$ by $(***)$, i.e. $\sup_n \varphi_n \in \mathscr{E}$. The proof of 6.5 implies $O \in \mathscr{E}$. This completes the proof.

Now we are ready to show that each of the above conditions implies iterativeness.

Proposition 18.12. If $\mathscr{S}$ is $(*)$-complete, then it is μA_2-iterative.

Proof. Let $\Gamma: \mathscr{F}^{n+1} \to \mathscr{F}$ be inductive and $\theta_1, \ldots, \theta_n \in \mathscr{F}$. Then the mapping $\Gamma^* = \lambda\theta.\Gamma(\theta_1, \ldots, \theta_n, \theta)$ is monotonic and normal since Γ preserves types by 18.7. It follows from 18.1, 18.9 that $\mu\theta.\Gamma^*(\theta)$ exists and is a member of all normal segments closed under Γ^*. Therefore, the axiom μA_2 holds, which completes the proof.

Proposition 18.13. If $\mathscr{S}$ is $(**)$-complete, then it is μA_3-iterative.

The proof follows that of 18.12, making use of 18.10 instead of 18.9.

Proposition 18.14. If $\mathscr{S}$ is $(***)$-complete, then it is μA_3-iterative.

Proof. Let $\Gamma: \mathscr{F}^{n+1} \to \mathscr{F}$ be inductive and $\theta_1, \ldots, \theta_n \in \mathscr{F}$. Then the mapping $\Gamma^* = \lambda\theta.\Gamma(\theta_1, \ldots, \theta_n, \theta)$ is $\mathscr{T}$-continuous by 18.8 and normal by 18.7. It follows from 18.2, 18.11 that $\mu\theta.\Gamma^*(\theta)$ exists and is a member of all regular segments closed under Γ^*. The proof is complete.

The next five statements make it possible to transfer iterativeness from a given space to a related one; the last three of them are analogues to certain assertions of Skordev [1980].

Proposition 18.15. Let $\mathscr{S}$ be an OS, $\mathscr{S}_1$ be a subspace and suppose that $\mathscr{S}, \mathscr{T}$ satisfy condition $(*)$ (condition $(**)$). Let $\mathscr{F}_1$ be closed under least upper bounds of well ordered subsets of $\mathscr{F}_{1,a}$ for all $a \in \mathscr{T}$. Then $\mathscr{S}_1, \mathscr{T}$

satisfy $(*)$ (respectively, $(**)$) with a type function $t \upharpoonright \mathscr{F}_1$, and $\mathscr{S}_1$ is a subspace of $\mathscr{S}$ as an IOS.

This follows immediately.

Proposition 18.16. Let $\mathscr{S}$ be an OS, let $\mathscr{S}_1$ be a subspace and suppose that $\mathscr{S}, \mathscr{T}$ satisfy $(***)$. Let $O \in \mathscr{F}_1$ and let $\mathscr{F}_1$ be closed under least upper bounds of chains in $\mathscr{F}_{1,a}$ for all $a \in \mathscr{T}$. Then $\mathscr{S}_1, \mathscr{T}$ satisfy $(***)$ and $\mathscr{S}_1$ is a subspace of $\mathscr{S}$ as an IOS.

This follows immediately.

Proposition 18.17. Let $\mathscr{S}, \mathscr{T}$ satisfy $(*)$ and let $\mathscr{S}_1$ be obtained from $\mathscr{S}$ by introducing a new partial order $\leq_1$ such that whenever $a \in \mathscr{T}$, $\mathscr{H}$ is a subset of $\mathscr{F}_a$ well ordered with respect to $\leq$ and $\varphi = \sup \mathscr{H}$ with respect to $\leq$, then $\varphi = \sup \mathscr{H}$ with respect to $\leq_1$. Then $\mathscr{S}_1$ is μA_2-iterative and $\mathscr{S}, \mathscr{S}_1$ have identical operations $\langle \ \ \rangle, [\ \ ]$.

Proof. Notions related to the partial order $\leq_1$ will be subscripted, e.g. $\sup_1$, segment$_1$, $\mu_1 \theta . \Gamma(\theta)$ etc.

Let us prove first that whenever $\Gamma: \mathscr{F} \to \mathscr{F}$ is monotonic and normal, then $\mu\theta . \Gamma(\theta)$ is a member of all normal segments$_1$ closed under Γ. Given such a segment$_1$

$$\mathscr{E} = \{\theta / \theta\psi \leq_1 \tau \quad \text{for all } \langle \psi, \tau \rangle \in \mathscr{A}\},$$

$\mathscr{A} \subseteq \mathscr{F}^2$, take the sequence $\{\theta_\xi\}$ assigned to Γ by the proof of 18.1. It suffices to show that $\theta_\xi \in \mathscr{E}$ for all ξ.

Suppose that $\theta_\eta \in \mathscr{E}$ for all $\eta < \xi$. Then $\Gamma(\theta_\eta) \in \mathscr{E}$ for all $\eta < \xi$, hence $\Gamma(\theta_\eta)\psi \leq_1 \tau$ for all $\eta < \xi$ and all $\langle \psi, \tau \rangle \in \mathscr{A}$. We get $\theta_\xi \psi = \sup \{\Gamma(\theta_\eta)\psi / \eta < \xi\}$ by $(*)$, hence $\theta_\xi \psi = \sup_1 \{\Gamma(\theta_\eta)\psi / \eta < \xi\}$, which implies $\theta_\xi \psi \leq_1 \tau$ for all $\langle \psi, \tau \rangle \in \mathscr{A}$, i.e. $\theta_\xi \in \mathscr{E}$. Therefore, $\theta_\xi \in \mathscr{E}$ for all ξ, hence $\mu\theta . \Gamma(\theta) \in \mathscr{E}$.

If $\Gamma(\tau) \leq_1 \tau$, then $\mu\theta . \Gamma(\theta) \in \mathscr{E} = \{\theta / \theta \leq_1 \tau\}$. Therefore, $\mu_1 \theta . \Gamma(\theta) = \mu\theta . \Gamma(\theta)$, and an inductive argument on the construction shows that a mapping is inductive$_1$ iff it is inductive. Taking 18.7 into account, we conclude that $\mathscr{S}_1$ is μA_2-iterative. This completes the proof.

Notice that whenever $\varphi \leq \psi$, then $\varphi \leq_1 \psi$ since $\psi = \sup \{\varphi, \psi\}$.

Proposition 18.18. Let $\mathscr{S}, \mathscr{T}$ satisfy $(**)$ and let $\mathscr{S}$, $\mathscr{S}_1$ be related as in the previous statement. Then $\mathscr{S}_1$ is μA_3-iterative and $\mathscr{S}, \mathscr{S}_1$ have identical operations $\langle \ \ \rangle, [\ \ ]$.

The proof follows that of 18.17.

Propositon 18.19. Let $\mathscr{S}, \mathscr{T}$ satisfy $(***)$ and let $\mathscr{S}_1$ be obtained from $\mathscr{S}$ by introducing a new partial order $\leq_1$ such that whenever $a \in \mathscr{T}, \{\varphi_n\}$ is a chain in $\mathscr{F}_a$ and $\varphi = \sup_n \varphi_n$, then $\varphi = \sup_{n1} \varphi_n$. Then $\mathscr{S}_1$ is μA_3-iterative and $\mathscr{S}, \mathscr{S}_1$ have identical operations $\langle \ \ \rangle, [\ \ ]$.

Proof. Following the proof of 18.17, we show first that whenever $\Gamma: \mathscr{F} \to \mathscr{F}$ is $\mathscr{T}$-continuous and normal, then $\mu\theta . \Gamma(\theta)$ is a member of all regular segments$_1$ $\mathscr{E}$ closed under Γ; we have $O \in \mathscr{E}$ since $\varphi \leq \psi$ always implies $\varphi \leq_1 \psi$.

Then we show that the inductive$_1$ mappings are exactly the inductive ones, which will complete the proof by 18.7, 18.8.

Assume that $\langle\!\langle\ \ \rangle\!\rangle$ is a t-operation over $\mathscr{F}$ in the sense of chapter 10. (Mappings Σ_2, Σ_3 satisfying the equalities (2), (3) of chapter 10 need not be assumed to be given since their existence will follow by 10.9*, 10.10*.) The following sufficient conditions are designed to ensure that $\mathscr{S}$, $\langle\!\langle\ \ \rangle\!\rangle$ satisfy a tμ-axion. Namely, take (t$*$) (respectively, (t$**$)) to be condition ($*$) (condition ($**$)) plus the additional assumption that $\langle\!\langle\ \ \rangle\!\rangle$ preserves types and is continuous with respect to least upper bounds of well ordered subsets of $\mathscr{F}_a$ for all a. Take (t$***$) to be ($***$) plus the additional assumption that $\langle\!\langle\ \ \rangle\!\rangle$ preserves types and is $\mathscr{T}$-continuous, and $\langle\!\langle O\rangle\!\rangle = O$. (Notice that (t$*$), (t$**$) impose heavier constraints on the operation $\langle\!\langle\ \ \rangle\!\rangle$ than (t$***$) does.)

The spaces and t-operations of exercise 10.1 satisfy both (t$**$)$_0$, (t$***$)$_0$, where the subscripts indicate that $\mathscr{T}$ is a singleton.

Proposition 18.20. If $\mathscr{S}, \langle\!\langle\ \ \rangle\!\rangle, \mathscr{T}$ satisfy condition (t$*$) (condition (t$**$)), then $\mathscr{S}, \langle\!\langle\ \ \rangle\!\rangle$ satisfy axiom tμA$_2$ (respectively, tμA$_3$).

Proof. $\mathscr{S}$ is μA$_2$-iterative by 18.12 (respectively, μA$_3$-iterative by 18.13). All t-inductive mappings are monotonic and preserve types. All t-simple segments satisfy the property established in 18.9. Therefore, $\mathscr{S}$, $\langle\!\langle\ \ \rangle\!\rangle$ satisfy axiom tμA by 18.1. The proof is complete.

It often happens that all t-simple segments are normal by 10.18. Then one need not bother to verify that $\langle\!\langle\ \ \rangle\!\rangle$ is continuous with respect to least upper bounds of well ordered subsets of $\mathscr{F}_a$, all a.

Proposition 18.21. Let $\mathscr{S}, \mathscr{T}$ satisfy condition ($*$) (condition ($**$)), let $\langle\!\langle\ \ \rangle\!\rangle$ be monotonic and preserve types and let all the t-simple segments be normal (respectively regular). Then $\mathscr{S}, \langle\!\langle\ \ \rangle\!\rangle$ satisfy axiom tμA$_2$(tμA$_3$).

Proof. $\mathscr{S}$ is μA$_2$-iterative by 18.12 (μA$_3$-iterative by 18.13). All t-inductive mappings are monotonic and preserve types; hence the validity of tμA follows by 18.9 (18.10) and 18.1.

Proposition 18.22. If $\mathscr{S}, \langle\!\langle\ \ \rangle\!\rangle, \mathscr{T}$ satisfy (t$***$), then $\mathscr{S}, \langle\!\langle\ \ \rangle\!\rangle$ satisfy the axiom tμA$_3$.

This follows from 18.2 since all t-inductive mappings preserve types and are $\mathscr{T}$-continuous in this case, while all t-simple initial segments satisfy the property of 18.11.

t-Analogues to 18.15–18.19 can be established as well.

EXERCISES TO CHAPTER 18

Exercise 18.1. Let $(L, R) \leq I$ hold in $\mathscr{S}$. Show that Π is continuous with respect to least upper bounds of nonempty subsets of $\mathscr{F}$. In particular, the requirement $(I, \varphi) = \sup_n (I, \varphi_n)$ in ($***$) could be dropped.

Hint. Use 16.2 and exercise 16.3.

Exercise 18.2. Let $(L,R)\geq I$ hold in $\mathscr{S}$. Show that whenever $\mathscr{H}\subseteq\mathscr{F}$ and $\varphi=\sup\mathscr{H}$, then $L\varphi=\sup L\mathscr{H}$ and $R\varphi=\sup R\mathscr{H}$. In particular, (*) and (**) will be equivalent.

Hint. Use 16.2 and exercise 16.4.

The equality $(L,R)=I$ is valid in examples 3.1, 3.2, while $(L,R)<I$ in example 4.3. As far as examples 4.7, 4.8 are concerned, everything depends on the splitting scheme.

Exercise 18.3. Let $\mathscr{S}$ be (*)-complete ((**)-complete, (***)-complete). Show that $\mathscr{S}_a=(\mathscr{F}_a, I, \Pi\restriction\mathscr{F}_a^2, L, R)$ is a $(*)_0$-complete (respectively, $(**)_0$-complete, $(***)_0$-complete) OS for all $a\in\mathscr{T}$.

Exercise 18.4. Let $\mathscr{S}$ be (*)-complete and Π $\mathscr{T}$-continuous. Observing that $\Delta(\varphi,\psi)=\mu\theta.(\varphi,\theta\psi)$ by 6.33, show that this least fixed point is reached at level ω and Δ is $\mathscr{T}$-continuous in its first argument. In particular, $\langle\ \rangle$ is $\mathscr{T}$-continuous.

Hint. Show that the mapping $\lambda\varphi\theta.(\varphi,\theta\psi)$ is $\mathscr{T}$-continuous, then use the proofs of 18.2, 18.7, 18.8.

Exercise 18.5. Let $\mathscr{S}$ be $(***)_0$-complete or $\mathscr{S}$ be $(*)_0$-complete and Π continuous. Prove that $\mathscr{S}$ admits a continuous collection operation. Assuming that $\mathscr{S}_1$ is obtained from $\mathscr{S}$ by modifying its partial order, show that $\mathscr{S}_1$ also admits a collection operation.

Hint. $\mathscr{S}$ satisfies condition (c*) of exercise 11.1. The same operation *Co* does for $\mathscr{S}_1$.

Exercise 18.6. Let $\mathscr{S}$ be an OS, $\mathscr{S}_1$ a subspace, let $\mathscr{S}$ be (***)-complete and suppose that $\mathscr{F}_1$ be closed under $\langle\ \rangle$, $[\ \ ]$. Prove that $\mathscr{S}_1$ is μA_3-iterative and is a subspace of $\mathscr{S}$ as an IOS.

Hint. Show first that μA_3 holds for mappings $\Gamma:\mathscr{F}_1\to\mathscr{F}_1$ recursive in $\mathscr{F}_1$, observing that $\mu_1\theta.\Gamma(\theta)=\sup_n\Gamma^n(O)$. Thus we obtain a First Recursion Theorem for $\mathscr{S}_1$, which in turn implies that μA_3 holds for arbitrary inductive mappings.

A version of the last exercise for spaces with t-operations can also be given.

Exercise 18.7. Let $\langle\!\langle\ \rangle\!\rangle$ be a storing operation which preserves types. Show that in order to ensure the validity of the stronger axiom tμA of exercise 10.9 it suffices to require in (*) and (**) that $x\varphi=\sup x\mathscr{H}$ for all $x\in\mathscr{L}$, while condition (t***) needs no amendments.

Exercise 18.8. Construct a 5-tuple $\mathscr{S}$ to meet μA_3 and all the axioms of OS but A_2.

Hint. Take $\mathscr{S}$ to be a $(**)_0$-complete OS with a pairing scheme modified as suggested by the hint to exercise 4.3. Then $\mathscr{S}$ will be $(**)_0$-complete, hence μA_3-iterative since the proof of 18.13 makes no use of A_2.

Exercise 18.9. Construct a 5-tuple $\mathscr{S}$ to meet μA_3 and all the axioms of OS but A_3.

Hint. Take $\mathscr{S}$ to be a $(***)_0$-complete OS with a pairing scheme modified as in the hint to exercise 4.3. The new pairing operation $\lambda\varphi\psi.\varphi$ is continuous, hence $\mathscr{S}$ is μA_3-iterative by the proof of 18.14.

While the OS of example 4.2 satisfies (£) but not (££), the following exercise shows that (£) is also independent.

Exercise 18.10. Construct an OS to satisfy (££) but not (£).

Hint. Take a μA_3-iterative OS $\mathscr{S}$ in which $\langle I \rangle = I$, then take $\mathscr{F}_1 = \{\varphi/\varphi$ is prime recursive$\}$. The OS $\mathscr{S}_1 = (\mathscr{F}_1, I, \Pi \restriction \mathscr{F}_1^2, L, R)$ satisfies (££) since $\mathscr{F}_1$ is closed under $[\quad]$. Assuming that the operation $\langle\quad\rangle_1$ over $\mathscr{F}_1$ satisfies (£), get $\langle R \rangle = \langle R \rangle_1 \in \mathscr{F}_1$ contrary to exercise 6.14***.

CHAPTER 19

Constructing operative spaces

This chapter presents several standard constructions which produce OS from pairing spaces or partially ordered semigroups. In particular, consecutive spaces will be constructed.

Often a pairing space can be augmented with multiplication to become an OS.

Proposition 19.1. Let $\mathscr{S} = (\mathscr{F}, \Pi, L', R')$ be a pairing space, let $\circ : \mathscr{F}^2 \to \mathscr{F}$ be monotonic and associative and let $I \in \mathscr{F}$ be a unit such that $L' = \lambda\theta . L'(I)\theta$, $R' = \lambda\theta . R'(I)\theta$. Let $\circ$ be right distributive with respect to Π, i.e. $(\varphi, \psi)\chi = (\varphi\chi, \psi\chi)$ for all φ, ψ, χ. Then $\mathscr{S}_1 = (\mathscr{F}, I, \Pi, L'(I), R'(I))$ is an OS.

This follows from the relevant definitions.

Of course, the above statement gives no general construction since multiplication can sometimes be introduced in various ways. Once such an operation is singled out, 19.1 just says what has to be verified in order to that the structure at hand be an OS.

Proposition 19.2. Let $\mathscr{S}, \mathscr{S}_1$ be the same as in 19.1 and let $\mathscr{S}$ be $\mathscr{T}$-complete (strongly $\mathscr{T}$-complete). Let $\circ$ preserve types and suppose that whenever $a \in \mathscr{T}$, $\mathscr{H}$ is a well ordered subset of $\mathscr{F}_a$ and $\varphi = \sup \mathscr{H}$ in $\mathscr{F}$, then $\varphi\psi = \sup(\mathscr{H}\psi)$ for all ψ. Then $\mathscr{S}_1$ is $(*)$-complete ($(**)$-complete).

This follows from the corresponding definitions.

Proposition 19.3. Let $\mathscr{S}, \mathscr{S}_1$ be the same as in 19.1 and let $\mathscr{S}$ be continuously $\mathscr{T}$-complete. Let $\circ$ preserve types and be $\mathscr{T}$-continuous and suppose that O satisfies the equality $O\psi = O$ for all ψ. Then $\mathscr{S}_1$ is $(***)$-complete.

This follows from the corresponding definitions.

On the other hand, IOS can be obtained by augmenting certain partially ordered semigroups with pairing schemes.

Proposition 19.4. Let $\mathscr{F}$ be a partially ordered semigroup with a unit I and suppose that every subset $\mathscr{H}$ of $\mathscr{F}$ has a least upper bound φ such that $\varphi\psi = \sup(\mathscr{H}\psi)$ and $\psi\varphi = \sup \psi\mathscr{H}$ for all ψ. Let $L, L_1, R, R_1 \in \mathscr{F}$, $LL_1 = RR_1 = I$ and $LR_1 = RL_1 = O$, where $O = \sup\varnothing$. Take $(\varphi, \psi) = \sup\{L_1\varphi, R_1\psi\}$. Then $\mathscr{S} = (\mathscr{F}, I, \Pi, L, R)$ is a $(**)_0$, $(***)_0$-complete OS.

Proof. Suppose that $\varnothing \subset \mathscr{H}_1, \mathscr{H}_2 \subseteq \mathscr{F}$, $\varphi = \sup \mathscr{H}_1$ and $\psi = \sup \mathscr{H}_2$.

Then

$$\begin{aligned}(\varphi,\psi) &= \sup\{L_1\varphi, R_1\psi\}\\ &= \sup\{\sup\{L_1\theta/\theta\in\mathscr{H}_1\},\quad \sup\{R_1\tau/\tau\in\mathscr{H}_2\}\}\\ &= \sup\{\sup\{L_1\theta, R_1\tau\}/\theta\in\mathscr{H}_1 \& \tau\in\mathscr{H}_2\}\\ &= \sup\{(\theta,\tau)/\theta\in\mathscr{H}_1 \& \tau\in\mathscr{H}_2\};\end{aligned}$$

hence Π is continuous with respect to least upper bounds of nonempty sets. In particular, Π is continuous and certainly monotonic. It follows that

$$\begin{aligned}&(\varphi,\psi)\psi = \sup\{L_1\varphi, R_1\psi\}\chi = \sup\{L_1\varphi\chi, R_1\psi\chi\} = (\varphi\chi,\psi\chi)\\ &L(\varphi,\psi) = L\sup\{L_1\varphi, R_1\psi\} = \sup\{LL_1\varphi, LR_1\psi\} = \sup\{\varphi, O\} = \varphi,\\ &R(\varphi,\psi) = \sup\{RL_1\varphi, RR_1\psi\} = \psi;\end{aligned}$$

hence $\mathscr{S}$ is a $(**)_0$-complete OS. Moreover, $O \leq \psi$ and $O\psi = (\sup\varnothing)\circ\psi = \sup\varnothing = O$ for all ψ, hence $\mathscr{S}$ is also $(***)_0$-complete. The proof is complete.

Notice that in this space we have also $\psi O = O$ for all ψ.

Proposition 19.5. (A typed version of 19.4.) Let $\mathscr{F}$ be a partially ordered semigroup with unit I such that $\circ$ preserves types and whenever $a\in\mathscr{T}$, $\mathscr{H}\subseteq\mathscr{F}_a$, there exists a $\varphi\in\mathscr{F}_a$ such that $\varphi\psi = \sup(\mathscr{H}\psi)$, $\psi\varphi = \sup\psi\mathscr{H}$ for all ψ. Let L, L_1, R, R_1, Π be as in 19.4. Then $\mathscr{S} = (\mathscr{F}, I, \Pi, L, R)$ is a $(**)$, $(***)$-complete OS.

The proof repeats that of 19.4. If $\mathscr{H}\subseteq\mathscr{F}$ is finite, $\mathscr{H}\subseteq\mathscr{F}_a$ for a certain a; hence $\sup\mathscr{H}$ exists. Therefore, the operation Π is correctly introduced.

It is worth mentioning that the construction of 19.4, 19.5 applies equally well to the dual semigroup $(\mathscr{F}, \lambda\varphi\psi.\psi\varphi)$.

Proposition 19.6. Let $\mathscr{S}$ be constructed by 19.4 or 19.5. Then $\langle\varphi\rangle = \sup_n R_1^n L_1\varphi L R^n$, $\Delta(\varphi,\psi) = \sup_n R_1^n L_1\varphi\psi^n$ and $[\varphi] = \sup_n (R_1\varphi)^n L_1$. The element $U = \sup\{L, R\}$ satisfies condition (1) of exercise 7.10. The above characterizations of $\langle\ \rangle$, Δ, $[\ \ ]$ take place for all iterative subspaces of $\mathscr{S}$.

Proof. Proposition 6.33 gives by the proof of 18.2 that $\Delta(\varphi,\psi) = \sup_n\theta_n$, where $\theta_0 = O$, $\theta_{n+1} = (\varphi, \theta_n\psi)$. Supposing $\theta_n = \sup_{i<n} R_1^i L_1\varphi\psi^i$, one gets

$$\theta_{n+1} = \sup\{L_1\varphi, R_1\theta_n\psi\} = \sup_{i<n+1} R_1^i L_1\varphi\psi^i,$$

hence $\theta_n = \sup_{i<n} R_1^i L_1\varphi\psi^i$ for all n. Therefore, $\Delta(\varphi,\psi) = \sup_n R_1^n L_1\varphi\psi^n$. In particular, $\langle\varphi\rangle = \sup_n R_1^n L_1\varphi L R^n$ by 6.32.

It follows similarly that $[\varphi] = \sup_n\theta_n$, where $\theta_0 = O$ and $\theta_{n+1} = (I, \varphi\theta_n)$. One gets by induction that $\theta_n = \sup_{i<n}(R_1\varphi)^i L_1$ for all n, hence $[\varphi] = \sup_n(R_1\varphi)^n L_1$.

The element U satisfies condition (2) of exercise 7.10 since $\varphi U\psi = \varphi\sup\{L, R\}\psi = \sup\{\varphi L\psi, \varphi R\psi\}$ for all φ, ψ.

Suppose that $\mathscr{S}_1 = (\mathscr{F}_1, I, \Pi\restriction\mathscr{F}_1^2, L, R)$ is an iterative subspace of $\mathscr{S}$. Then $L_1, R_1\in\mathscr{F}_1$ since $L_1 = \sup\{L_1, R_1 O\} = (I, O)$ and $R_1 = (O, I)$. Therefore, the expressions for $\langle \quad \rangle, \Delta, [\quad]$ obtained above still make sense. The proof is complete.

The following construction modifies that of 16.5.

Proposition 19.7. Let $\mathscr{S} = (\mathscr{F}, \Pi, L', R')$ be a pairing space (continuous pairing space). Take $\mathscr{F}' = \{\varphi' / \varphi' : \mathscr{F} \to \mathscr{F} \,\&\, \varphi'$ is monotonic (continuous)$\}$, $I' = \lambda\theta.\theta$, $\varphi' \le \psi'$ iff $\forall\theta(\varphi'(\theta) \le \psi'(\theta))$, $\varphi'\psi' = \lambda\theta.\varphi'(\psi'(\theta))$ and $\Pi'(\varphi', \psi') = \lambda\theta.(\varphi'(\theta), \psi'(\theta))$. Then $\mathscr{S}' = (\mathscr{F}', I', \Pi', L', R')$ is an OS.

Proof. If $\varphi', \psi' \in \mathscr{F}'$, then $\varphi'\psi' \in \mathscr{F}'$ since composition of mappings preserves monotonicity (continuity). It follows from the proof of 12.1 that $\mathscr{F}'$ is a partially ordered semigroup with unit I'. The monotonicity (respectively continuity) of L', R' implies $L', R' \in \mathscr{F}'$. If $\varphi', \psi' \in \mathscr{F}'$, then $(\varphi', \psi') \in \mathscr{F}'$ by the monotonicity (continuity) of Π. Moreover, Π' is monotonic since Π is. The distributive law is verified as in 12.1, while

$$L'(\varphi', \psi') = \lambda\theta.L'((\varphi'(\theta), \psi'(\theta))) = \lambda\theta.\varphi'(\theta) = \varphi'$$

and similarly $R'(\varphi', \psi') = \psi'$; hence $\mathscr{S}'$ is an OS. The proof is complete.

Notice that in the continuous version the operation multiplication and pairing of $\mathscr{S}'$ are continuous as well.

The following statement reaffirms the construction of 12.1.

Proposition 19.8. Let $\mathscr{S} = (\mathscr{F}, I, \Pi, L, R)$ be an OS and $\mathscr{S}' = (\mathscr{F}', I', \Pi', L', R')$ be obtained from the pairing space $(\mathscr{F}, \Pi, \bar{L}, \bar{R})$ by means of 19.7. Then $\mathscr{S}, \mathscr{S}'$ are consecutive OS.

Proof. We recall that $\bar{\varphi} = \lambda\theta.\varphi\theta$, $Id = \lambda\theta.I$ and $Ml = \lambda\theta.L\theta R\theta$. It follows that $\bar{\mathscr{F}} \subseteq \mathscr{F}'$ and $Id, Ml \in \mathscr{F}'$; hence the spaces $\mathscr{S}, \mathscr{S}'$ are consecutive.

The following statement modifies 16.11.

Proposition 19.9. Let $\mathscr{S} = (\mathscr{F}, \Pi, L', R')$ be a $\mathscr{T}$-CPS ($\mathscr{T}$-SCPS), $\mathscr{F}' = \{\varphi' / \varphi' : \mathscr{F} \to \mathscr{F} \,\&\, \varphi'$ is monotonic and normal$\}$, $I', \le, \circ, \Pi'$ be introduced as in 19.7 and $\mathscr{T}', t'$ the same as in 16.11. Then $\mathscr{S}' = (\mathscr{F}', I', \Pi', L', R')$ is an OS and $\mathscr{S}', \mathscr{T}'$ satisfy $(*)$ (respectively $(**)$). In particular, if $\mathscr{S}$ is a CPS (SCPS), then $\mathscr{F}'$ consists of all the monotonic unary mappings over $\mathscr{F}$ and $\mathscr{S}'$ is $(*)_0$-complete ($(**)_0$-complete).

Proof. Suppose that $\varphi', \psi' \in \mathscr{F}'$, $a \in t'(\varphi')$ and $b \in t'(\psi')$. Let $c \ge a, b$ and $\theta \in \mathscr{F}_c$. The mapping ψ' is normal and $b \le c$, hence $\psi'(\theta) \in \mathscr{F}_c$. The mapping φ' is normal and $a \le c$, hence $\varphi'(\psi'(\theta)) \in \mathscr{F}_c$, i.e. $\varphi'\psi'(\theta) \in \mathscr{F}_c$. Therefore, $\varphi'\psi'$ is a monotonic and normal mapping, hence $\varphi'\psi' \in \mathscr{F}'$. Whenever $\varphi', \psi' \in \mathscr{F}'_{a'}$, then the above argument gives $a' \subseteq t'(\varphi'\psi')$, hence $\varphi'\psi' \in \mathscr{F}'_{a'}$. We conclude that the multiplication of $\mathscr{F}'$ preserves types. It follows similarly that $\mathscr{F}'$ is closed under Π' and Π' preserves types, using the fact that Π preserves types.

It is immediate that I' preserves types. Observing that L', R' preserve types, we get

$$t'(I') = t'(L') = t'(R') = \mathscr{T} \le a'$$

for all $a' \in \mathscr{T}'$. In particular, I', L', R' are monotonic and normal, hence $I', L', R' \in \mathscr{F}'$. Using 19.7, it follows that $\mathscr{S}'$ is an OS.

Let $a' \in \mathscr{T}'$ and $\mathscr{H}'$ be a well ordered subset of $\mathscr{F}'_{a'}$. Then we take $\varphi' = \lambda\theta.\sup\{\theta'(\theta)/\theta' \in \mathscr{H}'\}$ as in the proof of 16.11 and show that φ' is monotonic and normal and also $a' \subseteq t'(\varphi')$, hence $\varphi' \in \mathscr{F}'_{a'}$. It follows that

$$\varphi'\psi'(\theta) = \varphi'(\psi'(\theta)) = \sup\{\theta'(\psi'(\theta))/\theta' \in \mathscr{H}'\} = \sup\{\theta'\psi'(\theta)/\theta' \in \mathscr{H}'\}$$

for all ψ', θ, hence $\varphi'\psi' = \sup(\mathscr{H}'\psi')$ for all ψ'. Therefore, $\mathscr{S}', \mathscr{T}'$ satisfy $(*)$.

Suppose that $\mathscr{S}$ is strongly $\mathscr{T}$-complete. Then

$$\begin{aligned} L'\varphi'(\theta) = L'(\varphi'(\theta)) &= L'(\sup\{\theta'(\theta)/\theta' \in \mathscr{H}'\}) = \sup\{L'(\theta'(\theta))/\theta' \in \mathscr{H}'\} \\ &= \sup\{L'\theta'(\theta)/\theta' \in \mathscr{H}'\} \end{aligned}$$

for all θ, hence $L'\varphi' = \sup L'\mathscr{H}'$. Similarly, $R'\varphi' = \sup R'\mathscr{H}'$, hence $\mathscr{S}', \mathscr{T}'$ satisfy $(**)$. The proof is complete.

Proposition 19.10. Let $\mathscr{S} = (\mathscr{F}, \Pi, L', R')$ be a $\mathscr{T}$-CCPS, $\mathscr{F}' = \{\varphi'/\varphi' : \mathscr{F} \to \mathscr{F}\ \&\ \varphi'$ is $\mathscr{T}$-continuous and normal$\}$, $I', \leq, \circ, \Pi'$ be as in 19.7 and $\mathscr{T}', t'$ those of 16.11. Then $\mathscr{S}' = (\mathscr{F}', I', \Pi', L', R')$ is an OS and $\mathscr{S}', \mathscr{T}'$ satisfy $(***)$. In particular, $\mathscr{F}'$ consists of all continuous unary mappings over $\mathscr{F}$ and $\mathscr{S}'$ is $(***)_0$-complete, provided $\mathscr{S}$ is a CCPS.

Proof. Suppose that $\varphi', \psi' \in \mathscr{F}', a \in \mathscr{T}, \{\varphi_n\}$ is a chain in $\mathscr{F}_a$ and $\varphi = \sup_n \varphi_n$. The mapping ψ' is $\mathscr{T}$-continuous; hence $\psi'(\varphi) = \sup_n \psi'(\varphi_n)$. Let $b \in t'(\psi')$ and $c \geq a, b$. Then $\{\psi'(\varphi_n)\}$ is a chain in $\mathscr{F}_c$ since $b \leq c$ and $\varphi_n \in \mathscr{F}_c$ for all n. The mapping φ' is $\mathscr{T}$-continuous; hence $\varphi'(\psi'(\varphi)) = \sup_n \varphi'(\psi'(\varphi_n))$, i.e. $\varphi'\psi'(\varphi) = \sup_n \varphi'\psi'(\varphi_n)$. Therefore $\varphi'\psi'$ is a $\mathscr{T}$-continuous mapping. The proof of 19.9 implies that $\varphi'\psi'$ is normal, hence $\varphi'\psi' \in \mathscr{F}'$. The proof of 19.9 implies also that the multiplication of $\mathscr{F}'$ preserves types. It follows similarly that $\mathscr{F}'$ is closed under Π' and Π' preserves types on account of the fact that Π is $\mathscr{T}$-continuous and preserves types.

The mappings I', L', R' are $\mathscr{T}$-continuous and preserve types, hence $I, L', R' \in \mathscr{F}'$ and $t'(I'), t'(L'), t'(R') \leq a'$ for all $a' \in \mathscr{T}'$. Using 19.7, it follows that $\mathscr{S}'$ is an OS.

Suppose that $a' \in \mathscr{T}'$ and $\{\varphi'_n\}$ is a chain in $\mathscr{F}'_{a'}$. Let $\theta \in \mathscr{F}$, $a \in a'$ and $b \geq a, t(\theta)$. It follows that $\varphi'_n \in \mathscr{F}'_{a'}$; hence $a \in t'(\varphi'_n)$ and $\varphi'_n(\theta) \in \mathscr{F}_b$ for all n. Therefore, $\{\varphi'_n(\theta)\}$ is a chain in $\mathscr{F}_b$; hence there is a $\tau_\theta \in \mathscr{F}_b$ such that $\tau_\theta = \sup_n \varphi'_n(\theta)$. Take $\varphi' = \lambda\theta.\tau_\theta$. Then whenever $c \in \mathscr{T}$, $\{\varphi_n\}$ is a chain in $\mathscr{F}_c$ and $\varphi = \sup_n \varphi_n$, we obtain from the $\mathscr{T}$-continuity of φ'_n,

$$\begin{aligned} \varphi'(\varphi) = \sup_n \varphi'_n(\varphi) &= \sup_n \varphi'_n\left(\sup_m \varphi_m\right) = \sup_n \sup_m \varphi'_n(\varphi_m) \\ &= \sup_m \sup_n \varphi'_n(\varphi_m) = \sup_m \varphi'(\varphi_m); \end{aligned}$$

hence φ' is $\mathscr{T}$-continuous. It is also normal and $a' \subseteq t'(\varphi')$; hence $\varphi' \in \mathscr{F}'_{a'}$.

Using the $\mathscr{T}$-continuity of Π, we get

$$(I', \varphi')(\theta) = (\theta, \varphi'(\theta)) = \left(\theta, \sup_n \varphi'_n(\theta)\right) = \sup_n (\theta, \varphi'_n(\theta)) = \sup_n (I, \varphi'_n)(\theta)$$

for all θ; hence $(I', \varphi') = \sup_n (I', \varphi'_n)$.

If $\psi' \in \mathscr{F}'$, then

$$\varphi'\psi'(\theta) = \varphi'(\psi'(\theta)) = \sup_n \varphi'_n(\psi'(\theta)) = \sup_n \varphi'_n\psi'(\theta)$$

for all θ, hence $\varphi'\psi' = \sup_n(\varphi'_n\psi')$. It follows from the $\mathscr{T}$-continuity of ψ' that

$$\psi'\varphi'(\theta) = \psi'\left(\sup_n \varphi'_n(\theta)\right) = \sup_n \psi'(\varphi'_n(\theta)) = \sup_n \psi\varphi'_n(\theta)$$

for all θ; hence $\psi'\varphi' = \sup_n \psi'\varphi'_n$.

The mapping $O' = \check{O}$ is continuous and preserves types; hence $O' \in \mathscr{F}'$ and $t'(O') \leq a'$ for all a'. It follows that $O' \leq \psi'$ and $O'\psi' = \lambda\theta.O'(\psi'(\theta)) = \lambda\theta.O = O'$ for all ψ'. Therefore, $\mathscr{S}', \mathscr{T}'$ satisfy (***), which completes the proof.

The following two statements present our standard constructions yielding consecutive IOS. They will also be used in the next chapter to construct hierarchies of IOS.

Proposition 19.11. Let $\mathscr{S}$ be a (*)-complete ((**)-complete) OS and let $\mathscr{S}'$ be obtained from $(\mathscr{F}, \Pi, \bar{L}, \bar{R})$ by 19.9. Then $\mathscr{S}$, $\mathscr{S}'$ are consecutive IOS.

Proof. $(\mathscr{F}, \Pi, \bar{L}, \bar{R})$ is a $\mathscr{T}$-CPS ($\mathscr{T}$-SCPS) by 18.5, hence 19.9 may be applied. The spaces $\mathscr{S}, \mathscr{S}'$ are μA_2-iterative by 18.12 (respectively, μA_3-iterative by 18.13). All $\bar{\varphi}$ are monotonic and normal and so are Id, Ml; hence $\bar{\mathscr{F}} \subseteq \mathscr{F}'$, $Id, Ml \in \mathscr{F}'$. Therefore $\mathscr{S}, \mathscr{S}'$ are consecutive OS. In order to show that they are consecutive IOS it suffices to show that $\langle \bar{I} \rangle = \langle I' \rangle$. As mentioned in chapter 12 however, this equality follows by the μA_2-iterativeness of $\mathscr{S}$. The proof is complete.

Proposition 19.12. Let $\mathscr{S}$ be (***)-complete and let $\mathscr{S}'$ be obtained from $(\mathscr{F}, \Pi, \bar{L}, \bar{R})$ by 19.10. Then $\mathscr{S}, \mathscr{S}'$ are consecutive IOS.

Proof. $(\mathscr{F}, \Pi, \bar{L}, \bar{R})$ is a $\mathscr{T}$-CCPS by 18.6. Both $\mathscr{S}, \mathscr{S}'$ are μA_3-iterative by 18.14. The mappings $\bar{\varphi}, Id, Ml$ are $\mathscr{T}$-continuous and normal, hence $\bar{\mathscr{F}} \subseteq \mathscr{F}'$, $Id, Ml \in \mathscr{F}'$, which implies that $\mathscr{S}, \mathscr{S}'$ are consecutive OS. The equality $\langle \bar{I} \rangle = \langle I' \rangle$ also holds, hence $\mathscr{S}, \mathscr{S}'$ are consecutive IOS. The proof is complete.

Proposition 19.13. Let $\mathscr{S}'$ be obtained from $\mathscr{S}$ by 19.11. Then the monotonic mappings Q_σ of chapter 13 are members of $\mathscr{F}'$ and so are the mappings $Q_{\sigma,\mathscr{C}}$, provided L and R have an upper bound U.

This follows immediately, from the normality of these mappings.

If $\mathscr{S}'$ is obtained from $\mathscr{S}$ by 19.12 however, one should not expect that $Q_\sigma, Q_{\sigma,\mathscr{C}} \in \mathscr{F}'$ for arbitrary $\sigma, \mathscr{C}$. A counterexample will be given in the exercises. This indicates that the monotonic version is richer and perhaps, more interesting, than the continuous one.

The following statement shows that consecutive spaces constructed by 19.11 admit transfers.

Proposition 19.14. Let $\mathscr{S}, \mathscr{T}$ satisfy condition (*) (condition (**)), let $\mathscr{S}', \mathscr{T}'$ be obtained from $\mathscr{S}, \mathscr{T}$ by 19.11 and let $\mathscr{S}'', \mathscr{T}''$ be obtained from $\mathscr{S}', \mathscr{T}'$ by

19.11 again. Then

$$Tf = \lambda\varphi''.\lambda\theta'.\lambda\theta.\varphi''(\theta'(L'\check{\theta}, I'))(R\theta)$$

is a correctly introduced operation over $\mathscr{F}''$ and the axiom tfμA_2 (respectively, tfμA_3) is valid. If $\mathscr{S}'''$ is obtained from $\mathscr{S}''$ by 19.11, then $Tf \in \mathscr{F}'''$.

Proof. Suppose that $\varphi'' \in \mathscr{F}'', \theta' \in \mathscr{F}', a' \in t''(\varphi'')$ and $b' \geq a', t'(\theta')$. Let $b \in b'$ and $\theta \in \mathscr{F}_b$. Then $t'(\check{\theta}) \leq \check{b} = \{a/b \leq a\}$ and $t'(\theta') \leq b' \leq \check{b}$; hence $\theta'(L'\check{\theta}, I') \in \mathscr{F}'_{\check{b}}$. It follows that $\varphi''(\theta'(L'\check{\theta}, I')) \in \mathscr{F}'_{\check{b}}$ since $a' \leq \check{b}$. Noting that $R\theta \in \mathscr{F}_b$ and $b \in \check{b}$, we get $\varphi''(\theta'(L'\check{\theta}, I'))(R\theta) \in \mathscr{F}_b$. Therefore, the monotonic mapping $Tf(\varphi'')(\theta') = \lambda\theta.\varphi''(\theta'(L'\check{\theta}, I'))(R\theta)$ is normal, and hence in $\mathscr{F}'$.

Suppose that $\varphi'' \in \mathscr{F}''$ and $a' \in t''(\varphi'')$. If $\theta' \in \mathscr{F}'_{a'}$, then the above argument gives $t'(Tf(\varphi'')(\theta')) \leq a'$, i.e. $Tf(\varphi'')(\theta') \in \mathscr{F}'_{a'}$. Therefore, the monotonic mapping $Tf(\varphi'')$ over $\mathscr{F}'$ is normal; hence $Tf(\varphi'') \in \mathscr{F}''$. We conclude that $Tf: \mathscr{F}'' \to \mathscr{F}''$. Moreover, we get $Tf(\mathscr{F}''_{a''}) \subseteq \mathscr{F}''_{a''}$ for all a'', hence Tf preserves types. Propositions 10.18, 18.21 ensure that $\mathscr{S}'', Tf$ satisfy tfμA_2 (respectively tfμA_3).

Finally, if $\mathscr{S}'''$ is obtained from $\mathscr{S}''$ by 19.11, then $Tf \in \mathscr{F}'''$ since Tf is monotonic and normal. This completes the proof.

One last remark. In order to consider the transfer operation over $\mathscr{F}''$, the space $\mathscr{S}$ need not be an OS. It suffices to assume $\mathscr{S}$ a pairing space and replace throughout chapter 14 $L\theta, R\theta$ by $L'(\theta), R'(\theta)$.

EXERCISES TO CHAPTER 19

Exercise 19.1. Let $\mathscr{S} = (\mathscr{F}, I, \Pi, L, R)$ be an OS with a least element O such that $O\psi = O$ for all ψ. Take $\mathscr{F}' = \{\varphi'/\varnothing \subset \varphi' \subseteq \mathscr{F}\}$, $\varphi' \leq \psi'$ iff $\forall \varphi \in \varphi' \exists \psi \in \psi'(\varphi \leq \psi)$, $\varphi'\psi' = \{\varphi\psi/\varphi \in \varphi' \,\&\, \psi \in \psi\}$, $\Pi'(\varphi', \psi') = \Pi(\varphi', \psi')$, $I' = \{I\}$, $L' = \{L\}$ and $R' = \{R\}$. Show that $\mathscr{S}' = (\mathscr{F}', I', \Pi', L', R')$ is a $(**)_0$, $(***)_0$-complete OS and the element $U' = \{L, R\}$ satisfies condition (1) of exercise 7.10. Give explicit characterizations of the operations $\langle\ \ \rangle, \Delta, [\ \ ]$ of $\mathscr{S}'$.

Hint. Use exercise 16.1 and propositions 19.1–19.3.

Remarks. The necessity of factorization can be avoided by modifying $\mathscr{F}'$ as suggested in the remarks to exercise 16.1. The advantages and drawbacks of this construction are the same as in the case of pairing spaces. One always gets an iterative space, even if the given one is not; e.g. the subspace of an arbitrary IOS consisting of all elements polynomial in O, I is not iterative by exercise 6.14***. The given OS $\mathscr{S}$ is isomorphic with the subspace of $\mathscr{S}'$ based on $\{\{\varphi\}/\varphi \in \mathscr{F}\}$. If, however, $\mathscr{S}$ is iterative, this isomorphism may fail to agree with the operations $\langle\ \ \rangle, [\ \ ]$ of $\mathscr{S}$ and $\mathscr{S}'$.

Exercise 19.2. Let N be a nonempty set and $\mathscr{S}_s$ be an IOS for all $s \in N$. Take $\mathscr{F} = \times_{s \in N} \mathscr{F}_s$, $\varphi \leqslant \psi$ iff $\forall s(\varphi(s) \leq \psi(s))$, $\varphi\psi = \lambda s.\varphi(s)\psi(s)$, $(\varphi, \psi) = \lambda s.\Pi_s(\varphi(s), \psi(s))$, $I = \lambda s.I_s$, $L = \lambda s.L_s$ and $R = \lambda s.R_s$. Show that $\mathscr{S} = (\mathscr{F}, I, \Pi, L, R)$ is an IOS and whenever $\mathscr{S}_s, \mathscr{T}_s$ satisfy one of $(*)$, $(**)$, $(***)$ for all s, then $\mathscr{S}, \mathscr{T}$ satisfy the same condition, where $\mathscr{T}$ is the type set of exercise 16.2.

Exercise 19.3. Let $\mathscr{S}$ be an IOS, $\mathscr{F}^{\cup}$ a set and ${}^{\cup}: \mathscr{F} \to \mathscr{F}^{\cup}$ a bijection. Take

$\varphi^{\cup} \leq^{\cup} \psi^{\cup}$ iff $\varphi \leq \psi$, $\varphi^{\cup} \circ^{\cup} \psi^{\cup} = (\varphi\psi)^{\cup}$ etc. Show that $\mathscr{S}^{\cup} = (\mathscr{F}^{\cup}, I^{\cup}, \Pi^{\cup}, L^{\cup}, R^{\cup})$ is an IOS isomorphic with $\mathscr{S}$.

Remarks. It also follows that t-operations over $\mathscr{F}$ are transformed by $^{\cup}$ into t-operations over $\mathscr{F}^{\cup}$ etc. Despite its seeming triviality this construction has interesting applications, for if $\mathscr{F}^{\cup} = \mathscr{F}$, then the operation $^{\cup}$ need not necessarily be expressible by the initial IOS-operations of $\mathscr{S}$.

Exercise 19.4. Let $\mathscr{S}$ be the IOS of example 4.8, $\mathscr{F}^{\cup} = \mathscr{F}$ $\varphi^{\cup} = \varphi^{-1} = \{(t,s)/(s,t) \in \varphi\}$ for all φ, and $\mathscr{S}^{\cup}$ be obtained from $\mathscr{S}$ by exercise 19.3. Prove that $\varphi \leq^{\cup} \psi$ iff $\varphi \leq \psi$, $\varphi \circ^{\cup} \psi = \psi\varphi$, $\Pi^{\cup}(\varphi, \psi) = \varphi L \cup \psi R$, $\langle\ \ \rangle^{\cup} = \langle\ \ \rangle$ and $[\ \ ]^{\cup}(\varphi) = \bigcup_n L(\varphi R)^n$.

Hint. Use exercise 5.2 and the equalities $L^{\cup} = L_1$, $R^{\cup} = R_1$.

Another natural bijection in the above example is $\varphi^{\cup} = \neg\varphi = M^2 \backslash \varphi$.

The following exercise shows that the axiom μA_2 does not imply μA_3.

Exercise 19.5. Let $\mathscr{S}$ be the CPS of exercise 16.7 and let the OS $\mathscr{S}'$ be obtained from it by 19.9. Prove that $\mathscr{S}'$ satisfies $(*)_0$ but not μA_3.

Hint. Take $\rho'(\theta)(3s)$, $\rho'(\theta)(3s+2) = \theta(3s)$ and $\rho'(\theta)(3s+1) = \theta(3s+1)$. Then $\rho' \in \mathscr{F}'$ and whenever $\operatorname{Dom}\psi \subset \omega$, then $L'(\rho'((\varphi, \psi))) = \varphi$.

Consider the mapping $\lambda\theta'.\rho'(I', \theta')$ and the regular segment $\mathscr{E}' = \{\theta'/L'\theta' \leq I'\}$. Suppose that $\theta' \in \mathscr{E}'$ and $\theta \in \mathscr{F}$.

If $\operatorname{Dom}\theta'(\theta) \subset \omega$, then $L'(\rho'((\theta, \theta'(\theta)))) = \theta$.

If $\operatorname{Dom}\theta'(\theta) = \omega$, then $L'(\theta'(\theta)) = \omega^2$ and $L'\theta' \leq I'$ implies $L'(\theta'(\theta)) \leq \theta$; hence $\theta = \omega^2$. Therefore, $L'(\rho'((\theta, \theta'(\theta)))) \leq \theta$. We get $L'\rho'(I', \theta')(\theta) \leq \theta$ for all θ, hence $\mathscr{E}'$ is closed under $\lambda\theta'.\rho'(I', \theta')$.

Take $\sigma' = \mu\theta'.\rho'(I', \theta')$. It follows easily that $\operatorname{Dom}\sigma'(I) = \omega$, where $I = \lambda s.s$, hence $L'(\sigma'(I)) = \omega^2 \not\leq I = I'(I)$, Therefore, $\sigma' \notin \mathscr{E}'$.

Exercise 19.6. Let $\mathscr{S}'$ be obtained from $\mathscr{S}$ by 19.12. Show that there are $\sigma \in \mathscr{F}$ such that $A_\sigma \notin \mathscr{F}'$.

Hint. Take $\varphi_0 = O$, $\varphi_{n+1} = (I, \varphi_n)$ and $\sigma = [I] = \sup_n \varphi_n$. Show that $\varphi_n < \varphi_{n+1}$; hence $\varphi_n < \sigma$ and $A_\sigma(\varphi_n) = O$ for all n, while $A_\sigma(\sigma) = L$.

Exercise 19.7. Let $\mathscr{S}''$, Tf be the same as in 19.14. Show that they meet the stronger tμ-axiom of exercise 10.9.

Hint. Use exercise 18.7.

Exercise 19.8. Let $\mathscr{S}$, $\mathscr{T}$ satisfy $(***)$, let $\mathscr{S}'$, $\mathscr{T}'$ be obtained from $\mathscr{S}$, $\mathscr{T}$ by 19.12 and let $\mathscr{S}''$, $\mathscr{T}''$ be obtained from $\mathscr{S}'$, $\mathscr{T}'$ again by 19.12. Prove that $\mathscr{S}''$ admits the operation Tf and $\mathscr{S}''$, Tf satisfy tfμA_3 and the stronger axiom of exercise 10.9. Assuming that $\mathscr{S}'''$ is obtained from $\mathscr{S}''$ by 19.12, show that $Tf \in \mathscr{F}'''$.

Hint. Following the proof of 19.14, show in addition that for all φ'', θ' the mapping $Tf(\varphi'')(\theta')$ over $\mathscr{F}$ is $\mathscr{T}$-continuous, while the mapping $Tf(\varphi'')$ over $\mathscr{F}'$ is $\mathscr{T}'$-continuous. (Do not forget that θ' is $\mathscr{T}$-continuous and φ'' is $\mathscr{T}'$-continuous.) Finally, the mapping Tf itself is $\mathscr{T}''$-continuous and preserves types. Use 18.22 and exercise 18.7.

CHAPTER 20

Constructing hierarchies of operative spaces

Given an OS $\mathscr{S}$ and a type set $\mathscr{T}$ satisfying $(*)$ or $(**)$, we construct in this chapter pairs $\mathscr{S}_\xi$, $\mathscr{T}_\xi$ satisfying the same sufficient condition for all ξ, so that $\{\mathscr{S}_\xi\}$ is a monotonic hierarchy based on $\mathscr{S}$ and $\{\mathscr{S}_\xi\}$ admits transfers. If $\xi = \xi_1 + 1$, then the semigroup $\mathscr{F}_\xi$ consists of all the monotonic and normal mappings over $\mathscr{F}_{\xi_1}$, while if ξ is a limit, then all the preceding spaces $\mathscr{S}_\eta$, $\eta < \xi$ are brought together to form $\mathscr{S}_\xi$. A continuous version of this construction is considered in the exercises.

Proposition 20.1. Let $\mathscr{S}$ be $(*)$-complete ($(**)$-complete). Then a monotonic hierarchy $\{\mathscr{S}_\xi\}$ based on $\mathscr{S}$ can be constructed such that $\mathscr{S}_\xi$ is $(*)$-complete (respectively, $(**)$-complete) for all ξ.

Proof. The proof is by transfinite induction on ξ and closely follows that of 17.2. We first construct a hierarchy of type sets $\{\mathscr{T}_\xi\}$ based on $\mathscr{T}$ by 17.1.

Take $\mathscr{S}_0 = \mathscr{S}$ and $t_0 = t$.

Suppose that $\xi > 0$ and for all $\eta < \xi$ an OS $\mathscr{S}_\eta$ and a type function $t_\eta : \mathscr{F}_\eta \to \mathscr{T}_\eta$ have been constructed such that $\mathscr{S}_\eta$, $\mathscr{T}_\eta$ satisfy $(*)$ (respectively, $(**)$). If $\eta = \eta_1 + 1 < \xi$, let $\mathscr{S}_\eta$ be obtained from $\mathscr{S}_{\eta_1}$ by 19.11 and identify the subspace $\overline{\mathscr{S}_{\eta_1}}$ of $\mathscr{S}_\eta$ with $\mathscr{S}_{\eta_1}$. If $\eta < \xi$ is a limit ordinal, then let $\mathscr{F}_\eta = \bigcup_{\zeta < \eta} \mathscr{F}_\zeta$ and $\mathscr{S}_\zeta$ be a subspace of $\mathscr{S}_\eta$, whenever $\zeta < \eta$. If $\varphi \in \bigcup_{\eta < \xi} \mathscr{F}_\eta$, then the ordinal $r(\varphi) = \min\{\eta / \varphi \in \mathscr{F}_\eta\}$ is the rank of φ as usual. If $\eta < \xi$ and $\zeta = r(\varphi) < \eta$, then let $t_\zeta(\varphi)$ be represented in $\mathscr{T}_\eta$ by $t_\eta(\varphi)$. This is the inductive hypothesis.

A familiar notation. If $\eta + 1 < \xi$, $r(\varphi) \leq \eta + 1$, $\psi \in \mathscr{F}_\eta$ and φ_1 is the member of $\mathscr{F}_{\eta+1}$ identified with φ, then $\varphi(\psi)_\eta$ will stand for $\varphi_1(\psi)$ as in chapter 15. Observe that $\varphi(\psi)_\eta = \varphi(\psi)$, if $r(\varphi) = \eta + 1$, while $\varphi(\psi)_\eta = \varphi\psi$, if $r(\varphi) < \eta + 1$. Moreover, $\chi \in \mathscr{F}_{\eta+1}$, $\varphi \leqslant \chi$ imply $\varphi(\psi)_\eta \leqslant \chi(\psi)_\eta$.

1. $\xi = \xi_1 + 1$.

Obtain $\mathscr{S}_\xi, t_\xi$ from $\mathscr{S}_{\xi_1}, t_{\xi_1}$ by 19.11 (i.e., by the construction of 19.9) and identify $\overline{\mathscr{S}_{\xi_1}}$ with $\mathscr{S}_{\xi_1}$. This can be done since the IOS $\mathscr{S}_{\xi_1}$, $\overline{\mathscr{S}_{\xi_1}}$ are isomorphic by 12.24.

If $\varphi \in \mathscr{F}_\xi$ and $\zeta = r(\varphi) < \xi$, then the argument adduced at the corresponding point of the proof of 17.2 implies that $t_\xi(\varphi)$ represents $t_\zeta(\varphi)$ in $\mathscr{T}_\xi$. Therefore, the induction clause holds for ξ.

2. ξ is a limit ordinal.

Take $\mathscr{F}_\xi = \bigcup_{\eta<\xi}\mathscr{F}_\eta$ and $t_\xi(\varphi) = (\zeta, t_\zeta(\varphi))$ for $\varphi \in \mathscr{F}_\xi$, where $\zeta = r(\varphi)$. The proof of 17.1 implies that $t_\xi(\varphi) \in \mathscr{T}_\xi$ and $t_\zeta(\varphi)$ is represented in $\mathscr{T}_\xi$ by $(\zeta, t_\zeta(\varphi))$, i.e. by $t_\xi(\varphi)$.

If $\varphi, \psi \in \mathscr{F}_\xi$, then take $\varphi \leq_\xi \psi$ iff $\varphi \leq_\zeta \psi$, $\varphi \circ_\xi \psi = \varphi \circ_\zeta \psi$ and $(\varphi, \psi)_\xi = (\varphi, \psi)_\zeta$, where $\zeta = \max\{r(\varphi), r(\psi)\}$. It follows from the induction hypothesis that if $\zeta \leq \eta < \xi$, then $\varphi \leq_\xi \psi$ iff $\varphi \leq_\eta \psi$, $\varphi \circ_\xi \psi = \varphi \circ_\eta \psi$ and $(\varphi, \psi)_\xi = (\varphi, \psi)_\eta$, so we may write simply $\varphi \leq \psi, \varphi\psi, (\varphi, \psi)$. The 5-tuple $\mathscr{S}_\xi = (\mathscr{F}_\xi, I, \Pi_\xi, L, R)$ is an OS since $\mathscr{S}_\eta$ is for all $\eta < \xi$.

Our next aim is to verify that $\mathscr{S}_\xi$, $\mathscr{T}_\xi$ satisfy condition $(*)$ (condition $(**)$).

Suppose that $t_\xi(\varphi) = (\eta, t_\eta(\varphi))$, $t_\xi(\psi) = (\zeta, t_\zeta(\psi))$, $(\xi_1, a) \in \mathscr{T}_\xi$ and $t_\xi(\varphi), t_\xi(\psi) \leq (\xi_1, a)$. Then $\eta, \zeta \leq \xi_1$, and $t_{\xi_1}(\varphi)$, $t_{\xi_1}(\psi)$ represent $t_\eta(\varphi)$, $t_\zeta(\psi)$ in $\mathscr{F}_{\xi_1}$ and $t_{\xi_1}(\varphi)$, $t_{\xi_1}(\psi) \leq a$. It follows that $r(\varphi), r(\psi) \leq \xi_1$, hence $\varphi, \psi \in \mathscr{F}_{\xi_1}$, which implies $\varphi\psi \in \mathscr{F}_{\xi_1}$ and $t_{\xi_1}(\varphi\psi) \leq a$ since $\circ_{\xi_1}$ preserves types. If $\xi_2 = r(\varphi\psi)$, then $\xi_2 \leq \xi_1$, $t_\xi(\varphi\psi) = (\xi_2, t_{\xi_2}(\varphi\psi))$ and $t_{\xi_1}(\varphi\psi)$ represents $t_{\xi_2}(\varphi\psi)$ in $\mathscr{T}_{\xi_1}$; hence $t_\xi(\varphi\psi) \leq (\xi_1, a)$. Therefore, the operation $\circ_\xi$ preserves types. Similarly, Π_ξ preserves types.

Suppose that $a = (\eta, b) \in \mathscr{T}_\xi$ and $\mathscr{H}$ is a well ordered subset of $\mathscr{F}_{\xi,a}$. Then $t_\xi(\theta) \leq a$; hence $\theta \in \mathscr{F}_\eta$ and $t_\eta(\theta) \leq b$ for all $\theta \in \mathscr{H}$. Therefore, $\mathscr{H}$ is a well ordered subset of $\mathscr{F}_{\eta,b}$. It follows from the inductive hypothesis that there is a $\varphi \in \mathscr{F}_{\eta,b}$ such that $\varphi\psi = \sup(\mathscr{H}\psi)$ in $\mathscr{F}_\eta$ for all $\psi \in \mathscr{F}_\eta$. We get $t_\xi(\varphi) \leq a$ from condition (ξ) of 17.1; hence $\varphi \in \mathscr{F}_{\xi,a}$. To complete the proof we shall prove by transfinite induction that $\varphi\psi = \sup(\mathscr{H}\psi)$ in $\mathscr{F}_\zeta$ for all $\psi \in \mathscr{F}_\zeta$, $\eta < \zeta \leq \xi$. In the case of condition $(**)$ the equalities $L\varphi = \sup L\mathscr{H}$, $R\varphi = \sup R\mathscr{H}$ will also be established in $\mathscr{F}_\zeta$.

a. Let $\zeta = \zeta_1 + 1$, $\eta < \zeta < \xi$ and suppose the assertion holds for ζ_1.

It is immediate that $\theta\psi \leq_\zeta \varphi\psi$ for all $\theta \in \mathscr{H}, \psi \in \mathscr{F}_\zeta$. On the other hand, suppose that $\psi, \tau \in \mathscr{F}_\zeta$ and $\theta\psi \leq_\zeta \tau$ for all $\theta \in \mathscr{H}$. If $\sigma \in \mathscr{F}_{\zeta_1}$, then $(\varphi\psi)(\sigma)_{\zeta_1} \leq_{\zeta_1} \tau(\sigma)_{\zeta_1}$, hence $\theta \circ_{\zeta_1} \psi(\sigma)_{\zeta_1} \leq_{\zeta_1} \tau(\sigma)_{\zeta_1}$. The inductive hypothesis implies that $\varphi \circ_{\zeta_1} \psi(\sigma)_{\zeta_1} = \sup(\mathscr{H} \circ_{\zeta_1} \psi(\sigma)_{\zeta_1})$ in $\mathscr{F}_{\zeta_1}$, hence $\varphi \circ_{\zeta_1} \psi(\sigma)_{\zeta_1} \leq_{\zeta_1} \tau(\sigma)_{\zeta_1}$, i.e. $(\varphi\psi)(\sigma)_{\zeta_1} \leq_{\zeta_1} \tau(\sigma)_{\zeta_1}$. This holds for all $\sigma \in \mathscr{F}_{\zeta_1}$, hence $\varphi\psi \leq_\zeta \tau$. Therefore, $\varphi\psi = \sup(\mathscr{H}\psi)$ in $\mathscr{F}_\zeta$.

Suppose that $\mathscr{S}_{\zeta_1}$, $\mathscr{T}_{\zeta_1}$ satisfy $(**)$. Let $\tau \in \mathscr{T}_\zeta$ and $L\theta \leq_\zeta \tau$ for all $\theta \in \mathscr{H}$. If $\sigma \in \mathscr{F}_{\zeta_1}$, then $(L\theta)(\sigma)_{\zeta_1} \leq_{\zeta_1} \tau(\sigma)_{\zeta_1}$, hence $L\theta\sigma \leq_{\zeta_1} \tau(\sigma)_{\zeta_1}$ for all $\theta \in \mathscr{H}$. The inductive hypothesis implies $L\varphi\sigma = \sup(L\mathscr{H}\sigma)$ in $\mathscr{F}_{\zeta_1}$; hence $L\varphi\sigma \leq_{\zeta_1} \tau(\sigma)_{\zeta_1}$, i.e. $(L\varphi)(\sigma)_{\zeta_1} \leq_{\zeta_1} \tau(\sigma)_{\zeta_1}$. This holds for all $\sigma \in \mathscr{F}_{\zeta_1}$; hence $L\varphi \leq_\zeta \tau$. Therefore, $L\varphi = \sup L\mathscr{H}$ in $\mathscr{F}_\zeta$ and similarly $R\varphi = \sup R\mathscr{H}$ in $\mathscr{F}_\zeta$.

b. Let ζ be a limit ordinal, $\eta < \zeta \leq \xi$ and suppose the assertion holds for all $\eta_1 < \zeta$.

It is immediate that $\theta\psi \leq_\zeta \varphi\psi$ for all $\theta \in \mathscr{H}$, $\psi \in \mathscr{F}_\zeta$. Suppose that $\psi, \tau \in \mathscr{F}_\zeta$ and $\theta\psi \leq_\zeta \tau$ for all $\theta \in \mathscr{H}$. Taking $\eta_1 = \max\{r(\psi), r(\tau), \eta\}$, we get $\eta_1 < \zeta, \psi$, $\tau \in \mathscr{F}_{\eta_1}$, $\mathscr{H} \subseteq \mathscr{F}_{\eta_1}$ and $\theta\psi \leq_{\eta_1} \tau$ for all $\theta \in \mathscr{H}$. The inductive hypothesis gives $\varphi\psi = \sup(\mathscr{H}\psi)$ in $\mathscr{F}_{\eta_1}$, hence $\varphi\psi \leq_{\eta_1} \tau$. It follows that $\varphi\psi \leq_\zeta \tau$; hence $\varphi\psi = \sup(\mathscr{H}\psi)$ in $\mathscr{F}_\zeta$.

Suppose that $\mathscr{S}_{\eta_1}$, $\mathscr{T}_{\eta_1}$ satisfy $(**)$ for all $\eta_1 < \zeta$. Let $\tau \in \mathscr{F}_\zeta$ and $L\theta \leq_\zeta \tau$ for all $\theta \in \mathscr{H}$. Taking $\eta_1 = \max\{r(\tau), \eta\}$, we get $\eta_1 < \zeta, L, \tau \in \mathscr{F}_{\eta_1}$, $\mathscr{H} \subseteq \mathscr{F}_{\eta_1}$ and

$L\theta \leq_{\eta_1} \tau$ for all $\theta \in \mathscr{H}$. The inductive hypothesis gives $L\varphi = \sup L\mathscr{H}$ in $\mathscr{F}_{\eta_1}$; hence $L\varphi \leq_{\eta_1} \tau$ which implies $L\varphi \leq_\zeta \tau$. Therefore, $L\varphi = \sup L\mathscr{H}$ in $\mathscr{F}_\zeta$ and similarly $R\varphi = \sup R\mathscr{H}$ in $\mathscr{F}_\zeta$. We conclude that $\mathscr{S}_\xi$, $\mathscr{T}_\xi$ satisfy $(*)$ (respectively, $(**)$), while it is immediate that $\mathscr{S}_\eta$ is a subspace of $\mathscr{S}_\xi$, provided $\eta < \xi$. The proof is complete.

Proposition 20.2. Let $\mathscr{S}$ be $(*)$-complete ($(**)$-complete) and let $\{\mathscr{S}_\xi\}$ be the monotonic hierarchy based on $\mathscr{S}$ constructed in 20.1. Then $\{\mathscr{S}_\xi\}$ admits transfers, i.e. for all ξ the element $Tf_{\xi+2} \in \mathscr{F}_{\xi+3}$ is a transfer operation over $\mathscr{F}_{\xi+2}$ satisfying the axiom tfμA$_2$ (respectively, tfμA$_3$). The stronger axiom of exercise 10.9 is also valid.

This follows from 19.14 and exercise 19.7.

Proposition 20.3. Let $\{\mathscr{S}_\xi\}$ be the hierarchy constructed by 20.1. Then for all ξ there are members $Q_{\sigma,\xi}$ (and $Q_{\sigma,\mathscr{C},\xi}$, provided L, R have an upper bound U) of $\mathscr{F}_{\xi+1}$ which are the mappings Q_σ (and $Q_{\sigma,\mathscr{C}}$) corresponding to $\mathscr{S}_\xi$.

This follows from 19.13.

Suppose that $\mathscr{S}$ is $(*)_0$-complete or $(**)_0$-complete and a hierarchy $\{\mathscr{S}_\xi\}$ based on $\mathscr{S}$ is constructed by 20.1. Then all $\mathscr{S}_n, n < \omega$ satisfy the same sufficient condition and types arise beyond ω, exactly as they did in hierarchies of pairing spaces. (Cf. exercises 17.1–17.3.)

EXERCISES TO CHAPTER 20

Exercise 20.1. Let $\mathscr{S}$ be a $(***)$-complete OS. Construct a monotonic hierarchy $\{\mathscr{S}_\xi\}$ based on $\mathscr{S}$.

Hint. Follow the proof of 20.1, substituting $(***)$ for $(*)$ and 19.12 for 19.11 in the induction clause.

Remarks. Hierarchies constructed this way are more appropriately called *continuous* since for all ξ the semigroup $\mathscr{F}_{\xi+1}$ consists of $\mathscr{T}_\xi$-continuous mappings over $\mathscr{F}_\xi$. Exercise 19.8 ensures that all continuous hierarchies admit transfers, while exercise 19.6 shows that proposition 20.3 does not hold for continuous hierarchies.

PART E

First order examples and applications

CHAPTER 21

Fuzzy relations

In this chapter three general families of examples (i.e., models) of IOS are introduced, from which a number of more particular ones are obtained in the next three chapters. The initial IOS-operations are explicitly characterized. A $\langle\ \ \rangle$-elimination theorem is used to show that translation in these examples can be implemented by means of an extra counter. Certain specific t-operations are also studied.

Assume that M is a nonempty set, f_1, f_2 is a splitting scheme for it, E is a complete lattice with at least two distinct members and the distributive law

$$\inf\{a, \sup E_1\} = \sup\{\inf\{a,b\}/b\in E_1\}$$

holds for all $a\in E$, $E_1 \subseteq E$. (A *complete lattice* is a partially ordered set all of whose subsets have least upper bounds; the existence of greatest lower bounds follows, for $\inf E_1 = \sup\{a\backslash\forall b\in E_1(a\le b)\}$.) Writing $\bot$, $\top$ respectively for $\sup\varnothing$, $\sup E$, we have $\bot = \inf E$, $\top = \inf\varnothing$ and $\bot < \top$.

The pairing space of 16.4 is the starting point for the following example which corresponds to example 17 in Skordev [1980], chapter 3. If the reader is not interested in fuzzy relations, then she or he may assume $E = \{\bot, \top\}$ and interpret $\varphi(s,t) = \top$, $\varphi(s,t) = \bot$ respectively as $(s,t)\in\varphi$, $(s,t)\notin\varphi$ (or vice versa), so that the semigroup $\mathscr{F}$ below will then consist of ordinary relations.

Proposition 21.1 (Example 21.1). Take $\mathscr{F} = \{\varphi/\varphi: M^2 \to E\}$ (all the binary E-valued *fuzzy relations* over M), $\varphi \le \psi$ iff $\forall st(\varphi(s,t) \le \psi(s,t))$, $I(s,s) = \top$ and $I(s,t) = \bot$ otherwise, $\varphi\psi(s,t) = \sup_r \inf\{\varphi(s,r), \psi(r,t)\}$, $(\varphi,\psi)(f_1(s),t) = \varphi(s,t)$, $(\varphi,\psi)(f_2(s),t) = \psi(s,t)$ and $(\varphi,\psi)(s,t) = \bot$ otherwise, $L = \lambda st.I(f_1(s),t)$ and $R = \lambda st.I(f_2(s),t)$. Then $\mathscr{S} = (\mathscr{F}, I, \Pi, L, R)$ is a $(**)_0$, $(***)_0$-complete OS.

Proof. We have

$$\begin{aligned}
\varphi(\psi\chi)(s,t) &= \sup_r \inf\{\varphi(s,r), \psi\chi(r,t)\}\\
&= \sup_r \inf\{\varphi(s,r), \sup_u \inf\{\psi(r,u), \chi(u,t)\}\}\\
&= \sup_r \sup_u \inf\{\varphi(s,r), \psi(r,u), \chi(u,t)\}\\
&= \sup_u \sup_r \inf\{\varphi(s,r), \psi(r,u), \chi(u,t)\}
\end{aligned}$$

$$= \sup_u \inf\{\sup_r \inf\{\varphi(s,r), \psi(r,u)\}, \chi(u,t)\}$$

$$= \sup_u \inf\{\varphi\psi(s,u), \chi(u,t)\} = (\varphi\psi)\chi(s,t)$$

for all s, t, hence $\varphi(\psi\chi) = (\varphi\psi)\chi$. We also get $\varphi I = \lambda st.\sup_r \inf\{\varphi(s,r), I(r,t)\} = \varphi$ and similarly $I\varphi = \varphi$. Therefore, $\mathscr{F}$ is a semigroup with unit I.

All the subsets of $\mathscr{F}$ have least upper bounds, namely

$$\sup \mathscr{H} = \lambda st.\sup\{\theta(s,t)/\theta \in \mathscr{H}\}.$$

Whenever $\mathscr{H} \subseteq \mathscr{F}$, $\varphi = \sup \mathscr{H}$ and $\psi \in \mathscr{F}$, then

$$\varphi\psi(s,t) = \sup_r \inf\{\varphi(s,r), \psi(r,t)\}$$

$$= \sup_r \inf\{\sup\{\theta(s,r)/\theta \in \mathscr{H}\}, \psi(r,t)\}$$

$$= \sup_r \sup\{\inf\{\theta(s,r), \psi(r,t)\}/\theta \in \mathscr{H}\}$$

$$= \sup\{\sup_r \inf\{\theta(s,r), \psi(r,t)\}/\theta \in \mathscr{H}\}$$

$$= \sup\{\theta\psi(s,t)/\theta \in \mathscr{H}\}$$

for all s, t; hence $\varphi\psi = \sup(\mathscr{H}\psi)$, while the equalities

$$\psi\varphi(s,t) = \sup_r \inf\{\psi(s,r), \varphi(r,t)\}$$

$$= \sup_r \inf\{\psi(s,r), \sup\{\theta(r,t)/\theta \in \mathscr{H}\}\}$$

$$= \sup_r \sup\{\inf\{\psi(s,r), \theta(r,t)\}/\theta \in \mathscr{H}\}$$

$$= \sup\{\psi\theta(s,t)/\theta \in \mathscr{H}\}$$

imply $\psi\varphi = \sup \psi\mathscr{H}$. In particular, multiplication is monotonic.

Take $L_1 = \lambda st.I(f_1(t), s)$ and $R_1 = \lambda st.I(f_2(t), s)$. It follows that $LL_1 = RR_1 = I$ and $LR_1 = RL_1 = O = \lambda st.\bot$ using the fact that f_1, f_2 is a splitting scheme for M. We get $(\varphi, \psi) = \sup\{L_1\varphi, R_1\psi\}$; hence $\mathscr{S}$ is a $(**)_0$, $(***)_0$-complete OS by 19.4. The proof is complete.

The above IOS can be obtained in another way. Namely, take the pairing space of 16.4 with $N = M$ and augment it with multiplication, then use 16.10, 19.1–19.3.

In order to further sharpen the characterizations given in 19.6, consider the sets $f_2^n(f_1(M))$, $n \geq 0$. It follows that whenever $i \neq j$, then $f_2^i(f_1(M)) \cap f_2^j(f_1(M)) = \varnothing$. Therefore, for every $s \in M$ there is at most one n (if any) such that $s \in f_2^n(f_1(M))$.

Proposition 21.2. Let $\mathscr{S}$ be the IOS of example 21.1 or a subspace of it. Then

$$\langle\varphi\rangle(f_2^n(f_1(s)), f_2^n(f_1(t))) = \varphi(s,t)$$

and $\langle\varphi\rangle(s,t)=\perp$ otherwise,

$$\Delta(\varphi,\psi)(f_2^n(f_1(s)),t)=\varphi\psi^n(s,t)$$

and $\Delta(\varphi,\psi)(s,t)=\perp$ otherwise,

$$[\varphi](s,t)=\sup\{\inf_{i<n}\varphi(f_2^{-1}(r_i),r_{i+1})/n\in\omega\,\&\,r_0=s$$
$$\&\,r_0,\ldots,r_{n-1}\in f_2(M)\,\&\,r_n=f_1(t)\}.$$

Proof. We have $R_1\varphi(f_2(s),t)=\varphi(s,t)$ and $R_1\varphi(s,t)=\perp$ otherwise, while $L_1\varphi(f_1(s),t)=\varphi(s,t)$ and $L_1\varphi(s,t)=\perp$ otherwise. Proposition 19.6 gives $\Delta(\varphi,\psi)=\sup_n R_1^n L_1\varphi\psi^n$, hence $\Delta(\varphi,\psi)(f_2^n(f_1(s)),t)=\varphi\psi^n(s,t)$ and $\Delta(\varphi,\psi)(s,t)=\perp$ otherwise. In particular, $\langle\varphi\rangle(f_2^n(f_1(s)),t)=\varphi LR^n(s,t)$ and $\langle\varphi\rangle(s,t)=\perp$ otherwise. It also follows that $\varphi LR^n(s,f_2^n(f_1(t)))=\varphi(s,t)$ and $\varphi LR^n(s,t)=\perp$ otherwise, hence $\langle\varphi\rangle(f_2^n(f_1(s)),f_2^n(f_1(t)))=\varphi(s,t)$ and $\langle\varphi\rangle(s,t)=\perp$ otherwise.

Proposition 19.6 gives $[\varphi]=\sup_n(R_1\varphi)^n L_1$, hence $[\varphi](s,t)=\sup_n(R_1\varphi)^n L_1(s,t)$ for all s,t. As already mentioned, $R_1\varphi(f_2(s),t)=\varphi(s,t)$ and $R_1\varphi(s,t)=\perp$ otherwise. Using this and the equality $\varphi L_1(s,t)=\varphi(s,f_1(t))$, we get the desired characterization of [ ]. The proof is complete.

The following example corresponds to the pairing space of 16.4 again, this time with $N\times M$ playing the role of N.

Proposition 21.3 (Example 21.2). Let $N\neq\varnothing$. Take $\mathscr{F}=\{\varphi/\varphi:M\times N\times M\to E\}$, $\varphi\leq\psi$ iff $\varphi(s,x,t)\leq\psi(s,x,t)$ for all $s,t\in M$ and $x\in N$, $\varphi\psi(s,x,t)=\sup_r\inf\{\varphi(s,x,r),\psi(r,x,t)\}$, $(\varphi,\psi)(f_1(s),x,t)=\varphi(s,x,t)$, $(\varphi,\psi)(f_2(s),x,t)=\psi(s,x,t)$ and $(\varphi,\psi)(s,x,t)=\perp$ otherwise, $I(s,x,s)=\top$ and $I(s,x,t)=\perp$ otherwise, $L=\lambda sxt.I(f_1(s),x,t)$ and $R=\lambda sxt.I(f_2(s),x,t)$. Then $\mathscr{S}=(\mathscr{F},I,\Pi,L,R)$ is a $(**)_0$, $(***)_0$-complete OS.

The proof repeats that of 21.1 with members of N treated as parameters.

Proposition 21.4. Let $\mathscr{S}$ be the IOS of example 21.2 or a subspace of it. Then

$$\langle\varphi\rangle(f_2^n(f_1(s)),x,f_2^n(f_1(t)))=\varphi(s,x,t),\Delta(\varphi,\psi)(f_2^n(f_1(s)),x,t)=\varphi\psi^n(s,x,t)$$

and $\langle\varphi\rangle(s,x,t)$, $\Delta(\varphi,\psi)(s,x,t)=\perp$ otherwise,

$$[\varphi](s,x,t)=\sup\{\inf_{i<n}\varphi(f_2^{-1}(r_i),x,r_{i+1})/n\in\omega\,\&\,r_0=s$$
$$\&\,r_0,\ldots,r_{n-1}\in f_2(M)\,\&\,r_n=f_1(t)\}.$$

The proof repeats that of 21.2.

The following two statements establish a close two-way connection between examples 21.1, 21.2.

Proposition 21.5. Let $\mathscr{S}_1,\mathscr{S}_2$ be the IOS respectively of examples 21.1, 21.2 based on the same sets M, splitting schemes and lattices E. Then $\mathscr{S}_1$ is isomorphic with the subspace $\tilde{\mathscr{S}}_1$ of $\mathscr{S}_2$ based on $\tilde{\mathscr{F}}_1=\{\tilde{\varphi}/\varphi\in\mathscr{F}_1\}$, where

$\tilde{\varphi} = \lambda sxt.\varphi(s,t)$. The spaces $\mathscr{S}_1$ and $\mathscr{S}_2$ are isomorphic whenever N is a singleton.

The proof is immediate.

Proposition 21.6. Let $\mathscr{S}_2$ be the IOS of example 21.2 and let $\mathscr{S}_1$ be obtained from example 21.1 by taking $M \times N$ for M. (Observe that whenever f_1, f_2 is a splitting scheme for M, then $\lambda sx.(f_1(s),x)$, $\lambda sx.(f_2(s),x)$ is a splitting scheme for $M \times N$.) Then $\mathscr{S}_2$ is isomorphic with the subspace $\tilde{\mathscr{S}}_2$ of $\mathscr{S}_1$ based on $\tilde{\mathscr{F}}_2 = \{\tilde{\varphi}/\varphi \in \mathscr{F}_2\}$, where $\tilde{\varphi}(s,x,t,x) = \varphi(s,x,t)$ and $\tilde{\varphi}(s,x,t,y) = \perp$ otherwise.

This isomorphism is quite immediate, too. For instance,

$$\begin{aligned}\tilde{\varphi}\tilde{\psi}(s,x,t,x) &= \sup_{r,z} \inf\{\tilde{\varphi}(s,x,r,z), \tilde{\psi}(r,z,t,x)\}\\ &= \sup_{r} \inf\{\tilde{\varphi}(s,x,r,x), \tilde{\psi}(r,x,t,x)\}\\ &= \sup_{r} \inf\{\varphi(s,x,r), \psi(r,x,t)\} = \varphi\psi(s,x,t)\end{aligned}$$

and $\tilde{\varphi}\tilde{\psi}(s,x,t,y) = \perp$ whenever $y \neq x$, hence $\tilde{\varphi}\tilde{\psi} = (\varphi\psi)^{\sim}$.

Following Skordev [1980], binary fuzzy relations can be intuitively interpreted as semantical counterparts of programs processed by a computer (or a man). Members of M are data and $\varphi(s,t) = d$ means that, given an input s, a part t is produced among the outputs; the extent of accuracy is evaluated by d. If $\varphi(s,t) = \top$, then t is fully produced, while $\varphi(s,t) = \perp$ would mean that nothing of t appears in the outputs at all. The initial IOS-operations correspond to certain natural program constructs, while $\varphi \leq \psi$ means that 'ψ is better than φ'.

The above interpretation is not the only possible one; in fact, there are as many as four pairwise dual interpretations. (Compare with the dual spaces in exercise 19.4.) First, inputs and outputs may be interchanged, with $\varphi(s,t) = d$ meaning that t is processed by φ and a part of s appears among the results. Secondly, the partial ordering of E could be viewed in another way. Namely, whenever $\varphi(s,t) = d$, $\varphi_1(s,t) = d_1$ and $d \leq d_1$, then the former computation may be regarded as providing more information than the latter, so that $\varphi(s,t) = \perp$ would mean that the result, respectively t or s, is produced entirely. Thus the motto of the corresponding two interpretations is 'The less information the better', which seems to make some sense nowadays.

Example 21.2 involves parameters, i.e. additional sources of information which is not liable to change. Examples 21.1, 21.2 can be slightly generalized by introducing a kind of complexity measure for data processing. For instance, if s is processed into t for time p_1 and the time needed to process t into r is p_2, then it is reasonable to assume that s is processed into r for time $p_1 + p_2$. If, however, p_1 and p_2 are bites of storage needed for the corresponding computations, then it would be reasonable to assume that $p_1 + p_2$ stands for the maximum rather than the sum of p_1, p_2. Of course, one may measure both these quantities by means of pairs of time and space. This suggests that complexity of data processing could be measured by members of certain semigroups as done in Skordev [1980]. A 'complexity' version of example 21.1 follows; example 21.2 may be modified similarly.

Proposition 21.7 (Example 21.3). Let S be a semigroup with a zero 0. Take $\mathscr{F}=\{\varphi/\varphi:M\times S\times M\to E\}$, $\varphi\le\psi$ iff $\varphi(s,p,t)\le\psi(s,p,t)$ for all $s,t\in M$ and $p\in S$, $\varphi\psi(s,p,t)=\sup_{r,p_1+p_2=p}\inf\{\varphi(s,p_1,r),\ \psi(r,p_2,t)\}$, $(\varphi,\psi)(f_1(s),p,t)=\varphi(s,p,t)$, $(\varphi,\psi)(f_2(s),p,t)=\psi(s,p,t)$ and $(\varphi,\psi)(s,p,t)=\bot$ otherwise, $I(s,0,s)=\top$ and $I(s,p,t)=\bot$ otherwise, $L=\lambda spt.I(f_1(s),p,t)$ and $R=\lambda spt.I(f_2(s),p,t)$. Then $\mathscr{S}=(\mathscr{F},I,\Pi,L,R)$ is a $(**)_0$, $(***)_0$-complete OS.

Proof. The verifications of 21.1 may be repeated once again, *mutatis mutandis*. We have:

$$\varphi(\psi\chi)(s,p,t)=\sup_{r,p_1+p_2=p}\inf\{\varphi(s,p_1,r),\psi\chi(r,p_2,t)\}$$

$$=\sup_{r,p_1+p_2=p}\inf\{\varphi(s,p_1,r),\ \sup_{u,p_3+p_4=p_2}\inf\{\psi(r,p_3,u),\chi(u,p_4,t)\}\}$$

$$=\sup_{r,p_1+p_2=p}\ \sup_{u,p_3+p_4=p_2}\inf\{\varphi(s,p_1,r),\inf\{\psi(r,p_3,u),\chi(u,p_4,t)\}\}$$

$$=\sup_{r,u,p_1+p_3+p_4=p}\inf\{\varphi(s,p_1,r),\psi(r,p_3,u),\chi(u,p_4,t)\}$$

$$=\sup_{u,p_2+p_4=p}\ \sup_{r,p_1+p_3=p_2}\inf\{\inf\{\varphi(s,p_1,r),\psi(r,p_3,u)\},\chi(u,p_4,t)\}$$

$$=\sup_{u,p_2+p_4=p}\inf\{\varphi\psi(s,p_2,u),\chi(u,p_4,t)\}=(\varphi\psi)\chi(s,p,t)$$

for all s,p,t; hence $\varphi(\psi\chi)=(\varphi\psi)\chi$. Taking $L_1=\lambda spt.I(f_1(t),p,s)$ and $R_1=\lambda spt.I(f_2(t),p,s)$, we get $LL_1=RR_1=I$, $LR_1=RL_1=O$ and $(\varphi,\psi)=\sup\{L_1\varphi,R_1\psi\}$. Any subset $\mathscr{H}$ of $\mathscr{F}$ has a least upper bound $\sup\mathscr{H}=\lambda spt.\sup\{\theta(s,p,t)/\theta\in\mathscr{H}\}$ and multiplication is continuous with respect to least upper bounds, which completes the proof by 19.4.

Proposition 21.8. Let $\mathscr{S}$ be the IOS of example 21.3 or a subspace of it. Then

$$\langle\varphi\rangle(f_2^n(f_1(s)),p,f_2^n(f_1(t)))=\varphi(s,p,t),$$
$$\Delta(\varphi,\psi)(f_2^n(f_1(s)),p,t)=\varphi\psi^n(s,p,t)$$

and $\langle\varphi\rangle(s,p,t)$, $\Delta(\varphi,\psi)(s,p,t)=\bot$ otherwise,

$$[\varphi](s,p,t)=\sup\{\inf_{i<n}\varphi(f_2^{-1}(r_i),p_i,r_{i+1})/n\in\omega\,\&\,r_0=s\,\&\,p_0+\cdots+p_{n-1}=p$$
$$\&\,r_0,\ldots,r_{n-1}\in f_2(M)\,\&\,r_n=f_1(t)\}.$$

The proof follows that of 21.2.

Proposition 21.9. Let $\mathscr{S}_1,\mathscr{S}_3$ be the IOS respectively of examples 21.1, 21.3. Then $\mathscr{S}_1$ is isomorphic to the subspace $\tilde{\mathscr{S}}_1$ of $\mathscr{S}_3$ based on $\tilde{\mathscr{F}}_1=\{\tilde{\varphi}/\varphi\in\mathscr{F}_1\}$, where $\tilde{\varphi}(s,0,t)=\varphi(s,t)$ and $\tilde{\varphi}(s,p,t)=\bot$ otherwise. The spaces $\mathscr{S}_1$ and $\mathscr{S}_3$ are isomorphic whenever $S=\{0\}$.

The proof is immediate.

An interesting semigroup along with those suggested above is $S = \cup_n M^n$, where $0 = \Lambda$ and the semigroup operation is concatenation. Members of S are regarded as computation traces, partial ones in general, for I, L, R leaves no traces.

One can obtain more particular IOS from examples 21.1–21.3 by taking specific sets M augmented with splitting schemes; several options were suggested in the comments to 16.3. On the other hand, specific lattices E can be taken. For instance, any finite distributive lattice would do. The sets $\{\eta/\eta \leq \xi\}$ and the intervals $[0,1]$, $[0,\infty]$ with their ordinary total orderings can also play the role of E. Some standard first order examples will be obtained this way in the next three chapters with $E = \{\perp, T\}$. (We recall that first order examples (i.e. models) are those composed of function-like elements, while higher order examples are those composed of operator-like elements.)

As far as first order examples are concerned, while the pairing scheme Π, L, R involves a splitting scheme usually realized by a *counter*, the operation $\langle\ \ \rangle$ can be implemented by means of an additional counter. To show that one needs the following Translation Elimination Theorem, a theoretical enclave in this applied part of the book.

Proposition 21.10. Let $\mathscr{S} = (\mathscr{F}, I, \Pi, L, R)$ be an IOS and $W, W_1, W_2 \in \mathscr{F}$ such that $W_1 W = L$, $W_2 W = R$, $W(L,R) = W$, $LW = W(L^2, LR)$ and $RW = W(RL, R^2)$. Let

$$\mathscr{C} = \{\sigma \in \mathscr{F} / W_i \sigma = \sigma W_i, \quad i = 1,2\},$$

assuming that $L, R, (L,R)$, $\langle I \rangle \in \mathscr{C}$. Then φ is recursive in $\{W, W_1, W_2\} \cup \mathscr{B}$ iff φ is prime recursive in $\{W, W_1, W_2\} \cup \mathscr{B}$, provided $\mathscr{B} \subseteq \mathscr{C}$.

Proof. We have $W_i(L,R) = (W_i L, W_i R)$ and the more general equalities

$$W_i(\varphi,\psi) = (W_i\varphi, W_i\psi),$$
$$LW(\varphi,\psi) = W(L\varphi, L\psi), RW(\varphi,\psi) = W(R\varphi, R\psi)$$

hold for $i = 1,2$ and all $\varphi, \psi \in \mathscr{F}$.

The set $\mathscr{C}$ is obviously closed under $\circ, \Pi$. In order to show that it is closed under $\langle\ \ \rangle$, suppose $\sigma \in \mathscr{C}$. Then the equalities

$$\sigma L W_i = W_i \sigma L = L W_i \langle \sigma \rangle,$$
$$R W_i = W_i R, \quad W_i \langle \sigma \rangle R = R W_i \langle \sigma \rangle, \quad i = 1,2$$

imply $\langle \sigma \rangle W_i = \langle I \rangle W_i \langle \sigma \rangle$ by 6.20. Therefore,

$$W_i \langle \sigma \rangle = W_i \langle I \rangle \langle \sigma \rangle = \langle I \rangle W_i \langle \sigma \rangle = \langle \sigma \rangle W_i, i = 1,2;$$

hence $\langle \sigma \rangle \in \mathscr{C}$.

Our next aim is to prove that $\langle \sigma \rangle$ is prime recursive in W, W_1, W_2, σ for all $\sigma \in \mathscr{C}$, while $\langle W \rangle$, $\langle W_1 \rangle$, $\langle W_2 \rangle$ are prime recursive in W, W_1, W_2.

Using 6.13, we get

$$[W_2] W_2 = R[(W_2 L, W_2 R)] = R[W_2(L,R)] = R[W_2].$$

Writing W_0 for $W(L, R^2)$, it follows that

$$R^2[W_0] = RW_0[W_0] = W_0(RL, R^2)[W_0] = W_0(R, R^2[W_0]),$$

hence $R[W_0]R \leq R^2[W_0]$ by 6.11. Conversely,

$$\begin{aligned} W_0(I, W(I, R[W_0]R)) &= W(I, RW(I, R[W_0]R)) = W(I, W(R, R^2[W_0]R)) \\ &= W(I, W_0[W_0]R) \end{aligned}$$

gives $R[W_0] \leq W(I, R[W_0]R)$ by 6.11, hence

$$R^2[W_0] \leq RW(I, R[W_0]R) = W(R, R^2[W_0]R) = W_0[W_0]R = R[W_0]R.$$

Therefore, $R^2[W_0] = R[W_0]R$.

Using the above equalities, we get for $\sigma \in \mathscr{C}$

$$\begin{aligned} LW_1[W_2]\sigma LW_0[W_0] &= W_1\sigma LW_0[W_0] = \sigma LW_1W_0[W_0] \\ &= \sigma L^2[W_0] = \sigma L \end{aligned}$$

and

$$\begin{aligned} &RW_1[W_2]\sigma LW_0[W_0] \\ &\quad = W_1R[W_2]\sigma LW_0[W_0] = W_1[W_2]W_2\sigma LW_0[W_0] \\ &\quad = W_1[W_2]\sigma LW_2W_0[W_0] = W_1[W_2]\sigma LR^2[W_0] \\ &\quad = W_1[W_2]\sigma LR[W_0]R = W_1[W_2]\sigma LW_0[W_0]R, \end{aligned}$$

which implies $\langle\sigma\rangle = \langle I\rangle W_1[W_2]\sigma LR[W_0]$ by 6.20. This equality is to be simplified by deleting the multiplier $\langle I\rangle$.

On the one hand, $R[W_2] = [W_2]W_2$ and

$$(L[W_2], [W_2]W_2) = (I, R[W_2]) = [W_2]$$

imply $\langle I\rangle[W_2] \leq [W_2]$ by (£). On the other hand,

$$(I, W_2\langle I\rangle[W_2]) = (I, \langle I\rangle W_2[W_2]) = (L, \langle I\rangle R)[W_2] = \langle I\rangle[W_2]$$

implies $[W_2] \leq \langle I\rangle[W_2]$ by (££). Therefore,

$$\begin{aligned} \langle\sigma\rangle &= \langle I\rangle W_1[W_2]\sigma LR[W_0] = W_1\langle I\rangle[W_2]\sigma LR[W_0] \\ &= W_1[W_2]\sigma LR[W_0] \end{aligned}$$

for all $\sigma \in \mathscr{C}$. In particular, $\langle I\rangle$ is prime recursive in W, W_1, W_2.

The equalities $W_1L = LW_1$, $W_1R = RW_1$ give $\langle W_1\rangle = \langle I\rangle W_1$ by 6.20 and similarly $\langle W_2\rangle = \langle I\rangle W_2$.

The equalities

$$\begin{aligned} &LW(\langle L\rangle, \langle R\rangle) = W(L\langle L\rangle, L\langle R\rangle) = W(L^2, RL) = W(L, R)L = WL, \\ &RW(\langle L\rangle, \langle R\rangle) = W(R\langle L\rangle, R\langle R\rangle) = W(\langle L\rangle, \langle R\rangle)R \end{aligned}$$

imply $\langle W\rangle = \langle I\rangle W(\langle L\rangle, \langle R\rangle)$ by 6.20.

Suppose that $\mathscr{B} \subseteq \mathscr{C}$ and φ is recursive in $\{W, W_1, W_2\} \cup \mathscr{B}$. Then φ is prime recursive in $\mathscr{B}_1 = \{\langle B\rangle, \langle\langle L\rangle\rangle, \langle\langle A\rangle\rangle, \langle W\rangle, \langle W_1\rangle, \langle W_2\rangle\} \cup \langle\mathscr{B}\rangle$ by 7.11. Noting that B, $\langle L\rangle$, $\langle A\rangle \in \mathscr{C}$, we conclude that all the members of

$\mathscr{B}_1$ are prime recursive in $\{W, W_1, W_2\} \cup \mathscr{B}$, hence so is φ. This completes the proof.

The elements W, W_1, W_2 seem to express axiomatically the presence of an extra counter besides that implied by Π, L, R. Example 21.1 illustrates this, while other first order examples can be treated similarly.

Proposition 21.11. Let $\mathscr{S}_0 = (\mathscr{F}_0, I_0, \Pi_0, L_0, R_0)$ be the IOS of example 21.1 and let $\mathscr{S} = (\mathscr{F}, I, \Pi, L, R)$ be obtained from the same example by taking $M \times \omega$ for M, i.e. by adding a new counter. To each $\varphi_0 \in \mathscr{F}_0$ assign an element $\tilde{\varphi}_0 \in \mathscr{F}$ such that $\tilde{\varphi}_0(s,n,t,n) = \varphi_0(s,t)$ and $\tilde{\varphi}_0(s,m,t,n) = \bot$ otherwise. Take $W_1(s,n,s,2n)$, $W_2(s,n,s,2n+1) = \top$ and $W_1(s,m,t,n)$, $W_2(s,m,t,n) = \bot$ otherwise, $W(s,2m,t,n) = L(s,m,t,n)$ and $W(s,2m+1,t,n) = R(s,m,t,n)$. Let $\varphi \in \mathscr{F}$, $\mathscr{B}_0 \subseteq \mathscr{F}_0$. Then φ is recursive in $\{W, W_1, W_2\} \cup \tilde{\mathscr{B}_0}$ iff φ is prime recursive in $\{W, W_1, W_2\} \cup \tilde{\mathscr{B}_0}$. If $\varphi_0 \in \mathscr{F}_0$ is recursive in $\mathscr{B}_0$, then $\tilde{\varphi_0}$ is prime recursive in $\{W, W_1, W_2\} \cup \tilde{\mathscr{B}_0}$.

Proof. The elements W, W_1, W_2 satisfy the corresponding equalities of 21.10 and all $\tilde{\psi}_0$ commute with W_1, W_2. Therefore, 21.10 implies that φ is recursive in $\{W, W_1, W_2\} \cup \tilde{\mathscr{B}_0}$ iff it is prime recursive in the same set.

It follows from 21.5, 21.6 that $\tilde{\mathscr{S}}_0 = (\tilde{\mathscr{F}}_0, I, \Pi \restriction \tilde{\mathscr{F}}_0^2, L, R)$ is a subspace of $\mathscr{S}$ isomorphic to $\mathscr{S}_0$. Whenever $\varphi_0 \in \mathscr{F}_0$ is recursive in $\mathscr{B}_0$, then $\tilde{\varphi}_0$ is recursive in $\tilde{\mathscr{B}_0}$, hence $\tilde{\varphi_0}$ is prime recursive in $\{W, W_1, W_2\} \cup \tilde{\mathscr{B}_0}$. The proof is complete.

The IOS of examples 21.1–21.3 have some nice properties not shared by spaces considered in chapter 25 below. Least fixed points of inductive mappings are reached at level ω according to the proofs of 18.2, 18.14. Collection operations can be introduced by exercise 18.5, while 19.6 implies that there are elements U satisfying condition (1) of exercise 7.10. In the case of example 21.1 $Co\{\varphi_n\}(f_2^n(f_1(s)), t) = \varphi_n(s,t)$ and $Co\{\varphi_n\}(s,t) = \bot$ otherwise, $U(s, f_1(s))$, $U(s, f_2(s)) = \top$ and $U(s,t) = \bot$ otherwise. Observe that $U(\varphi, \psi) = \sup\{\varphi, \psi\}$ for all φ, ψ.

The following two statements introduce certain t-operations in example 21.1, while examples 21.2, 21.3 are treated similarly.

Proposition 21.12. Let $\mathscr{S}$ be the IOS of example 21.1 and $\mathrm{In}(\varphi) = \varphi^{-1} = \lambda st.\varphi(t,s)$ (*inversion*). Then $\langle\!\langle\ \rangle\!\rangle = \lambda\varphi.(\varphi, In(\varphi))$ is a t-operation with a corresponding set of functional elements $\mathscr{B}_0 = \{U\}$ such that $\mathscr{S}, \langle\!\langle\ \rangle\!\rangle$ satisfy the axiom $\mathrm{t}\mu\mathrm{A}_3$.

Proof. While our intention is to add In to the initial IOS-operations, we take $\langle\!\langle\ \rangle\!\rangle$ instead in order to ensure that φ is recovered from $\langle\!\langle \varphi \rangle\!\rangle$ recursively. We have $\varphi = L\langle\!\langle \varphi \rangle\!\rangle$ and $\varphi^{-1} = R\langle\!\langle \varphi \rangle\!\rangle$. Using the easy equalities $L_1^{-1} = L$, $R_1^{-1} = R$, $(\varphi\psi)^{-1} = \psi^{-1}\varphi^{-1}$, $(\varphi^{-1})^{-1} = \varphi$ and $(\sup\{\varphi, \psi\})^{-1} = \sup\{\varphi^{-1}, \psi^{-1}\}$, we get

$$\begin{aligned}\langle\!\langle \varphi\psi \rangle\!\rangle &= (\varphi\psi, (\varphi\psi)^{-1}) = (\varphi\psi, \psi^{-1}\varphi^{-1})\\ &= (L\langle\!\langle \varphi \rangle\!\rangle L\langle\!\langle \psi \rangle\!\rangle, R\langle\!\langle \psi \rangle\!\rangle R\langle\!\langle \varphi \rangle\!\rangle),\end{aligned}$$

$$\langle\!\langle(\varphi,\psi)\rangle\!\rangle = ((\varphi,\psi),(\sup\{L_1\varphi,R_1\psi\})^{-1}) = ((\varphi,\psi),\sup\{\varphi^{-1}L,\psi^{-1}R\})$$
$$= ((L\langle\!\langle\varphi\rangle\!\rangle, L\langle\!\langle\psi\rangle\!\rangle), U(R\langle\!\langle\varphi\rangle\!\rangle L, R\langle\!\langle\psi\rangle\!\rangle R)),$$
$$\langle\!\langle\langle\!\langle\varphi\rangle\!\rangle\rangle\!\rangle = (\langle\!\langle\varphi\rangle\!\rangle,(\sup\{L_1\varphi,R_1\varphi^{-1}\})^{-1})$$
$$= (\langle\!\langle\varphi\rangle\!\rangle, U(\varphi^{-1}L,\varphi R)) = (\langle\!\langle\varphi\rangle\!\rangle, U(R\langle\!\langle\varphi\rangle\!\rangle L, L\langle\!\langle\varphi\rangle\!\rangle R)).$$

The operation In is continuous with respect to least upper bounds of arbitrary sets, hence so is $\langle\!\langle\ \rangle\!\rangle$. Therefore, $\mathscr{S}$, $\langle\!\langle\ \rangle\!\rangle$ satisfy both conditions $(t{*}{*})_0$ and $(t{*}{*}{*})_0$, hence $t\mu A_3$ is valid by 18.20 or 18.22. Using 10.9*, 10.10*, we get that $\langle\!\langle\ \rangle\!\rangle$ is a t-operation, which completes the proof.

Proposition 21.13. Let $\mathscr{S}$ be the IOS of example 21.1, $\varnothing \subset M_0 \subseteq M$, let $J: M_0 \times M \to M$ is injective and $J(M_0^2) \subseteq M_0$. Take

$$St(\varphi)(J(s,t),J(s,r)) = \varphi(t,r)$$

and $St(\varphi)(s,t) = \bot$ otherwise, and $\langle\!\langle\ \rangle\!\rangle = \lambda\varphi.(\varphi, St(\varphi))$. Then St is a storing operation and $\mathscr{S}$, $\langle\!\langle\ \rangle\!\rangle$ satisfy the axiom $t\mu A_3$.

Proof. Take $\mathscr{L} = M_0^{\sim} = \{\tilde{s}/s \in M_0\}$, where $\tilde{s} = \lambda tr.I(J(s,t),r)$, $K_0(J(s,f_i(t)), f_i(J(s,t))) = \top$, $i = 1,2$, and $K_0(s,t) = \bot$ otherwise. Then

$$\tilde{s}K_0(f_1(t),r) = I(f_1(J(s,t)),r) = \tilde{s}L(t,r) = (\tilde{s}L,\tilde{s}R)(f_1(t),r)$$

and $\tilde{s}K_0(f_2(t),r) = \tilde{s}R(t,r) = (\tilde{s}L,\tilde{s}R)(f_2(t),r)$, while $\tilde{s}K_0(t,r)$, $(\tilde{s}L,\tilde{s}R)(t,r) = \bot$ otherwise. Therefore, $\tilde{s}K_0 = (\tilde{s}L,\tilde{s}R)$.

Taking $K_1(J(s,J(t,r)), J(J(s,t),r)) = \top$, $K_1(s,t) = \bot$ otherwise, and $K_2 = K_1^{-1}$, one easily obtains $\tilde{s}\tilde{t}K_1 = J(t,s)^{\sim}$ and $J(t,s)^{\sim}K_2 = \tilde{s}\tilde{t}$.

The equality $\tilde{s}St(\varphi) = \varphi\tilde{s}$ follows from the definition of St, while $\forall\tilde{s}(\tilde{s}\varphi \leq \tilde{s}\psi)$ says exactly that $St(I)\varphi \leq St(I)\psi$; hence the axiom (\$) holds by exercise 10.7. Therefore, St is a storing operation, which implies by 10.17 that $\langle\!\langle\ \rangle\!\rangle$ is a t-operation. The latter is obviously continuous with respect to least upper bounds of arbitrary sets, hence $\mathscr{S}$, $\langle\!\langle\ \rangle\!\rangle$ satisfy conditions $(t{*}{*})_0$, $(t{*}{*}{*})_0$, each of which ensures the validity of the axiom $t\mu A_3$. This completes the proof.

Remark. To avoid trivialities M_0 is assumed not a singleton, hence $\mathrm{Card}(M_0) \geq \omega$ by $J(M_0^2) \subseteq M_0$. If $M_0 = M$, then we have a pairing function for M as in exercise 10.1. If $M_0 \subset M$, then we have a *restricted pairing function* for M as in Moschovakis [1971].

Restricted pairing functions appear naturally when Cartesian products are considered. For instance, let $J_0: N_0^2 \to N_0$ be injective, let N be a nonempty set and let $M = N_0 \times N$. Take a fixed $r_0 \in N$, $M_0 = N_0 \times \{r_0\}$ and $J((s,r_0),(t,r)) = (J_0(s,t),r)$ for all $s,t \in N_0$ and $r \in N$. Then M, M_0 and J will satisfy the assumptions of 21.13. (If a restricted pairing function for N_0 is given, then one nevertheless gets a restricted pairing function for $M = N_0 \times N$ this way.) A possible splitting scheme for this particular M is $f_1 = \lambda sr.(J_0(s_1,s),r)$, $f_2 = \lambda sr.(J_0(s_2,s),r)$ with certain fixed $s_1 \neq s_2 \in N_0$. If another splitting scheme for M has already been chosen, then simple transition between the corresponding pairing schemes for $\mathscr{F}$ are provided by exercises 4.6, 7.2.

It should be stressed that St is itself a t-operation by 10.18 since $I = K_3K_4$ for any $K_3 \in M_0^{\sim}$ and $K_4 = K_3^{-1}$ or $K_4(J(s,t),t) = \top$ and $K_4(s,t) = \bot$ otherwise.

The notions of t-recursiveness which correspond to the operations $\langle\!\langle\ \rangle\!\rangle$ of 21.12, 21.13 are called respectively in-*recursiveness*, st-*recursiveness*. The following assertion shows that these t-operations can be combined into a single t-operation, which allows both *In*, *St* to be added to the initial IOS-operations. The notion of t-recursiveness thus introduced is called st, in-*recursiveness*.

Proposition 21.14. Let $\mathscr{S}$ be the IOS of example 21.1 and let $\langle\!\langle\ \rangle\!\rangle$, $\langle\!\langle\ \rangle\!\rangle^*$ be introduced respectively by 21.12, 21.13. Then $\lambda\varphi.\langle\!\langle\langle\!\langle\varphi\rangle\!\rangle^*\rangle\!\rangle$ is a t-operation satisfying the axiom $\mathrm{t}\mu\mathrm{A}_3$ and so is $\lambda\varphi.\langle\!\langle\langle\!\langle\varphi\rangle\!\rangle\rangle\!\rangle^*$.

Proof. We have

$$St(\varphi^{-1})(J(s,t),J(s,r))=\varphi^{-1}(t,r)=\varphi(r,t)=St(\varphi)^{-1}(J(s,t),J(s,r))$$

and $St(\varphi^{-1})(s,t)$, $St(\varphi)^{-1}(s,t)=\bot$ otherwise, hence $St(\varphi^{-1})=St(\varphi)^{-1}$. Using this equality, 10.12 and 10.13, $\langle\!\langle\langle\!\langle\varphi\rangle\!\rangle\rangle\!\rangle^*$ is expressed by $\langle\!\langle\langle\!\langle\varphi\rangle\!\rangle^*\rangle\!\rangle$ and vice versa. Therefore, both $\lambda\varphi.\langle\!\langle\langle\!\langle\varphi\rangle\!\rangle^*\rangle\!\rangle$ and $\lambda\varphi.\langle\!\langle\langle\!\langle\varphi\rangle\!\rangle\rangle\!\rangle^*$ are t-operations by exercise 10.6. They are continuous with respect to least upper bounds, which implies $\mathrm{t}\mu\mathrm{A}_3$ by 18.22. The proof is complete.

The last three statements show that, leaving c-recursiveness to one side, there are at least four different concepts of effective computability for E-valued fuzzy relations: recursiveness, in-recursiveness, st-recursiveness and st, in-recursiveness. Of course, they are effective only if the splitting and pairing schemes involved are effective. A number of properties of these notions follow directly from the general theory developed in chapters 7–10.

It will be shown in chapter 24 that the notion of st-recursiveness with $M_0=M$ generalizes the prime computability of Moschovakis. Adding U to the initial elements, one gets a natural broader concept of effective computability which in a sense generalizes Friedman's computability by effectively definitional schemes. The replacement of U by $U^*=\lambda st.\top$ results in a broadest notion which generalizes the search computability of Moschovakis and which, as will be shown in the exercises below, extends in-recursiveness as well.

EXERCISES TO CHAPTER 21

Exercise 21.1. Let $\mathscr{S}$ be the IOS of example 21.1 or 21.2, 21.3. Show that $(L,R)\le I$, while $(L,R)=I$ iff $f_1(M)\cup f_2(M)=M$, and $\langle I\rangle=I$ iff $\bigcup_n f_2^n(f_1(M))=M$.

Hint. Use exercise 16.5.

Comments. While $\bigcup_n f_2^n(f_1(M))=M$ implies $f_1(M)\cup f_2(M)=M$, the reverse implication fails in example 22.9.

Exercise 21.2 (Example 21.4). Let $\mathscr{S}_0=(\mathscr{F}_0,I,\Pi_0,L,R)$ be the IOS of example 21.1 and let $\mathscr{F}=\{\varphi/\forall str(\varphi(s,t),\varphi(s,r)\ne\bot\Rightarrow r=t)\}$. Show that $\mathscr{S}=(\mathscr{F},I,\Pi_0\restriction\mathscr{F}^2,L,R)$ is a $(**)_0$, $(***)_0$-complete subspace of $\mathscr{S}_0$ and if $\langle\!\langle\ \rangle\!\rangle$ is an operation over $\mathscr{F}_0$ introduced by 21.13, then $\langle\!\langle\ \rangle\!\rangle\restriction\mathscr{F}$ is a t-operation over $\mathscr{F}$ which meets $\mathrm{t}\mu\mathrm{A}_3$.

Hint. Use 18.15, 18.16 and a t-analog to 18.16. The elements K_0, K_1, K_2 of the proof of 21.13 are in $\mathscr{F}$.

Similar assertions hold for example 21.2, 21.3; in the latter case one may also take

$$\mathscr{F} = \{\varphi/\forall spp_1 tr(\varphi(s,p,t), \varphi(s,p_1,r) \neq \bot \Rightarrow p_1 = p \& r = t)\}.$$

Exercise 21.3. Let $\mathscr{S}$ be the IOS of example 21.3. Specify a semi group S and elements φ, ψ such that φ is polynomial in ψ but neither strictly polynomial in ψ nor strictly primitive in ψ.

Hint. Take $S = (\omega, +)$, $\psi(s,1,s) = \top$ and $\psi(s,p,t) = \bot$ otherwise, then take $\varphi = \psi^2$.

Exercise 21.4. Let $W, W_1, W_2, \mathscr{C}$ be the same as in 21.10, $\mathscr{C}_1 = \{\sigma/L\sigma = \sigma L \& R\sigma = \sigma R\}$ and $\mathscr{C}_2 = \{\sigma/L\sigma = \sigma B^2 \& R\sigma = \sigma A^2 \& \sigma(L,R) = \sigma\}$. Show that if $\mathscr{B} \subseteq \mathscr{C} \cup \mathscr{C}_1 \cup \mathscr{C}_2$ and $\mathscr{B}_1 \subseteq \mathscr{F}$, then φ is recursive in $\{W, W_1, W_2\} \cup \mathscr{B} \cup \mathscr{B}_1$ iff it is prime recursive in $\{W, W_1, W_2\} \cup \mathscr{B} \cup \langle \mathscr{B}_1 \rangle$.

Hint. Following the proof of 21.10, show that $\langle \sigma \rangle = \langle I \rangle \sigma$ for all $\sigma \in \mathscr{C}_1$, while $\langle \sigma \rangle = \langle I \rangle \sigma(\langle L \rangle, \langle R \rangle)$ for all $\sigma \in \mathscr{C}_2$.

Notice that $\mathscr{C}_1 L \subseteq \mathscr{C}_2$ and the elements U, V of exercise 7.10, 7.14, if any, are in $\mathscr{C}_2$.

Exercise 21.5. Let $\mathscr{S}_0, \mathscr{S}$ be the IOS of 21.11, let $\mathscr{B} \subseteq \mathscr{F}$, $\mathscr{B}^* \subseteq \mathscr{F}_0$ and let a $\psi^* \in \mathscr{B}^*$ correspond to each $\psi \in \mathscr{B}$ such that $\psi^*(\bar{m}(s), \bar{n}(t)) = \psi(s,m,t,n)$, and whenever t is not of the form $\bar{n}(r)$, then $\psi^*(\bar{m}(s), t) = \bot$, where $\bar{m}(s)$ stands for $f_2^m(f_1(s))$. Let $Z(s,m,s,(2m+1)\mathrm{sgm}) = \top$ and $Z(s,m,t,n) = \bot$ otherwise. Show that for every $\varphi \in \mathscr{F}$ recursive in $\{W, W_1, W_2, Z\} \cup \mathscr{B}$ there is a $\varphi^* \in \mathscr{F}_0$ recursive in $\mathscr{B}^*$ corresponding to φ as above.

Hint. Take $L^* = \langle L_0 \rangle$, $R^* = \langle R_0 \rangle$. There is a recursive element $\varphi_0 \in \mathscr{F}_0$ such that $\bar{n}\varphi_0 = \overline{2n}$; take $W_1^* = \varphi_0$. Similarly W_2^*, W^*, Z^* are constructed. If φ^*, ψ^* correspond to φ, ψ, take $(\varphi\psi)^* = \varphi^*\psi^*$, $(\varphi,\psi)^* = C_0(\varphi^*, \psi^*)$, $\langle \varphi \rangle^* = G_0 \langle \varphi^* \rangle G_0$ and $[\varphi]^* = C_0[\varphi^* C_0]$, where C_0, G_0 are the elements of $\mathscr{F}_0$ introduced in 6.35 and exercise 6.4.

Exercise 21.6. Let $\mathscr{S}_0, \mathscr{S}$ be the IOS of 21.11, $\varphi_0 \in \mathscr{F}_0$ and $\mathscr{B}_0 \subseteq \mathscr{F}_0$. Show that if $\tilde{\varphi}_0$ is recursive in $\{W, W_1, W_2, Z\} \cup \tilde{\mathscr{B}_0}$, then φ_0 is recursive in $\mathscr{B}_0$.

Hint. Take $\tilde{\mathscr{B}}_0^* = \langle \mathscr{B}_0 \rangle$ and get by exercise 21.5 an element $\tilde{\varphi}_0^*$ recursive in $\langle \mathscr{B}_0 \rangle$ to correspond to $\tilde{\varphi}_0$. Then φ_0 will be recursive in $\mathscr{B}_0$ since $\varphi_0 = L_0 \tilde{\varphi}_0^*(I_0, I_0)$.

Remarks. Combined with 21.11, the last two exercises show that translation is equivalent to counting. More generally, the storing operation in first order contexts is connected with computations on complex storages and can be implemented by stacking. The latter can be established by using the fact that $St(\sigma) = St(I)S_1\sigma S_2$, provided $xS_1S_2 = x$ and $xS_1\sigma = \sigma x S_1$ for all $x \in \mathscr{L}$. However, we shall not burden the exposition with further (though interesting) details.

Exercise 21.7. Show that each of the operations $\langle\!\langle\ \rangle\!\rangle$, St of 21.13 satisfies the assumptions of exercises 10.2–10.5. Take $K_5 = \lambda st.I(J(s,s),t)$, $K_4(J(s,t),t) = \top$, $K_6(J(s,J(t,r)), J(t,J(s,r))) = \top$ and $K_4(s,t)$, $K_6(s,t) = \bot$ otherwise, and show that the assumptions of exercise 10.9 are also satisfied.

Hint. Use exercises 10.8**, 18.7.

Remarks. According to exercise 21.7, $\langle\ \rangle$ is always expressible in terms of the operation St of 21.13. On the other hand, whenever M_0 is countable, then St can be expressed in terms of $\langle\ \rangle$ and st-recursiveness reduces to recursiveness. It follows easily from 10.20 that operations St based on wider sets M_0 are more powerful. In particular, all such operations can be expressed in terms of one which corresponds to $M_0 = M$.

Exercise 21.8. Let $M_0 = M$. Show that the operation $\lambda\varphi\psi.\inf\{\varphi,\psi\}$ can be expressed by St and certain elements.

Hint. Show that $\inf\{\varphi,\psi\} = K_5 St(\varphi) K_1^* St(\psi) K_5^{-1}$, where $K_1^*(J(s,t), J(t,s)) = \top$ and $K_1^*(s,t) = \bot$ otherwise.

Exercise 21.9. Let $M_0 = M$. Show that the operation In can be expressed in terms of St and certain elements.

Hint. Show that

$$\varphi^{-1} = K_5 St(U^* K_5) K_6 St(St(\varphi) K_5^{-1}) K_1^* K_4.$$

Example 21.5. Example 21.2 with $N = M^{n-1}$, $n > 0$. Therefore, $\mathscr{F} = \{\varphi/\varphi : M^{n+1} \to E\}$ consists of $n+1$-ary E-valued fuzzy relations.

Exercise 21.10. Let $\mathscr{S}$ be the IOS of example 21.5, let $J: M^2 \to M$ be injective and, writing $J(s_1,\ldots,s_{n+1})$ for $J(s_1, J(s_2,\ldots,s_{n+1}))$, $St(\varphi)(J(s_1,t_1,\ldots,t_n), s_2,\ldots,s_n, J(s_1,r,t_2,\ldots,t_n)) = \varphi(t_1,\ldots,t_n,r)$ and $St(\varphi)(s_1,\ldots,s_n,r) = \bot$ otherwise. Prove that St is a storing operation in the sense of exercise 10.10 and $\langle\!\langle\ \rangle\!\rangle = \lambda\varphi.(\varphi, St(\varphi))$ is a t-operation satisfying $t\mu A_3$.

Hint. Take $\mathscr{L} = \widetilde{M^n}$, where

$$(s_1,\ldots,s_n)\tilde{}(\varphi) = \lambda t_1\ldots t_n r.\varphi(J(s_1,t_1,\ldots,t_n), s_2,\ldots,s_n,r).$$

While the multiplication of $\mathscr{S}$ is a particular instance of composition of $n+1$-ary fuzzy relations, the operation St makes it possible to fully work out such compositions.

Exercise 21.11. Let $\mathscr{S}$ be the IOS of example 21.5. Show that the mapping

$$\lambda\varphi\psi_1\ldots\psi_n.\lambda s_1\ldots s_n t.\sup_{r_1,\ldots,r_n}\inf_{1\le i\le n}\{\psi_i(s_1,\ldots,s_n,r_i), \varphi(r_1,\ldots,r_n,t)\}$$

is st-recursive in certain elements.

Exercise 21.12 (Example 21.6). Let $\mathscr{S}_0 = (\mathscr{F}_0, I, \Pi_0, L, R)$ be the IOS of example 21.1, let $\lesssim$ be a partial quasi-ordering of M (i.e., a reflexive and transitive binary relation on M) and let $J: M^2 \dashrightarrow M$ be injective, such that $f_1(s) \approx f_2(s) \approx s$ (i.e., both $\lesssim$ and $\gtrsim$ hold). Suppose also that, if $J(s,r)\downarrow$ and

$t \lesssim r$, then $J(s,t)\downarrow$ and $J(s,t)$, $r \lesssim J(s,r)$, and if $J(s,J(t,r))\downarrow$, then $J(J(s,t),r)\downarrow$ and $J(s,J(t,r)) \approx J(J(s,t),r)$ for all $s,t,r \in M$. Take the subspace $\mathscr{S} = (\mathscr{F}, I, \Pi, L, R)$ of $\mathscr{S}_0$ based on $\mathscr{F} = \{\varphi / \forall st(\varphi(s,t) \neq \bot \Rightarrow s \lesssim t)\}$, and take $St(\varphi)(J(s,t), J(s,r)) = \varphi(t,r)$ and $St(\varphi)(s,t) = \bot$ otherwise. Show that St is a storing operation and $\langle\ \ \rangle = \lambda\varphi.(\varphi, St(\varphi))$ is a t-operation satisfying $t\mu A_3$.

Hint. Take $\mathscr{L} = \{\tilde{s}/\exists t(J(s,t)\downarrow)\}$, where $\tilde{s}(t,r) = \top$, if $r = J(s,t)$, and $\tilde{s}(t,r) = \bot$ otherwise. Follow the proof of 21.13.

Remark. Proposition 21.13 corresponds to $\lesssim\, = M^2$. It cannot now be claimed that $I \in \mathscr{L} \circ \mathscr{F}$, St is a t-operation and is sufficient to express $\langle\ \ \rangle$; here is an example illustrating this.

Exercise 21.13. Let $\mathscr{S}_0$ be the space of example 21.1 with $M_n^* = \omega \times \bigcup_{0<i\leq n} M^i$ taken for M, the splitting scheme operating on ω, $x \lesssim y$ iff $\mathrm{lh}(x) \leq lh(y)$, where $lh(k, s_1, \ldots, s_m) = m$, and let $J: M_n^{*2} \dashrightarrow M_n^*$ be introduced as suggested in the remark preceding 16.5 so that $J(x,y)\downarrow$ iff $lh(x) + lh(y) \leq n$. Show that $\lesssim$, J satisfy the assumptions of exercise 21.12.

If M_n^* is defined as $\omega \times \bigcup_{i \leq n} M^i$, then $I \in \mathscr{L} \circ \mathscr{F}$ and $\langle\ \ \rangle$ is expressible, but still no elements K_4, K_5 can be specified to meet the requirements of exercise 10.9.

CHAPTER 22

Number functions and relations

This chapter studies in detail some standard examples which consist of functions and relations on natural numbers or sequences of natural numbers. The classical notions of recursive enumerability, relative μ-recursiveness and relative partial recursiveness are accommodated, establishing another connection with Ordinary Recursion Theory in addition to representability.

The starting point is examples 21.1, 21.2 with $\{\bot, \top\}$ taken for E. While these spaces consist of relations, i.e. multiple-valued functions, the single-valued functions are shown to form respective subspaces.

Example 22.1. Example 21.1 with $E = \{\bot, \top\}$.

Proposition 22.1 (Example 22.2). Let $\mathscr{S}_0 = (\mathscr{F}_0, I, \Pi_0, L, R)$ be the IOS of example 22.1 and

$$\mathscr{F} = \{\varphi / \forall str(\varphi(s,t), \varphi(s,r) = \top \Rightarrow r = t)\}.$$

Then $\mathscr{S} = (\mathscr{F}, I, \Pi_0 \restriction \mathscr{F}^2, L, R)$ is a $(**)_0$, $(***)_0$-complete subspace of $\mathscr{S}_0$.

Proof. It is immediate that $I, L, R \in \mathscr{F}$ and it easily follows that $\mathscr{F}$ is closed under multiplication and pairing, hence $\mathscr{S}$ is a subspace of $\mathscr{S}_0$ as an OS.

Suppose that $\mathscr{H}$ is a well ordered subset of $\mathscr{F}$ and $\varphi = \sup \mathscr{H}$ in $\mathscr{F}_0$. If $\varphi(s,t), \varphi(s,r) = \top$, then there is a $\theta \in \mathscr{H}$ such that $\theta(s,t), \theta(s,r) = T$, hence $r = t$. Therefore, $\varphi \in \mathscr{F}$, which completes the proof by 18.15, 18.16.

Later, members of $\mathscr{F}_0$ will be regarded as binary relations by interpreting $\varphi(s,t) = \bot$, $\varphi(s,t) = \top$ respectively as $(s,t) \notin \varphi$, $(s,t) \in \varphi$. Therefore, $\mathscr{F} = \{\varphi / \varphi : M \dashrightarrow M\}$, $\varphi \leq \psi$ iff $\varphi \subseteq \psi$, $\varphi\psi = \lambda s.\psi(\varphi(s))$, $(\varphi, \psi)(f_1(s)) = \varphi(s)$, $(\varphi, \psi)(f_2(s)) = \psi(s)$ and $(\varphi, \psi)(s)\uparrow$ otherwise, $I = \lambda s.s$, $L = f_1$ and $R = f_2$. In other words, $\mathscr{S}$ is exactly the IOS of example 4.7, while $\mathscr{S}_0$ is the IOS of example 4.8.

Example 22.3. Example 21.2 with $E = \{\bot, \top\}$.

Proposition 22.2 (Example 22.4). Let $\mathscr{S}_0 = (\mathscr{F}_0, I, \Pi_0, L, R)$ be the IOS of example 22.3 and

$$\mathscr{F} = \{\varphi / \forall sxtr(\varphi(s,x,t), \varphi(s,x,r) = \top \Rightarrow r = t)\}.$$

Then $\mathscr{S} = (\mathscr{F}, I, \Pi_0 \upharpoonright \mathscr{F}^2, L, R)$ is a $(**)_0$, $(***)_0$-complete subspace of $\mathscr{S}_0$.

The proof repeats that of 22.1.

Members of $\mathscr{F}_0$ will be regarded as relations over $M \times N \times M$ or multiple-valued functions from $M \times N$ to M, writing $(s, x, t) \in \varphi$ or $t \in \varphi(s, x)$ for $\varphi(s, x, t) = \top$ etc. Therefore, $\mathscr{F} = \{\varphi / \varphi : M \times N \dashrightarrow M\}$, $\varphi \leq \psi$ iff $\varphi \subseteq \psi$, $\varphi\psi = \lambda sx.\psi(\varphi(s, x), x)$, $(\varphi, \psi)(f_1(s), x) = \varphi(s, x)$, $(\varphi, \psi)(f_2(s), x) = \psi(s, x)$ and $(\varphi, \psi)(s, x)\uparrow$ otherwise, $I = \lambda sx.s$, $L = \lambda sx.f_1(s)$ and $R = \lambda sx.f_2(s)$.

It is worth mentioning that example 22.2 can be obtained from example 21.4 by taking $E = \{\bot, \top\}$; similarly for example 22.4.

Example 22.5. Example 3.1, that is example 22.2 with $M = \omega$, $f_1 = \lambda s.2s$ and $f_2 = \lambda s.2s + 1$.

Notice that for each s there are unique k, l such that $s = 2^k(2l + 1) - 1 = f_2^k(f_1(l)) = \bar{k}(l)$. In other words, $\omega = \bigcup_n f_2^n(f_1(\omega))$, which implies $\langle I \rangle = (L, R) = I$ by exercise 21.1.

Proposition 21.2 provides the following explicit characterizations of the IOS-operations of example 22.5. Given a number s, then $s = \bar{k}(l)$ and

$$\langle \varphi \rangle(s) = f_2^k(f_1(\varphi(l))),$$
$$\Delta(\varphi, \psi)(s) = \psi^k(\varphi(l)),$$

while $[\varphi](s) = t$ iff there are $n, r_0, \ldots, r_n$ such that $r_0 = s$, $r_0, \ldots, r_{n-1}$ are odd, $r_{i+1} = \varphi((r_i - 1)/2)$ for all $i < n$, and $r_n = 2t$. That is,

$$[\varphi](s) = f_1^{-1}((f_2^{-1}\varphi)^n(s)),$$

where n is the least number m (if any) such that $(f_2^{-1}\varphi)^m(s)$ is even.

'Primitive recursive$_0$' will stand below for 'primitive recursive in the sense of Ordinary Recursion Theory' as opposed to the IOS-notion of primitive recursiveness.

The notions of primitive recursiveness$_0$ and μ-recursiveness for unary number theoretic functions turn out to be particular instances of primitive recursiveness and recursiveness in the space considered.

Proposition 22.3 (μ-Recursiveness Theorem). Let $\mathscr{S}$ be the IOS of example 22.5, $Z = \lambda s.R(s)\text{sg}s$ and $Z_0 = \lambda s.2^s - 1$. Then the following equivalences hold for all $\varphi \in \mathscr{F}$, $\mathscr{B} \subseteq \mathscr{F}$.

(1) φ is primitive recursive$_0$ (μ-recursive) in $\mathscr{B}$ iff φ is primitive recursive (recursive) in $\{Z_0\} \cup \mathscr{B}$.

(2) φ is μ-recursive in $\mathscr{B}$ iff φ is recursive in $\{Z\} \cup \mathscr{B}$.

Proof. The 'if'-parts of (1), (2) are immediate since L, R, Z, Z_0 are primitive recursive$_0$ functions and $\circ, \Pi, \Delta$ are primitive recursive$_0$ operations, while $[\ \ ]$ is a μ-recursive operation by the explicit characterizations of $\Delta, [\ \ ]$.

Let us show that for every $\varphi \in \mathscr{F}$ there is a unique $\tilde{\varphi} \in \mathscr{F}$ representing φ in the sense of chapter 8. Take $\tilde{\varphi}(s) = \tilde{\varphi}(\bar{k}(l)) = \overline{\varphi(k)}(l)$ by definition. Then $\bar{k}\tilde{\varphi} = \overline{\varphi(k)}$ for all k; hence $\tilde{\varphi}$ represents φ. Supposing that φ_1 also represents φ, we get $\varphi_1(s) = \bar{k}\varphi_1(l) = \bar{k}\tilde{\varphi}(l) = \tilde{\varphi}(s)$ for all s; hence $\varphi_1 = \tilde{\varphi}$.

If φ is primitive recursive$_0$ (μ-recursive) in $\mathscr{B}$, then 8.1 (respectively, 8.3***) implies that $\tilde{\varphi}$ is primitive recursive (recursive) in $\mathscr{B}^{\sim}$, hence it suffices to show that φ is primitive recursive in $Z_0, \tilde{\varphi}$, $\tilde{\psi}$ is primitive recursive in Z_0, ψ and Z_0 is recursive in Z.

Take $\psi_0 = \Delta(R, L)$, $\psi_1 = \mu\theta.(Z(I, \theta R), L)$ and $\psi_2 = \mu\theta.Z(I, \psi_1\theta R)$.

An easy induction on s implies that $\psi_0 = \lambda s.s + 1$. If $s = 2l$, then

$$\psi_0(s) = \bar{0}\psi_0(l) = R(l) = 2l + 1 = s + 1.$$

If $s = 2l + 1$, then

$$\psi_0(s) = R\psi_0(l) = \psi_0 L(l) = 2(l + 1) = s + 1.$$

It also follows by induction that $\psi_1 = \lambda s.s \dot{-} 1$. First,

$$\psi_1(0) = Z(I, \psi_1 R)(0) = (I, \psi_1 R)(0) = 0.$$

If $s = 2l + 1$, then

$$\psi_1(s) = (Z(I, \psi_1 R), L)(2l + 1) = L(l) = s - 1.$$

If $s = 2l + 2$, then

$$\psi_1(s) = Z(I, \psi_1 R)(l + 1) = (I, \psi_1 R)(2(l + 1) + 1) = \psi_1 R(l + 1) = R(l) = s - 1.$$

We get

$$\begin{aligned}\psi_2(0) &= Z(I, \psi_1\psi_2 R)(0) = I(0) = 0,\\ \psi_2(s + 1) &= (I, \psi_1\psi_2 R)(2(s + 1) + 1)\\ &= \psi_1\psi_2 R(s + 1) = \psi_2 R(s) = 2\psi_2(s) + 1,\end{aligned}$$

hence $\psi_2 = \lambda s.R^s(0) = \lambda s.\bar{s}(0) = Z_0$. The elements ψ_1, ψ_2 are recursive in Z by the proof of 6.39, hence so is Z_0.

Take $\psi_3 = \Delta(I, \psi_0)$ and

$$\rho = \langle Z_0 \rangle G \langle \psi_3 \psi Z_0 \rangle G \langle \psi_3 \rangle.$$

If $\psi(k) = t$, then

$$\begin{aligned}\bar{k}\rho(l) &= Z_0 \langle \bar{k} \rangle \langle \psi_3 \psi Z_0 \rangle G \langle \psi_3 \rangle (l) = Z_0 \langle \psi_0^k \psi Z_0 \rangle G \langle \psi_3 \rangle (l)\\ &= \bar{l} \langle \psi_0^k \psi Z_0 \rangle G \langle \psi_3 \rangle (0) = \psi_0^k \psi Z_0 \langle \bar{l} \psi_3 \rangle (0) = \psi Z_0 \langle \psi_0^l \rangle (k)\\ &= Z_0 \langle \psi_0^l \rangle (t) = \bar{t} \langle \psi_0^l \rangle (0) = \psi_0^l t(0) = \bar{t}(l)\end{aligned}$$

for all l, hence $\bar{k}\rho = \bar{t}$. If $\psi(k)\uparrow$, then similarly $\bar{k}\rho(l) = \psi Z_0 \langle \psi_0^l \rangle (k)\uparrow$ for all l, hence $\bar{k}\rho = O$. Therefore, $\tilde{\psi} = \rho$; hence $\tilde{\psi}$ is primitive recursive in Z_0, ψ.

Finally, if $\varphi(k) = t$, then $\bar{k}\tilde{\varphi} = \bar{t}$; hence

$$Z_0\tilde{\varphi}\psi_3(k) = \bar{k}\tilde{\varphi}\psi_3(0) = \bar{t}\psi_3(0) = \psi_0^t(0) = t.$$

If $\varphi(k)\uparrow$, then $\bar{k}\tilde{\varphi} = O$; hence

$$Z_0\tilde{\varphi}\psi_3(k) = \bar{k}\tilde{\varphi}\psi_3(0) = O\psi_3(0)\uparrow.$$

We get $\varphi = Z_0\tilde{\varphi}\psi_3$, hence φ is primitive recursive in $Z_0, \tilde{\varphi}$. The proof is complete.

It was necessary to add Z or Z_0 to the initial elements for, as will be shown in the exercises, neither of them is IOS-recursive.

Proposition 22.3 yields a natural way of generating the unary partial recursive functions; other unary bases are provided by 7.11 and exercise 7.3. For instance, φ is partial recursive iff

$$\varphi \in cl(L, \langle\!\langle L \rangle\!\rangle, \langle\!\langle A \rangle\!\rangle, \langle Z \rangle/\circ, [\ \ ]).$$

The following analogue to 22.3 for mappings is obtained by an immediate parametrization.

Proposition 22.4. Let $\mathscr{S}$ be the space of example 22.5, $\mathscr{B} \subseteq \mathscr{F}$ and $\Gamma: \mathscr{F}^n \to \mathscr{F}$, $n > 0$. Then the following equivalences hold.

(1) Γ is primitive recursive$_0$ (μ-recursive) in $\mathscr{B}$ iff Γ is primitive recursive (recursive) in $\{Z_0\} \cup \mathscr{B}$.

(2) Γ is μ-recursive in $\mathscr{B}$ iff Γ is recursive in $\{Z\} \cup \mathscr{B}$.

The following example corresponds to example 8 in Skordev [1980], chapter 2.

Example 22.6. Example 22.1 with set M and splitting scheme as in example 22.5. Therefore, $\mathscr{F} = \{\varphi/\varphi \subseteq \omega^2\} = \{\varphi/\varphi: \omega \to 2^\omega\}$, $\varphi \leq \psi$ iff $\varphi \subseteq \psi$ iff $\forall s(\varphi(s) \subseteq \psi(s))$, $\varphi\psi = \lambda s.\psi(\varphi(s)) = \lambda s. \cup \{\psi(r)/r \in \varphi(s)\}$,

$$(\varphi, \psi)(s) = \begin{cases} \varphi(s/2), & \text{if } s \text{ is even,} \\ \psi((s-1)/2), & \text{if } s \text{ is odd,} \end{cases}$$

$I = \lambda s.s$, $L = f_1$ and $R = f_2$.

The explicit characterizations of 21.2 now read as follows.

$$\langle \varphi \rangle(s) = f_2^k(f_1(\varphi(l))),$$
$$\Delta(\varphi, \psi)(s) = \psi^k(\varphi(l)),$$

where $s = \overline{k}(l)$,

$t \in [\varphi](s)$ iff there are $n, r_0, \ldots, r_n$ such that $r_0 = s, r_0, \ldots, r_{n-1}$ are odd, $r_{i+1} \in \varphi((r_i - 1)/2)$ for all $i < n$, r_n is even and $t = r_n/2$.

The following two Partial Recursiveness Theorems embody the notions of relative partial recursiveness for single-valued and multiple-valued functions. The latter is introduced by the definition given in the remarks to exercise 8.2, allowing f to be multiple-valued; alternatively, adding to the definition of relative μ-recursiveness a certain multiple-valued initial function, say $\lambda s.\{0, 1\}$ or $\lambda s.\omega$. We recall that the space of example 22.6 has an element satisfying the assumptions of exercise 7.9***, $U = \lambda s.\{2s, 2s+1\}$.

Proposition 22.5. Let $\mathscr{S}$ be the IOS of example 22.6 and $\mathscr{S}_1$ its subspace consisting of all the single-valued functions. (In other words, $\mathscr{S}_1$ is the space of example 22.5.) Let $\varphi \in \mathscr{F}_1$, $\mathscr{B} \subseteq \mathscr{F}_1$. Then

(1) φ is partial recursive in $\mathscr{B}$ iff φ is recursive in $\{Z, U\} \cup \mathscr{B}$.

Remark. Though $\varphi \in \mathscr{F}_1$ and $\mathscr{B} \subseteq \mathscr{F}_1$, the considerations could not be carried out within $\mathscr{S}_1$ since $U \in \mathscr{F} \setminus \mathscr{F}_1$.

Proof. It is the 'only-if'-part of (1) that needs proof.

As shown in the proof of 22.3, there is a $\tilde{\varphi} \in \mathscr{F}_1$ representing φ. If $\varphi_1 \in \mathscr{F}$ also represents φ, it follows as in the proof of 22.3 that $\varphi_1 = \tilde{\varphi}$. Moreover, φ is recursive in $Z, \tilde{\varphi}$ and $\tilde{\psi}$ is recursive in Z, ψ by the proof of 22.3.

If φ is partial recursive in $\mathscr{B}$, then $\tilde{\varphi}$ is recursive in $\{U\}\cup\mathscr{B}^{\sim}$ by exercise 8.2***, hence φ is recursive in $\{Z, U\}\cup\mathscr{B}$. The proof is complete.

Proposition 22.6. Let $\mathscr{S}$ be the space of example 22.6, $\varphi\in\mathscr{F}$ and $\mathscr{B}\subseteq\mathscr{F}$. Then the following equivalences hold:

(1) φ is partial recursive in $\mathscr{B}$ iff φ is recursive in $\{Z, U\}\cup\mathscr{B}$.

(2) φ is a recursively enumerable relation iff φ is recursive in Z, U.

Proof. To each $\varphi\in\mathscr{F}$ assign a single-valued φ^* such that $\varphi^*(\bar{k}(l))=k$, if $k\in\varphi(l)$, and $\varphi^*(\bar{k}(l))\uparrow$ otherwise. Then it follows that $\varphi^*=\langle\varphi\rangle\psi_0$, where $\psi_0(\bar{k}(k))=k$ and $\psi_0(s)\uparrow$ otherwise. The element ψ_0 is a partial recursive function, hence it is recursive in Z by 22.3. On the other hand, we get $\varphi=\sup_k\bar{k}\varphi^*=\bar{\omega}\varphi^*$, where $\bar{\omega}$ is the element considered in exercise 7.9***. Therefore, φ^* is recursive in Z, φ, while φ is recursive in U, φ^*. The equivalence (1) is deduced as follows.

$$\begin{aligned}
&\varphi \text{ is partial recursive in } \mathscr{B}\\
\text{iff}\quad &\varphi^* \text{ is partial recursive in } \mathscr{B}^*\\
\text{iff}\quad &\varphi^* \text{ is recursive in } \{Z, U\}\cup\mathscr{B}^* &&\text{(by 22.5)}\\
\text{iff}\quad &\varphi \text{ is recursive in } \{Z, U\}\cup\mathscr{B}.
\end{aligned}$$

Since a relation is *recursively enumerable* iff it is partial recursive as a multiple-valued function, (2) follows immediately by taking $\mathscr{B}=\varnothing$. This completes the proof.

Example 22.7. Example 22.4 with $N=M^{n-1}$, $n>0$. (It is a subspace of example 21.5 as well.)

Example 22.8. Example 22.7 with set M and splitting scheme as in example 22.5. Thus $\mathscr{F}=\{\varphi/\varphi:\omega^n\longrightarrow\omega\}$, $\varphi\le\psi$ iff $\varphi\subseteq\psi$, $\varphi\psi=\lambda s_1\ldots s_n.\psi(\varphi(s_1,\ldots, s_n), s_2,\ldots,s_n)$,

$$(\varphi,\psi)(s_1,\ldots,s_n)=\begin{cases}\varphi(s_1/2, s_2,\ldots,s_n), & \text{if } s_1 \text{ is even},\\ \psi(((s_1-1)/2), s_2,\ldots,s_n), & \text{if } s_1 \text{ is odd},\end{cases}$$

$I=\lambda s_1\ldots s_n.s_1$, $L=\lambda s_1\ldots s_n.2s_1$ and $R=\lambda s_1\ldots s_n.2s_1+1$.

Then for each s_1 there are unique k, l such that

$$s_1=f_2^k(f_1(l))=\bar{k}(l, s_2,\ldots,s_n)$$

for all $s_2,\ldots,s_n$, hence there are unique $k_1,\ldots,k_n,l$ such that $s_1=\overline{k_1}\ldots\overline{k_n}(l, s_2,\ldots,s_n)$ for all $s_2,\ldots,s_n$. This implies that for every $\varphi\in\mathscr{F}$ there is a unique $\tilde{\varphi}\in\mathscr{F}$ representing it in the sense of chapter 8, namely

$$\tilde{\varphi}(\overline{k_1}\ldots\overline{k_n}(l, s_2,\ldots,s_n), s_2,\ldots,s_n)=\overline{\varphi(k_1,\ldots,k_n)}(l, s_2,\ldots,s_n).$$

Proposition 21.4 gives the following explicit characterizations of $\langle\quad\rangle$, Δ, $[\quad]$, which show that the first two operations are primitive recursive$_0$, while the last one is μ-recursive.

$$\begin{aligned}
\langle\varphi\rangle(s_1,\ldots,s_n)&=f_2^k(f_1(\varphi(l, s_2,\ldots,s_n))),\\
\Delta(\varphi,\psi)(s_1,\ldots,s_n)&=\varphi\psi^k(l, s_2,\ldots,s_n),
\end{aligned}$$

where $s_1 = f_2^k(f_1(l))$,
$[\varphi](s_1,\ldots,s_n)=t$ iff there are $n, r_0,\ldots,r_n$ such that $r_0 = s_1, r_0,\ldots,r_{n-1}$ are odd, $r_{i+1} = \varphi((r_i-1)/2, s_2,\ldots,s_n)$ for all $i<n$, r_n is even and $t = r_n/2$.

The following statement describes the primitive recursiveness$_{\mathrm{O}}$ and μ-recursiveness for n-ary number functions in terms of the corresponding IOS-notions. Other characterizations may be obtained by making use of 7.11 and exercise 7.3.

Proposition 22.7. Let $\mathscr{S}$ be the IOS of example 22.8, $Z = \lambda s_1\ldots s_n.f_2(s_1)\mathrm{sg}s_1$, $Z_0 = \lambda s_1\ldots s_n.2^{s_1}-1$ and $I_i^n = \lambda s_1\ldots s_n.s_i$, $1<i\le n$. Then the following equivalences hold for all $\varphi\in\mathscr{F}$, $\mathscr{B}\subseteq\mathscr{F}$.

(1) φ is primitive recursive$_{\mathrm{O}}$ (μ-recursive) in $\mathscr{B}$ iff φ is primitive recursive (recursive) in $\{Z_0, I_2^n,\ldots,I_n^n\}\cup\mathscr{B}^{\sim}$.

(2) φ is μ-recursive in $\mathscr{B}$ iff φ is recursive in $\{Z, I_2^n,\ldots,I_n^n\}\cup\mathscr{B}^{\sim}$.

In particular,

(3) φ is primitive recursive$_{\mathrm{O}}$ (partial recursive) iff φ is primitive recursive (recursive) in $Z_0, I_2^n,\ldots,I_n^n$.

(4) φ is partial recursive iff φ is recursive in $Z, I_2^n,\ldots,I_n^n$.

Proof. The 'if'-parts of (1), (2) are immediate. In order to establish their 'only if'-parts it suffices to show that Z_0 is recursive in Z and φ is primitive recursive in $Z_0, I_2^n,\ldots,I_n^n, \tilde{\varphi}$. This completes the proof by making use of 8.1, 8.3***.

Proposition 21.5 implies that the space of example 22.5 is isomorphic to a subspace of the present IOS, where the isomorphism assigns to each unary function f the n-ary one $\lambda s_1\ldots s_n.f(s_1)$. Therefore, Z_0 is recursive in Z by the proof of 22.3.

Take ψ_0,ψ_3 to correspond to the elements ψ_0,ψ_3 of the proof of 22.3. Then ψ_0,ψ_3 are primitive recursive in Z_0 and $\psi_0 = \lambda s_1\ldots s_n.s_1+1$, $\bar{k}\psi_3 = \psi_0^k$.

Let G_{n-1} be the element constructed in 7.15 and

$$\psi_4 = Z_0\langle I_n^n Z_0\langle I_{n-1}^n Z_0\ldots\langle I_2^n Z_0\rangle\ldots\rangle\rangle G_{n-1}.$$

Then ψ_4 is primitive recursive in $Z_0, I_2^n,\ldots,I_n^n$ and

$$\psi_4(s_1,\ldots,s_n) = \bar{s_1}\ldots\bar{s_n}(0,s_2,\ldots,s_n).$$

If $\varphi(s_1,\ldots,s_n)=t$, then $\bar{s_1}\ldots\bar{s_n}\tilde{\varphi}=\bar{t}$, hence

$$\begin{aligned}\psi_4\tilde{\varphi}\psi_3(s_1,\ldots,s_n) &= \bar{s_1}\ldots\bar{s_n}\tilde{\varphi}\psi_3(0,s_2,\ldots,s_n) = \bar{t}\psi_3(0,s_2,\ldots,s_n)\\ &= \psi_0^t(0,s_2,\ldots,s_n) = t.\end{aligned}$$

If $\varphi(s_1,\ldots,s_n)\uparrow$, then $\bar{s_1}\ldots,\bar{s_n}\tilde{\varphi}=O$, hence

$$\psi_4\tilde{\varphi}\psi_3(s_1,\ldots,s_n) = \bar{s_1}\ldots\bar{s_n}\tilde{\varphi}\psi_3(0,s_2,\ldots,s_n) = O\psi_3(0,s_2,\ldots,s_n)\uparrow.$$

We get $\varphi = \psi_4\tilde{\varphi}\psi_3$; hence φ is primitive recursive in $Z_0, I_2^n,\ldots,I_n^n$, $\tilde{\varphi}$. The proof is complete.

The reader has undoubtedly noticed the peculiarities which distinguish (1), (2) from the corresponding equivalences of 22.3. We did not prove that φ was μ-recursive in $\mathscr{B}$ iff it was recursive in $\{Z, I_2^n,\ldots,I_n^n\}\cup\mathscr{B}$ because of our inability to work out full compositions unless $n=1$: all the arguments of a function except the first one were parameters taken into account but not affected. This obstacle will be overcome in the exercises below either by making use of a t-operation or by considering the space of functions $\varphi:\omega^n \longrightarrow \omega^n$ instead.

The following wider space consists of functions operating on arbitrary tuples of natural numbers.

Example 22.9. The IOS $\mathscr{S}$ of example 22.2 with $M=\omega^*=\bigcup_n \omega^n$, $f_1=\lambda x.(0,x), f_2(\Lambda)=\Lambda$ and $f_2(s,x)=(s+1,x)$.

If $x=(s_1,\ldots,s_n)$, then we write (s,x) for $(s,s_1,\ldots,s_n)$; Λ is the empty tuple, the only member of ω^0.

It follows that $\bar{s_1}\ldots\bar{s_n}(x)=(s_n,\ldots,s_1,x)$ for all $x\in\omega^*$. In particular, $\bar{s}(x)=(s,x)$ and $\bar{s}(\Lambda)=s$. We get $(\varphi,\psi)(0,x)=\varphi(x), (\varphi,\psi)(\Lambda)=\psi(\Lambda)$ and $(\varphi,\psi)(s+1,x)=\psi(s,x)$. Notice that $f_1(\omega^*)\cup f_2(\omega^*)=\omega^*$ but $\bigcup_n f_2^n(f_1(\omega^*))\neq\omega^*$, hence $(L,R)=I, \langle I\rangle\neq I$ by exercise 21.1. Proposition 21.2 gives $\langle\varphi\rangle(\Lambda)\uparrow$ and

$$\langle\varphi\rangle(s,x)=\bar{s}\langle\varphi\rangle(x)=\varphi\bar{s}(x)=(s,\varphi(x)),$$

while $[\varphi](x)=y$ iff there are $n,z_0,\ldots,z_n$ such that $z_0=x$, $z_i\in f_2(\omega^*)$ and $z_{i+1}=\varphi(f_2^{-1}(z_i))$ for all $i<n$, and $z_n=(0,y)$.

The following Stack Recursiveness Theorem formulates the notion of partial recursive stack function introduced in Germano and Maggiolo-Schettini [1976] in terms of operations close to the basic IOS-operations in this example.

Proposition 22.8. Let $\mathscr{S}$ be the IOS of example 22.9, $Z(\Lambda)=0$ and $Z(s,x)=(s+1,x)$. Then $\varphi\in\mathscr{F}$ is a partial recursive stack function iff φ is recursive in Z.

Proof. The *partial recursive stack functions* are generated from the initial ones $\mathbb{O},\mathbb{S},\mathbb{P}$ by means of multiplication, which is exactly the semigroup multiplication of $\mathscr{F}$, left cylindrification operator $\lambda\varphi.{}^c\varphi$ and repetition operator $\lambda\varphi.\varphi^\nabla$, where $\mathbb{O}=L$, $\mathbb{S}=Z$, $\mathbb{P}(\Lambda)=\Lambda$ and $\mathbb{P}(s,x)=(s\dot{-}1,x)$, ${}^c\varphi(\Lambda)=0$, ${}^c\varphi(s,x)=(s,\varphi(x))$, $\varphi^\nabla(x)=y$ iff there are $n,z_0,\ldots,z_n$ such that $z_0=x$, $z_i\in f_2(\omega^*)$ and $z_{i+1}=\varphi(z_i)$ for all $i<n$, and $z_n=(0,y)$.

It follows almost immediately that $\mathbb{P}=(L,I)$, ${}^c\varphi=Z(L,\langle\varphi\rangle)$ and $\varphi^\nabla=[R\varphi]$; hence all partial recursive stack functions are recursive in Z.

On the other hand, $R=\mathbb{S}\mathbb{O}^\nabla$. Notice that $R\mathbb{P}=I$ and $\mathbb{O}\varphi^\nabla=I$ for all φ.

Writing $\sigma_0,\sigma_1,\sigma_2,\sigma$ respectively for $({}^c\mathbb{P}\mathbb{P}\psi\mathbb{O}^2)^\nabla$, $(\mathbb{P}\varphi\mathbb{O})^\nabla$, $(\mathbb{P}^cR\sigma_0\sigma_1\mathbb{O}^2)^\nabla$, $\mathbb{S}\sigma_2(\psi\mathbb{O})^\nabla$, one gets

$$\begin{aligned}
\sigma(0,x)&=\sigma_2(\psi\mathbb{O})^\nabla(1,x)={}^cR\sigma_0\sigma_1\mathbb{O}(\psi\mathbb{O})^\nabla(0,x)=\sigma_0\sigma_1(0,R(x))\\
&=\sigma_1(R(x))=\varphi\mathbb{O}(\mathbb{P}\varphi\mathbb{O})^\nabla(x)=\varphi(x),\\
\sigma(\Lambda)&=\sigma_2(\psi\mathbb{O})^\nabla(0)=(\psi\mathbb{O})^\nabla(\Lambda)=\psi(\Lambda),\\
\sigma(s+1,x)&=\sigma_2(\psi\mathbb{O})^\nabla(s+2,x)\\
&={}^cR\sigma_0\sigma_1\mathbb{O}(\psi\mathbb{O})^\nabla(s+1,x)=\sigma_0\sigma_1(s+1,R(x))\\
&=\psi\mathbb{O}\sigma_1(s,x)=\psi(s,x);
\end{aligned}$$

hence $(\varphi,\psi)=\sigma$. It also follows that $\langle\varphi\rangle=\mathbb{S}(\mathbb{P}^\nabla,{}^c\varphi)$ and $[\varphi]=(\mathbb{P}\varphi)^\nabla$, which completes the proof.

EXERCISES TO CHAPTER 22

Exercise 22.1. Let $\mathscr{S}$ be the space of example 22.5. Show that the members $Z, Z_0, \lambda s.s \dot{-} 1$ and $\psi_0 = \lambda s.0$ of $\mathscr{F}$ are not IOS-recursive.

Hint. It suffices to show that whenever φ is recursive, then either $\exists n(L^n\varphi \in \mathscr{D})$ or $\psi_0\varphi = O$, for the functions in question do not have this property. The element φ has a normal form $\bar{1}[\sigma]$ for a certain primitive σ. Notice that $\forall n(\psi_0 = \psi_0 L^n)$, while $\forall \alpha \in \mathscr{D} \exists n(L^n \alpha\sigma \in \mathscr{D})$ by the hint to exercise 6.12. Using the unwinding method, show that either $\exists n(L^n\varphi \in \mathscr{D})$ or $\psi_0\varphi = O$. The segment to be used is of the form $\{\theta/\forall n(\tau_n\theta \leq O)\}$; though not regular, it contains $[\sigma]$ whenever closed under $\lambda\theta.(I, \sigma\theta)$.

Exercise 22.2. Let $\mathscr{S}$ be an IOS and let $f:\omega^n \to \omega$ be represented by a primitive recursive element φ. Show that f is a primitive recursive function.

Hint. Assume $n = 1$. Let $\mathscr{S}_0$ be the space of example 22.5 and let φ_0 be a primitive recursive member of $\mathscr{F}_0$ the construction of which repeats that of φ. It follows as in the hint to exercise 6.13 that for all s there are m and $\alpha \in \mathscr{D}$ such that $L^m\bar{s}\varphi = \alpha$. As a consequence of the IOS-axioms, this holds both in $\mathscr{S}$ and $\mathscr{S}_0$; hence $L^m\bar{s}\varphi_0 = \alpha$ as well. However, $\bar{s}\varphi = \overline{f(s)}$; hence $\alpha = L^m \overline{f(s)}$. Therefore, $\bar{s}\varphi_0(0) = \overline{f(s)}(0)$ for all s, hence $f = Z_0\varphi_0\psi_3$ with Z_0, ψ_3 as in the proof of 22.3.

Exercise 22.3. Let $\mathscr{S}$ be the space of example 22.8 and let

$$St(\varphi)(\bar{t}_1 \ldots \bar{t}_n(s_1), s_2, \ldots, s_n) = \overline{\varphi(t_1, \ldots, t_n)}\,\bar{t}_2 \ldots \bar{t}_n(s_1),$$

$\bar{k}(s)$ standing for $f_2^k(f_1(s))$. Show that St is a t-operation with a set of functionary elements $\mathscr{B}_0 = \{Z_0, I_2^n, \ldots, I_n^n\}$ or $\mathscr{B}_0 = \{Z, I_2^n, \ldots, I_n^n\}$ and the following equivalences hold for all $\varphi \in \mathscr{F}$, $\mathscr{B} \subseteq \mathscr{F}$.

(1) φ is primitive recursive$_0$ (μ-recursive) in $\mathscr{B}$ iff φ is primitive st-recursive (st-recursive) in $\mathscr{B}$, provided we take the former $\mathscr{B}_0$.

(2) φ is μ-recursive in $\mathscr{B}$ iff φ is st-recursive in $\mathscr{B}$, provided we take the latter $\mathscr{B}_0$.

Hint. Show by exercise 21.1 that St is a storing operation. Construct elements K_3, K_4 primitive recursive in Z_0, $I_2^n, \ldots, I_n^n$ such that $\varphi = K_3 St(\varphi) K_4$ for all φ. Make use of 22.7 and the equality $\tilde{\psi} = St(\psi)\Delta(I, I)^{n-1}$.

Example 22.10. Example 22.2 with $M = \omega^n, n > 0, f_1 = \lambda s_1 \ldots s_n.(2s_1, s_2, \ldots, s_n)$ and similarly for f_2.

Consider the elements $Z_0 = \lambda s_1 \ldots s_n.(2^{s_1} - 1, s_2, \ldots, s_n)$, $Z = \lambda s_1 \ldots s_n.((2s_1 + 1)\text{sg}s_1, s_2, \ldots, s_n)$ and $I_{n,i} = \lambda s_1 \ldots s_n.(s'_1, \ldots, s'_n)$, $1 < i \leq n$, where $s'_1 = s_i$, $s'_i = s_1$ and $s'_j = s_j$ otherwise. For all $\varphi \in \mathscr{F}$, $\mathscr{B} \subseteq \mathscr{F}$ let $\varphi^{(i)}$ stand for the i-th co-ordinate function of φ and $\mathscr{B}^{(i)}$ for the set of i-th co-ordinate functions of members of $\mathscr{B}$.

Exercise 22.4. Let $\mathscr{S}$ be the IOS of example 22.10. Prove that the following equivalences hold.

(1) φ is primitive recursive (recursive) in $\{Z_0, I_{n,2}, \ldots, I_{n,n}\} \cup \mathscr{B}$ iff $\varphi^{(1)}, \ldots, \varphi^{(n)}$ are primitive recursive$_0$ (μ-recursive) in $\bigcup_{i=1}^{n} \mathscr{B}^{(i)}$.

(2) φ is recursive in $\{Z, I_{n,2}, \ldots, I_{n,n}\} \cup \mathscr{B}$ iff $\varphi^{(1)}, \ldots, \varphi^{(n)}$ are μ-recursive in $\bigcup_{i=1}^{n} \mathscr{B}^{(i)}$.

Hint. Reduce (1), (2) to the corresponding equivalences of 22.3.

Exercise 22.5. Let φ be a n-ary number function and $\mathscr{B}$ a set of such functions, $\varphi^* = \lambda s_1 \ldots s_n.(\varphi(s_1, \ldots, s_n), s_2, \ldots, s_n)$ and let $\mathscr{S}$ be the IOS of example 22.10 so that $\varphi^* \in \mathscr{F}$, $\mathscr{B}^* \subseteq \mathscr{F}$. Prove the following equivalences.

(1) φ is primitive recursive$_0$ (μ-recursive) in $\mathscr{B}$ iff φ^* is primitive recursive (recursive) in $\{Z_0, I_{n,2}, \ldots, I_{n,n}\} \cup \mathscr{B}^*$.

(2) φ is μ-recursive in $\mathscr{B}$ iff φ^* is recursive in $\{Z, I_{n,2}, \ldots, I_{n,n}\} \cup \mathscr{B}^*$.

Example 22.11. Example 22.2 with $M = \{\ldots \varepsilon_1 \varepsilon_0 * \delta_0 \delta_1 \ldots / \forall n(\varepsilon_n, \delta_n \in \{0, 1\})\}$, $f_1 = \lambda x * y.x0 * y$ and $f_2 = \lambda x * y.x1 * y$.

Members of M can be regarded as records on an infinite tape, the asterisk indicating the scanned symbol. Let $S_1 = \lambda x\varepsilon * y.x * \varepsilon y$ and $S_2 = S_1^{-1}$. An element $\psi \in \mathscr{F}$ is *normal* iff

$$\forall xyx'y'(\psi(x * y) = x' * y' \Rightarrow \forall y''(\psi(x * y'') = x' * y''))$$

Exercise 22.6. Let $\mathscr{S}$ be the IOS of example 22.11 and let $\mathscr{B}$ consist of normal elements. Show that φ is recursive in $\{S_1, S_2\} \cup \mathscr{B}$ iff φ is prime recursive in $\{S_1, S_2\} \cup \mathscr{B}$.

Hint. Take $W = S_2$, $W_1 = LS_1$, $W_2 = RS_1$ and use 21.10.

CHAPTER 23

Functions computable by programs

While the previous chapter was devoted to number functions and relations, the present chapter (as well as the next) deals with the more general case of functions on arbitrary domains. Concepts of computability for such functions are provided in particular by certain kinds of abstract programs studied in Computer Science, e.g. program schemes augmented by counters and stacks, which correspond to the so-called finite algorithmic procedures with counting (Friedman [1971]) and stacking (Moldestad, Stoltenberg-Hansen and Tucker [1981]). The considerations below aim to show that those notions of computability are particular instances of relative IOS-recursiveness, with all the attendant consequences.

Let us start with the simplest program schemes, the unary ones. All the other kinds will be obtained in terms of these.

We have unary function symbols $f, f_1, f_2, \ldots$ and unary predicate symbols $P, P_1, P_2, \ldots$. A *unary scheme* consists of instructions of the form

$$l_i \text{ do } f \text{ go to } l_j,$$
$$l_i \text{ if } P \text{ then go to } l_j \text{ else go to } l_k$$

and one instruction l_m stop. We shall write simply $l_i f l_j$ and l_i if P then l_j else l_k. Each label l_i occurs just once as a prefix of an instruction; the instruction itself will often be referred to as l_i. Therefore, a scheme is a nonempty list $l_1, \ldots, l_m$ of instructions one of which, say l_1, is declared to be first.

Unary schemes are interpreted by means of unary bases to give interpreted schemes or *programs* in those bases. A *unary basis* is a tuple

$$B = (M, f_1, \ldots, f_u, P_1, \ldots, P_v),$$

where M is a nonempty set, $f_i: M \dashrightarrow M$, $0 \leq i \leq u$, and $P_i: M \dashrightarrow \{0, 1\}$, $0 \leq i \leq v$. Function and predicate symbols are interpreted respectively by functions and predicates from B. The same letters will denote function symbols and functions, respectively predicate symbols and predicates.

The execution of a program $\mathscr{P}$ employs one processing *register* r and goes in a fairly ordinary manner. It starts with the first instruction; $l_i f l_j$ replaces r by $f(r)$ and the computation continues with l_j, provided $f(r)\downarrow$; in the case of l_i if P then l_j else l_k the content of the register remains unaltered and the next instruction to be executed is either l_j or l_k, depending on whether $P(r) = 0$ or

$P(r) = 1$. If $f(r)\uparrow$ or $P(r)\uparrow$, then the computation does not terminate; another instance in which the computation process fails to terminate is that of an infinite computation. If the last instruction l_m stop is reached, then the process terminates and the content of r is the final result. Therefore, the program $\mathscr{P}$ computes a partial function $\varphi_{\mathscr{P}}: M \dashrightarrow M$.

If l_2 is taken as a first instruction, then another function φ_2 will be computed etc. If l_m is the first instruction, then the function computed will, of course, be $\varphi_m = I = \lambda r.r$.

Each program $\mathscr{P}$ has a corresponding *characteristic system*

$$\Gamma_i(\theta_1, \ldots, \theta_m) = \theta_i, \quad 1 \leq i \leq m, \tag{1}$$

the equalities of which are obtained directly from the instruction of $\mathscr{P}$. Namely, $l_i f l_j$ yields the equality $f\theta_j = \theta_i$, l_i if P then l_j else l_k yields $(P \to \theta_j, \theta_k) = \theta_i$ and l_m stop yields $I = \theta_m$, where $\varphi\psi = \lambda r.\psi(\varphi(r))$ and

$$(P \to \varphi, \psi)(r) = \begin{cases} \varphi(r), & \text{if } P(r) = 0, \\ \psi(r), & \text{if } P(r) = 1, \\ \uparrow, & \text{if } P(r)\uparrow. \end{cases}$$

We shall make use of the fact that $\varphi_{\mathscr{P}}, \varphi_2, \ldots, \varphi_m$ is the least solution of (1) (Mazurkievicz [1971]; cf. exercise 23.1). Alternatively, one may formally define the execution of a program $\mathscr{P}$ as the first component of the least solution to its characteristic system. (Why does such a solution exist?) Of course, the equalities of (1) can be replaced by inequalities.

A few specific classes of program schemes will be considered now. These are unary schemes with several interpreted function and predicate symbols, i.e., function and predicate constants. Correspondingly, the presence of such constants assumes interpretations by means of more specific unary bases.

Firstly, there are unary schemes with *counters*, which we shall call *C-schemes*. It suffices to have two counting registers since the work of a greater number of counters can be modelled by just two.

A *unary C-scheme* is a unary scheme in which some of the function constants L, R, K, L_2, R_2, K_2 and the predicate constants Ev, Ze, Ev_2, Ze_2 may eventually occur.

Given a unary basis $B = (M, f_1, \ldots, f_u, P_1, \ldots, P_v)$, C-schemes are interpreted by means of the modified unary basis

$$B^c = (\omega^2 \times M, f_1^c, \ldots, f_u^c, P_1^c, \ldots, P_v^c),$$

where

$$f_i^c = \lambda c_1 c_2 r.(c_1, c_2, f_i(r)), i = 1, \ldots, u,$$
$$P_i^c = \lambda c_1 c_2 r.P_i(r), i = 1, \ldots, v.$$

The interpretation of the constants does not depend on B:

$$L = \lambda c_1 c_2 r.(2c_1, c_2, r),$$
$$R = \lambda c_1 c_2 r.(2c_1 + 1, c_2, r),$$
$$K = \lambda c_1 c_2 r.([c_1/2], c_2, r),$$
$$Ev = \lambda c_1 c_2 r.\mathrm{rem}(c_1, 2),$$
$$Ze = \lambda c_1 c_2 r.\mathrm{sg}c_1$$

and similarly $L_2, R_2, K_2, Ev_2, Ze_2$ operate the second counter c_2. The predicates Ze, Ze_2 are not really needed for computations involving functions over M. They are added to ensure that all partial recursive functions can be computed in the counters, including the successor and predecessor functions traditionally used to operate counters. (Cf. 22.3 and exercise 22.1.)

The interpreted C-schemes are C-*programs* in the basis B. Such a program $\mathscr{P}$ computes a function $\varphi_{\mathscr{P}}:\omega^2 \times M \longrightarrow \omega^2 \times M$. (Usually the first counter c_1 or the register r are regarded as outputs,) A function $\varphi:\omega^2 \times M \longrightarrow \omega^2 \times M$ is said to be C-*computable in B* iff $\varphi = \varphi_{\mathscr{P}}$ for a certain C-program $\mathscr{P}$ in B.

In order to characterize C-computability we design an appropriate space.

Example 23.1. The IOS $\mathscr{S}$ of example 22.2 with $\omega^2 \times M, L, R$ playing the roles of M, f_1, f_2. Therefore.

$$\mathscr{F} = \{\varphi/\varphi:\omega^2 \times M \longrightarrow \omega^2 \times M\},$$
$$\varphi\psi = \lambda c_1 c_2 r.\psi(\varphi(c_1, c_2, r)),$$
$$(\varphi, \psi)(2c_1, c_2, r) = \varphi(c_1, c_2, r),$$
$$(\varphi, \psi)(2c_1 + 1, c_2, r) = \psi(c_1, c_2, r),$$

while proposition 21.2 implies that

$$\langle\varphi\rangle(\bar{n}(c_1), c_2, r) = (\bar{n}(c'_1), c'_2, r') \quad \text{iff} \quad \varphi(c_1, c_2, r) = (c'_1, c'_2, r'),$$

$\bar{n}$ standing in the sense of example 22.5,
$[\varphi](c_1, c_2, r) = (c'_1, c'_2, r')$ iff there are n, $(k_0, m_0, r_0), \ldots, (k_n, m_n, r_n)$ such that $(k_0, m_0, r_0) = (c_1, c_2, r)$, k_i is odd and $(k_{i+1}, m_{i+1}, r_{i+1}) = \varphi((k_i - 1)/2, m_i, r_i)$ for all $i < n$, and $(k_n, m_n, r_n) = (2c'_1, c'_2, r')$. Therefore, the first counter c_1 handles the pairing schemes Π, L, R. The second counter c_2 will be used to implement the operation translation as done in 21.11.

Now the functions we are interested in are members of $\mathscr{F}$. Predicates will also be presented by members of $\mathscr{F}$ assigning $\tilde{P} = (P \to L, R)$ to $P:\omega^2 \times M \longrightarrow \{0, 1\}$. The following statement characterizes C-computability in the terms of IOS-recursiveness and prime computability of Moschovakis.

Proposition 23.1 (C-Computability Theorem). Let $\mathscr{S}$ be the IOS of example 23.1, $\varphi \in \mathscr{F}$ and B be a unary basis the carrier of which is M. Then the following are equivalent.

(1) φ is C-computable in B.

(2) φ is prime recursive in

$$\mathscr{B} = \{L_2, R_2, K_2, f_1^c, \ldots, f_u^c, \tilde{Z}e, \tilde{E}v_2, \tilde{Z}e_2, \tilde{P}_1^c, \ldots, \tilde{P}_v^c\}.$$

(3) φ is recursive in $\mathscr{B}$.

(4) φ is prime computable in B, i.e., so are its three co-ordinate functions.

Proof. We shall verify (1)$\Leftrightarrow$(2) and (3)$\Rightarrow$(2), while (2)$\Rightarrow$(3) is obvious and (1)$\Leftrightarrow$(4) is established in Soskov [1983].

Let φ be C-computable in B. Then $\varphi = \varphi_1$, where $\varphi_1, \ldots, \varphi_m$ is the least solution to a certain system

(5) $\Gamma_i(\theta_1, \ldots, \theta_m) = \theta_i$, $1 \le i \le m$, such that each mapping Γ_i is of the form

$\lambda\theta.f\theta$ or $\lambda\theta\tau.(P\to\theta,\tau)$ or $\lambda\theta.I$. Noting the equalities

$$(P\to\varphi,\psi)=(P\to L,R)(\varphi,\psi)=\tilde{P}(\varphi,\psi),$$
$$(Ev\to\varphi,\psi)=(L\varphi,R\psi),\quad K=(I,I),$$

the system (5) can be rewritten as $\sigma_i(I,\theta_1,\ldots,\theta_m)=\theta_i$, $1\le i\le m$, with certain $\sigma_1,\ldots,\sigma_m$ polynomial in $\mathscr{B}$. The latter system is in turn equivalent to the single equality.

(6) $\sigma(I,\theta)=\theta$,

where $\sigma=(\sigma_1,\ldots,\sigma_m)$ is polynomial in $\mathscr{B}$. Now, the element $(\varphi_1,\ldots,\varphi_m)$ is a least solution to (6). (Cf. the proof of 9.16*.) This implies $(\varphi_1,\ldots,\varphi_m)=R[\sigma]$ by 6.11, hence $\varphi=\varphi_1=\bar{1}[\sigma]$ is prime recursive in $\mathscr{B}$. (We even have φ presented in a normal form.)

Conversely, all the members of $\mathscr{F}$ prime recursive in $\mathscr{B}$ are C-computable in B. The program l_1Ll_2, l_2 stop computes L, while similar programs compute $R, L_2, R_2, K_2, f^c_1,\ldots,f^c_u$. The program l_1 if Ze then l_2 else l_3, l_2Ll_4, l_3Rl_4, l_4 stop computes $\tilde{Z}e$; similar programs compute $\tilde{E}v_2, \tilde{Z}e_2, \tilde{P}^c_1,\ldots,\tilde{P}^c_v$.

Let a program $l_1,\ldots,l_m$ compute φ and a program $l_{m+1},\ldots,l_{m+k}$ compute ψ, assuming without loss of generality that no label occurs in both programs at the same time.

Replacing throughout the first program the label l_m by l_{m+1}, the program $l_1,\ldots,l_{m-1},l_{m+1},\ldots,l_{m+k}$ will compute $\varphi\psi$.

Replacing throughout the first program the label l_m by l_{m+k}, the program $l_0,l_1,\ldots.l_{m-1},l_{m+1},\ldots,l_{m+k+2}$ will compute (φ,ψ), where the new instructions are l_0 if Ev then l_{m+k+1} else l_{m+k+2}, $l_{m+k+1}Kl_1$ and $l_{m+k+2}Kl_{m+1}$.

Replacing throughout the instructions $l_1,\ldots,l_{m-1}$ the label l_m by l_0, the function $[\varphi]$ will be computed by the program $l_0,l_1,\ldots,l_{m+2}$ the new instructions of which are l_0 if Ev then l_{m+1} else l_{m+2}, $l_{m+1}Kl_m$ and $l_{m+2}Kl_1$. (While the last three assertions seem obvious, this does not mean that they should not be formally proved.)

Suppose now that φ is recursive in $\mathscr{B}$. The elements $W_1=L_2$, $W_2=R_2$ and $W=E\tilde{v}_2K_2$ satisfy the assumptions of 21.10. (Compare with 21.11.) The elements $f^c_1,\ldots,f^c_u$, $\tilde{Z}e$, $\tilde{P}^c_1,\ldots,\tilde{P}^c_v$ commute with W_1, W_2; the element K_2 commutes with L, R, while the elements $E\tilde{v}_2$, $Z\tilde{e}_2$ satisfy the equalities $L\sigma=\sigma B^2$, $R\sigma=\sigma A^2$. Therefore, φ is prime recursive in $\{W,W_1,W_2\}\cup\mathscr{B}$ by exercise 21.4; hence φ is prime recursive in $\mathscr{B}$. The proof is complete.

It is worth mentioning that the programs constructed when proving the implication (2)$\Rightarrow$(1) above are *structured*, i.e., obtained from simplest programs of the form $\mathscr{P}=l_1fl_2$, l_2 stop by means of consecutive compositions, branchings and loopings. Taken together, the proofs of (1)$\Rightarrow$(2) and (2)$\Rightarrow$(1) ensure that all the unary C-programs can be transformed into equivalent structured ones. (Notice that it suffices to have one counter in order to achieve this.) This seems to capture the essence of the structurization result of Böhm and Jacopini [1966] which will be revisited at the scheme level in chapter 26.

We proceed with the case of n-ary schemes augmented by counters. An *n-ary C-scheme* is a unary C-scheme whose function and predicate symbols are among the following:

$$I_i^{(j)},\quad 1 \le i \ne j \le n,$$
$$f^{(j)}_{1,i_1,\ldots,i_n}, f^{(j)}_{2,i_1,\ldots,i_n}, \ldots, 1 \le i_1, \ldots, i_n,\ j \le n,$$
$$P_{1,i_1,\ldots,i_n}, P_{2,i_1,\ldots,i_n}, \ldots, 1 \le i_1, \ldots, i_n \le n,$$

plus constants operating the counters.

A *n-ary basis* is a tuple $B_n = (M, f_1, \ldots, f_u, P_1, \ldots, P_v)$, where

$$f_i : M^{m_i} \longrightarrow M,\ 1 \le m_i \le n,$$
$$P_i : M^{k_i} \longrightarrow \{0,1\},\ 1 \le k_i \le n.$$

A n-ary C-scheme is now interpreted by a unary modification of B_n,

$$B'_n = (M^n, I_i^{(j)}, 1 \le i \ne j \le n, f^{(j)}_{1,i_1,\ldots,i_n}, \ldots, f^{(j)}_{u,i_1,\ldots,i_n},$$
$$P_{1,i_1,\ldots,i_n}, \ldots, P_{v,i_1,\ldots,i_n}, 1 \le i_1, \ldots, i_n, j \le n),$$

where

$$I_i^{(j)} = \lambda r_1 \ldots r_n.(r_1, \ldots, r_{j-1}, r_i, r_{j+1}, \ldots, r_n),$$
$$f^{(j)}_{i,i_1,\ldots,i_n} = \lambda r_1 \ldots r_n.(r_1, \ldots, r_{j-1}, f_i(r_{i_1}, \ldots, r_{i_{m_i}}), r_{j+1}, \ldots, r_n),$$
$$P_{i,i_1,\ldots,i_n} = \lambda r_1 \ldots r_n. P_i(r_{i_1}, \ldots, r_{i_{k_i}}).$$

Therefore, the *n-ary* C-*programs* in B_n are just unary C-programs in B'_n and their execution requires two counters and n registers. Consequently, a function $\varphi : \omega^2 \times M^n \longrightarrow \omega^2 \times M^n$ is C-*computable in* B_n iff φ is C-computable in the unary basis B'_n. This notion of C-computability is characterized in terms of IOS as follows.

Example 23.2. Example 23.1 with M replaced by M^n.

Proposition 23.2. Let $\mathscr{S}$ be the IOS of example 23.2, $\varphi \in \mathscr{F}$ and let B_n be a n-ary basis with carrier M. Then the following are equivalent.

(1) φ is C-computable in B_n.

(2) φ is prime recursive in

$$\mathscr{B}_n = \{L_2, R_2, K_2, I_i^{(j)c}, 1 \le i \ne j \le n, f^{(j)c}_{1,i_1,\ldots,i_n}, \ldots, f^{(j)c}_{u,i_1,\ldots,i_n},$$
$$\widetilde{Ze}, \widetilde{Ev}_2, \widetilde{Ze}_2, \widetilde{P^c}_{1,i_1,\ldots,i_n}, \ldots, \widetilde{P^c}_{v,i_1,\ldots,i_n}, 1 \le i_1, \ldots, i_n, j \le n\}.$$

(3) φ is recursive in $\mathscr{B}_n$.

This follows from 23.1.

If all the functions of B_n are unary, then C-computability in B_n is again equivalent to prime computability in B_n, as shown in Soskov [1983]. Otherwise this equivalence may fail since one needs a *stack* in order to compute arbitrary functions prime computable in B_n. Thus we come to the next kind of scheme to be considered, viz., schemes with counting and stacking, which we shall call CS-schemes.

The *n-ary* CS-*schemes* are unary C-schemes whose function symbols include the constants $Si_1, \ldots, Si_n$, $So_1, \ldots, So_n$ (*Si* for 'stack-in', *So* for 'stack-out'). An extra predicate constant may also be added to check on the emptiness of the stack. Given a n-ary basis $B_n=(M,f_1,\ldots,f_u,P_1,\ldots,P_v)$, such a scheme is interpreted by means of a unary modification of B_n,

$$B_n^s=(M^n\times M^*,Si_1,\ldots,Si_n,So_1,\ldots,So_n,f_1^s,\ldots,f_u^s,P_1^s,\ldots,P_v^s),$$

where $M^*=\bigcup_n M^n$ and, x ranging over M^*,

$$Si_i=\lambda r_1\ldots r_n x.(r_1,\ldots,r_n,r_i,x),$$
$$So_i(r_1,\ldots,r_n)=(r_1,\ldots,r_n),$$
$$So_i(r_1,\ldots,r_n,r,x)=(r_1,\ldots,r_{i-1},r,r_{i+1},\ldots,r_n,x)$$
$$f_i^s=\lambda r_1\ldots r_n x.(f_i(r_1,\ldots,r_{m_i}),r_2,\ldots,r_n,x),$$
$$P_i^s=\lambda r_1\ldots r_n x.P_i(r_1,\ldots,r_{k_i}).$$

Correspondingly, the *n-ary* CS-*programs* in B_n are exactly the unary C-programs in B_n^s and a function $\varphi:\omega^2\times M^n\times M^*\longrightarrow\omega^2\times M^n\times M^*$ is CS-*computable in* B_n iff it is C-computable in B_n^s.

Example 23.3. Example 23.1 with M replaced by $M^n\times M^*$.

Proposition 23.3. Let $\mathscr{S}$ be the IOS of example 23.3, $\varphi\in\mathscr{F}$ and let B_n be a n-ary basis with carrier M. Then the following are equivalent.

(1) φ is CS-computable in B_n.

(2) φ is prime recursive in

$$\mathscr{B}_n=\{L_2,R_2,K_2,Si_i^c,So_i^c,\ 1\le i\le n,f_1^{sc},\ldots,f_u^{sc},\widetilde{Ze},\widetilde{Ev}_2,\widetilde{Ze}_2,\widetilde{P_1^{sc}},\ldots,\widetilde{P_v^{sc}}\}.$$

(3) φ is recursive in $\mathscr{B}_n$.

This follows from 23.1.

Usually the content of the stack is regarded neither as an input not as an output. That is, one retrieves from the stack only what has been loaded there earlier and, on the other hand, programs can be composed in such way that data loaded in the stack is later retrieved.

Example 23.2 is isomorphic to the subspace of example 23.3, assigning $\varphi^s=\lambda c_1c_2r_1\ldots r_nx.(\varphi(c_1,c_2,r_1,\ldots,r_n),x)$ to $\varphi:\omega^2\times M^n\longrightarrow\omega^2\times M^n$. The functions $I_i^{(j)cs}$ may be computed by making use of the stack since $I_i^{(j)cs}=Si_i^cSo_j^c$. It also follows that $f_{i,i_1,\ldots,i_n}^{(j)cs}$, $(\widetilde{P_{i,i_1,\ldots,i_n}^c})^s$ can be obtained by multiplying f_i^{sc}, $\widetilde{P_i^{sc}}$, $Si_1,\ldots,Si_n$, $So_1,\ldots,So_n$; hence whenever φ is C-computable in B_n, φ^s is CS-computable in B_n. According to Moldestad, Stoltenberg-Hansen and Tucker [1981] the last implication can be reversed, provided all the functions of B_n are unary. The same holds whenever M is finite (I. Soskov).

Functions over $\omega^2\times M^n$ CS-computable in B_n are characterized as follows.

Proposition 23.4. Let B_n be a n-ary basis, let $\mathscr{B}_n$ correspond to it as in the previous statement and let $\varphi:\omega^2\times M^n\longrightarrow\omega^2\times M^n$. Then the following are equivalent.

(1) φ is CS-computable in B_n, i.e., so is φ^s.

(2) φ^s is prime recursive in $\mathscr{B}_n$.

(3) φ^s is recursive in $\mathscr{B}_n$.

(4) φ is prime computable in B_n, i.e., so are its co-ordinate functions.

Proof. The assertions (1), (2), (3) are equivalent by 23.3, while (1)$\Leftrightarrow$(4) is proved in Soskov [1983].

Certain kinds of recursive schemes and programs may also be considered. These are systems of equalities

$$(1) \qquad \Gamma_i(\theta_1, \ldots, \theta_m) = \theta_i,\ 1 \le i \le m,$$

with arbitrary mappings Γ_i constructed by the operations of multiplication and branching. In this way one gets *unary recursive* C-*schemes*, *n-ary recursive* C-*schemes* and *n-ary recursive* CS-*schemes*, depending on the function and predicate constants allowed. If the function and predicate symbols are interpreted, then we have *recursive programs*. The function $\varphi_{\mathscr{P}}$ computed by such a program $\mathscr{P}$ is by definition the first component of the least solution to the corresponding system of equalities. However, recursion can be eliminated since the First Recursion Theorem enables us to solve equations.

Proposition 23.5 (Recursion Elimination Theorem). Let $\mathscr{P}$ be a unary recursive C-program (respectively n-ary recursive C-program, n-ary recursive CS-program). Then there is a unary C-program (n-ary C-program, n-ary CS-program) $\mathscr{P}_1$ in the same basis such that $\varphi_{\mathscr{P}} = \varphi_{\mathscr{P}_1}$.

Proof. Following the proof of 23.1, we rewrite the system (1) as a system introduced by mappings polynomial in the corresponding set $\mathscr{B}$ (respectively $\mathscr{B}_n$). The latter system is then solved by making use of 9.16* and finally the implication (3)$\Rightarrow$(1) of 23.1 (respectively 23.2, 23.3) completes the proof.

In fact, the above recursion elimination (i.e., implementation) algorithm does not depend on interpretations, so the result holds for schemes as well. The case of unary C-schemes is considered in Greibach [1975], while it is well known in Computer Science that recursion can be implemented by stacking. It should be stressed that the availability of $I_i^{(j)}$ or Si_i, So_i determines the particular n-ary recursive schemes which one can actually write down. (Cf. exercise 23.4.)

In another development, one may consider *multiple-valued program schemes*. The definitions are the same but function and predicate symbols are interpreted by multiple-valued functions and predicates. Analogues of 23.1–23.5 hold with $\mathscr{S}$ obtained from example 22.1 rather than 22.2.

A natural multiple-valued function constant to be allowed is $U = \lambda c_1 c_2 r.\{(2c_1, c_2, r), (2c_1+1, c_2, r)\}$ or, equivalently, $\lambda c_1 c_2 r.\{(0, c_2, r), (1, c_2, r)\}$ or $\lambda c_1 c_2 r.\{(n, c_2, r)/n \in \omega\}$, or the predicate constant $\lambda c_1 c_2 r.\{0, 1\}$. The same effect would be achieved by lifting the restriction that each label in a scheme appear only once as a prefix. If the scheme consists of m instructions in which k distinct labels occur, $k \le m$, then the corresponding characteristic system is of the form

$$\Gamma_j(\theta_1, \ldots, \theta_k) \le \theta_{i_j}, \quad 1 \le j \le m,$$

where $1 \le i_1, \ldots, i_m \le k$. According to exercise 9.4 such systems may be solved by making use of U while, conversely, U may be computed by the program $l_1 L l_2$, $l_1 R l_2$, l_2 stop. In the case of unary C-schemes and n-ary CS-schemes the computability so obtained is equivalent to prime computability in B, U, respectively B_n, U. As pointed out in Soskov [1983], the latter is in essence computability in B or B_n by the effective definitional schemes of Friedman [1971], a natural analogue to relative partial recursiveness.

Finally, *infinitary schemes* may also be considered. These are infinite lists of instructions, one of them being l_m stop. Constants L_2, R_2, K_2, Ev_2 and Ze_2 are not needed since one counter proves sufficient now. The corresponding notions of infinitary C-computability and infinitary CS-computability can be characterized by means of (prime) c-recursiveness in appropriate IOS.

EXERCISES TO CHAPTER 23

Exercise 23.1. Let $\mathscr{P} = l_1,\ldots,l_m$ be a unary program, let $\varphi_{\mathscr{P}}$ be the function computed by $\mathscr{P}$ and let $\tau_1,\ldots,\tau_m$ be a solution to the characteristic system of $\mathscr{P}$. Show that $\varphi_{\mathscr{P}} \leq \tau_1$.

Hint. Let $\varphi_{\mathscr{P}}(s) = t$. Then there are $n, i_0,\ldots,i_n, r_0,\ldots,r_n$ such that $i_0 = 1$, $i_n = m$, $r_0 = s$, $r_n = t$ and for all $k < n$ either $l_{i_k} f l_{i_{k+1}} \in \mathscr{P}$ and $r_{k+1} = f(r_k)$ or if P then l_i else $l_j \in \mathscr{P}$, $r_{k+1} = r_k$ and either $P(r_k) = 0$, $i_{k+1} = i$ or $P(r_k) = 1$, $i_{k+1} = j$. Starting with $k = n$ show that $\tau_{i_k}(r_k) = t$ for all $k \leq n$. In particular, $\tau_1(s) = t$.

Example 23.4. Example 22.2 with $\omega \times M$ taken for M, $L = \lambda c_1 r.(2c_1, r)$ and $R = \lambda c_1 r.(2c_1 + 1, r)$.

Exercise 23.2. Let $\mathscr{S}$ be the IOS of example 23.4, let $B = (M, f_1,\ldots,f_u, P_1,\ldots,P_v)$ be a unary basis, let $\varphi \in \mathscr{F}$ and let $\varphi^*: \omega^2 \times M \longrightarrow \omega^2 \times M$ be such that $\varphi^*(c_1, c_2, r) = (c_1', c_2', r')$ iff $c_2' = c_2$ and $\varphi(c_1, r) = (c_1', r')$. Show that φ^* is C-computable in B iff φ is recursive in $\mathscr{B} = \{\lambda c_1 r.(c_1, f_i(r)), 1 \leq i \leq u, (\lambda c_1 r.\mathrm{sg} c_1)^{\sim}, (\lambda c_1 r.P_i(r))^{\sim}, 1 \leq i \leq v\}$.

Hint. Let $\mathscr{S}_1$ be the IOS of example 23.1. Then φ^* is C-computable in B iff it is recursive in a corresponding subset of $\mathscr{F}_1$ by 23.1. Using exercise 21.6, show that the latter is equivalent to φ being recursive in $\mathscr{B}$.

Exercise 23.3. Let $\mathscr{S}$ be the IOS of example 22.2, i.e. example 4.7. Show that φ is recursive in $\psi_1,\ldots,\psi_m$ iff $\lambda c_1 c_2 r.(c_1, c_2, \varphi(r))$ is C-computable in $B = (M, f_1, f_2, f_1^{-1}, f_2^{-1}, \psi_1,\ldots,\psi_m, Ev)$, where Ev is a predicate true of $f_1(M)$, false on $f_2(M)$ and undefined otherwise.

Hint. Let $\mathscr{S}_1$ be the IOS of example 22.2 with $\omega \times M$ taken for M, preserving the splitting scheme of M, and let $\mathscr{S}_2$ be the IOS of example 23.4. Apply exercise 21.6 to $\mathscr{S}, \mathscr{S}_1$, exercise 7.2 to $\mathscr{S}_1, \mathscr{S}_2$ and exercise 23.2 to $\mathscr{S}_2$.

Remark. By 23.1 φ is recursive in $\psi_1,\ldots,\psi_m$ iff φ is prime computable in $f_1, f_2, f_1^{-1}, f_2^{-1}, \psi_1,\ldots,\psi_m, Ev$. The relative recursiveness of example 4.8 is characterized in the same way since exercise 23.2, 23.3 can be restated for multiple-valued functions.

Exercise 23.4. Let $g, g_1, g_2: M \longrightarrow M$, $h: M^2 \longrightarrow M$ and $P: M \longrightarrow \{0,1\}$. Write recursive programs to compute the functions f introduced by recursion as follows.

a. $$f(r)=\begin{cases} g(r), & \text{if } P(r)=0,\\ g_2(f(g_1(r))), & \text{if } P(r)=1,\\ \uparrow, & \text{otherwise.}\end{cases}$$

b. $$f(r)=\begin{cases} g(r), & \text{if } P(r)=0,\\ h(f(g_1(r)),r), & \text{if } P(r)=1,\\ \uparrow, & \text{otherwise.}\end{cases}$$

c. $$f(r)=\begin{cases} g(r), & \text{if } P(r)=0,\\ h(f(g_1(r)),f(g_2(r))), & \text{if } P(r)=1,\\ \uparrow, & \text{otherwise.}\end{cases}$$

Hint. a. $(P\to g, g_1\theta g_2)=\theta$.
b. $(P^s\to g^s, Si_2Si_1So_2g_1^s\theta h^sSo_2)=\theta$.
c. $(P^s\to g^s, Si_2Si_1So_2g_2^s\theta Si_1Si_2So_1So_2g_1^s\theta h^sSo_2)=\theta$.

Remark. Superscripts c turn the last two binary recursive programs with stacking into binary recursive CS-programs which can be transformed into equivalent binary CS-programs by 23.5. More generally, this exercise shows how arbitrary polyadic recursive procedures can be transformed into equivalent recursive CS-schemes (cf. Skordev [1983]).

CHAPTER 24

Prime and search computability

The best known concepts of effectively computable functions over arbitrary domains are perhaps those of prime and search computability of Moschovakis. This chapter is aimed to show that they are particular instances of t-recursiveness.

We begin with some definitions from Moschovakis [1969] adduced in slightly modified notations.

Let M be an arbitrary set the members of which are not ordered pairs, $0 \notin M$ and

$$N^* = cl(N \cup \{0\}/\lambda st.(s,t))$$

whenever $N \subseteq M$. Natural numbers are introduced by taking $n+1=(0,n)$, while $\langle s_1,\ldots,s_n\rangle$ stands for $(n,(s_1,(s_2,\ldots,(s_n,0)\ldots)))$ and $L^*, R^*: M^* \to M^*$ such that $L^*(0)=R^*(0)=0$, $L^*(M)=R^*(M)=1$ and $L^*((s,t))=s$, $R^*((s,t))=t$. Object of interest are the n-ary partial multiple-valued functions over $M^*, \varphi: M^{*n} \to 2^{M^*}$.

The basic notion of prime computability is introduced via index schemes following the pattern of Kleene [1959]. Given a subset N of M^* and a list of functions $\psi_1,\ldots,\psi_l, \psi_i$ being n_i-ary, an n-ary function φ is *prime computable with constants derived from N in* $\psi_1,\ldots,\psi_l$, denoted $\varphi \in PC(N,\psi_1,\ldots,\psi_l)$, iff $\varphi = \lambda s_1\ldots s_n.\sigma(e,s_1,\ldots,s_n)$ for a certain index $e \in N^*$. The universal function σ is the least fixed point of the mapping $\Omega: \mathscr{F}_0 \to \mathscr{F}_0$, where $\mathscr{F}_0 = \{\varphi_0/\varphi_0: \bigcup_{n>0} M^{*n} \to 2^{M^*}\}$ and $\Omega(\varphi_0)$ is defined by the following clauses for all $\varphi_0 \in \mathscr{F}_0$.

0. If $t \in \psi_i(s_1,\ldots,s_{n_i})$, $1 \le i \le l$, then $t \in \Omega(\varphi_0)(\langle 0, n_i+n, i\rangle, s_1,\ldots,s_{n_i}, r_1,\ldots,r_n)$.
1. $t \in \Omega(\varphi_0)(\langle 1,n,t\rangle, s_1,\ldots,s_n)$.
2. $t \in \Omega(\varphi_0)(\langle 2,n+1\rangle, t, s_1,\ldots,s_n)$.
3. $(t,r) \in \Omega(\varphi_0)(\langle 3,n+2\rangle, t, r, s_1,\ldots,s_n)$.

4_0. $L^*(t) \in \Omega(\varphi_0)(\langle 4,n+1,0\rangle, t, s_1,\ldots,s_n)$.

4_1. $R^*(t) \in \Omega(\varphi_0)(\langle 4,n+1,1\rangle, t, s_1,\ldots,s_n)$.

5. If $t \in \varphi_0(e_1,r,s_1,\ldots,s_n)$ for a certain $r \in \varphi_0(e_2,s_1,\ldots,s_n)$, then $t \in \Omega(\varphi_0)(\langle 5,n,e_1,e_2\rangle, s_1,\ldots,s_n)$.
6. If $r \in M \cup \{0\}$ and $t \in \varphi_0(e_1,r,s_1,\ldots,s_n)$, then $t \in \Omega(\varphi_0)(\langle 6,n+1,e_1,e_2\rangle, r, s_1,\ldots,s_n)$.

If $t \in \varphi_0(e_2, u, v, s, r, s_1, \ldots, s_n)$ for certain $u \in \varphi_0(\langle 6, n+1, e_1, e_2 \rangle, s, s_1, \ldots, s_n)$ and $v \in \varphi_0(\langle 6, n+1, e_1, e_2 \rangle, r, s_1, \ldots, s_n)$, then $t \in \Omega(\varphi_0)(\langle 6, n+1, e_1, e_2 \rangle, (s, r), s_1, \ldots, s_n)$.

7. If $t \in \varphi_0(e, s_{j+1}, s_1, \ldots, s_j, s_{j+2}, \ldots, s_n)$, then $t \in \Omega(\varphi_0)(\langle 7, n, j, e \rangle, s_1, \ldots, s_n)$.

8. If $t \in \varphi_0(e, s_1, \ldots, s_n)$, then $t \in \Omega(\varphi_0)(\langle 8, n+m+1, n \rangle, e, s_1, \ldots, s_n, r_1, \ldots, r_m)$.

Notice that Ω is a continuous mapping, hence $\sigma = \sup_n \sigma_n$, where σ_0 is the nowhere defined function and $\sigma_{n+1} = \Omega(\sigma_n)$. Traditionally, $\{e\}(s_1, \ldots, s_n) \to t$ is written for $t \in \sigma(e, s_1, \ldots, s_n)$. If $\varphi \in PC(\varnothing, \psi_1, \ldots, \psi_l)$, then φ is *prime computable in* $\psi_1, \ldots, \psi_l$. Some of the functions $\psi_1, \ldots, \psi_l$ may be regarded as predicates, provided they are $\{0,1\}$-valued.

Let us design an appropriate IOS to express the notion of prime computability. It suffices to consider a space composed of unary functions over M^* since this set is closed under the prime computable pairing function $\lambda st.(s,t)$.

Example 24.1. The IOS $\mathscr{S} = (\mathscr{F}, I, \Pi, L, R)$ of example 22.1 (alias example 4.8) with M^* taken for $M, L = \lambda s.(0,s)$ and $R = \lambda s.(1,s)$. Then, $\mathscr{F} = \{\varphi/\varphi\colon M^* \to 2^{M^*}\}$, $I = \lambda s.s$, $\varphi \le \psi$ iff $\varphi \subseteq \psi$, $\varphi\psi = \lambda s.\psi(\varphi(s))$, $(\varphi, \psi)((0,s)) = \varphi(s)$, $(\varphi, \psi)((1,s)) = \psi(s)$ and $(\varphi, \psi)(s)\uparrow$ otherwise.

Proposition 21.2 characterizes the operations $\langle\ \rangle$, $[\ \ ]$ explicitly as follows.

$t \in \langle \varphi \rangle(s)$ iff there are n, s_0, t_0 such that $s = \bar{n}(s_0)$, $t = \bar{n}(t_0)$ and $t_0 \in \varphi(s_0)$.

$t \in [\varphi](s)$ iff there are $n, r_0, \ldots, r_n, t_0, \ldots, t_{n-1}$ such that $r_0 = s$, $r_i = (1, t_i)$ and $r_{i+1} = \varphi(t_i)$ for all $i < n$, while $r_n = (0, t)$.

A storing operation corresponds to the pairing function $J = \lambda st.(s,t)$ by 21.13. Namely,

$$St(\varphi)((s,t)) = (s, \varphi(t))$$

and $St(\varphi)(s)\uparrow$ otherwise. We broaden the notion of st-recursiveness by assuming the following elements $K_0^* - K_3^*$ initial:

$K_0^* = \lambda s.(s,s)$, $K_1^*((s,t)) = (t,s)$ and $K_1^*(s)\uparrow$ otherwise, $K_2^*(0)$, $K_3^*(M) = 0$ and $K_2^*(s)$, $K_3^*(s) = 1$ otherwise. Then the elements $K_0 - K_6$ of the proof of 21.13 and exercise 21.7 are expressed as follows. One may take $K_3 = L$, while

$$K_4 = K_1^* St(K_2^*) K_1^*(I, I), \qquad K_5 = K_0^*,$$
$$K_6 = K_0^* St(K_4 K_1^* K_4) K_1^* St^2(K_4),$$
$$K_0 = K_6(L, R), \qquad K_1 = St(K_1^*) K_6 K_1^*,$$
$$K_2 = K_1^* K_6 St(K_1^*).$$

Therefore, an element φ is now st-recursive in $\mathscr{B}$ iff $\varphi \in cl(\{L, R, K_0^* - K_3^*\} \cup \mathscr{B}/\circ, \Pi, \langle\ \rangle, [\ \ ], St)$.

We recall that the set $\mathscr{L}$ of St is $M^{\sim} = \{\tilde{s} = \lambda t.(s,t)/s \in M\}$ and assign to each $\varphi_0 \in \mathscr{F}_0$ a unary function $\varphi_0^{\sim} \in \mathscr{F}$ such that $t \in \varphi_0^{\sim}(s)$ iff $\exists n s_1 \ldots s_n (s = \langle s_1, \ldots, s_n \rangle \,\&\, t \in \varphi_0(s_1, \ldots, s_n))$.

Proposition 24.1. If $\phi \in PC(N, \psi_1, \ldots, \psi_l)$, then φ is st-recursive in $N^{\sim} \cup \{\tilde{\psi}_1, \ldots, \tilde{\psi}_l\}$.

Proof. There is an $e \in N^*$ such that $\varphi = \lambda s_1 \dots s_n.\sigma(e, s_1, \dots, s_n)$, hence $\tilde{\varphi} = \tilde{e} K_6 K_1^* St(L) K_1^* \tilde{\sigma}$. The element $\tilde{e}$ is st-recursive in $N^{\sim}$, hence it suffices to prove that $\tilde{\sigma}$ is st-recursive in $\tilde{\psi}_1, \dots, \tilde{\psi}_l$. For this purpose we construct a mapping $\Omega^{\sim}: \mathscr{F} \to \mathscr{F}$ st-recursive in $\tilde{\psi}_1, \dots, \tilde{\psi}_l$ such that $\tilde{\sigma} = \mu\theta.\Omega^{\sim}(\theta)$. Namely, take

$$\Omega^{\sim} = \lambda\theta.\rho((\tilde{\psi}_1, \dots, \tilde{\psi}_l, I), I, St(\theta)\rho_1\theta, \theta, St(St(\theta)K_1^* St(\theta))\rho_2\theta),$$

where

$$\rho(\langle\langle 0, n_i + n, i\rangle, s_1, \dots, s_{n_i}, r_1, \dots, r_n\rangle) = \overline{i0}(\langle s_1, \dots, s_{n_i}\rangle), \quad 1 \le i \le l,$$
$$\rho(\langle\langle 1, n, t\rangle, s_1, \dots, s_n\rangle) = \bar{1}(t),$$
$$\rho(\langle\langle 2, n+1\rangle, t, s_1, \dots, s_n\rangle) = \bar{1}(t),$$
$$\rho(\langle\langle 3, n+2\rangle, t, r, s_1, \dots, s_n\rangle) = \bar{1}((t, r)),$$
$$\rho(\langle\langle 4, n+1, 0\rangle, t, s_1, \dots, s_n\rangle) = \bar{1}(L^*(t)),$$
$$\rho(\langle\langle 4, n+1, 1\rangle, t, s_1, \dots, s_n\rangle) = \bar{1}(R^*(t)),$$
$$\rho(\langle\langle 5, n, e_1, e_2\rangle, s_1, \dots, s_n\rangle) = \bar{2}((\langle e_1, s_1, \dots, s_n\rangle, \langle e_2, s_1, \dots, s_n\rangle)),$$
$$\rho(\langle\langle 6, n+1, e_1, e_2\rangle, r, s_1, \dots, s_n\rangle) = \bar{3}(\langle e_1, r, s_1, \dots, s_n\rangle), \quad \text{if } r \in M \cup \{0\},$$
$$\rho(\langle\langle 6, n+1, e_1, e_2\rangle, (s, r), s_1, \dots, s_n\rangle) = R^4((\langle e_2, s, r, s_1, \dots, s_n\rangle,$$
$$(\langle\langle 6, n+1, e_1, e_2\rangle, s, s_1, \dots, s_n\rangle, \langle\langle 6, n+1, e_1, e_2\rangle, r, s_1, \dots, s_n\rangle))),$$
$$\rho(\langle\langle 7, n, j, e\rangle, s_1, \dots, s_n\rangle) = \bar{3}(\langle e, s_{j+1}, s_1, \dots, s_j, s_{j+2}, \dots, s_n\rangle),$$
$$\rho(\langle\langle 8, n+m+1, n\rangle, e, s_1, \dots, s_n, r_1, \dots, r_m\rangle) = \bar{3}(\langle e, s_1, \dots, s_n\rangle)$$

and $\rho(s)\uparrow$ otherwise,

$$\rho_1((\langle e, s_1, \dots, s_n\rangle, r)) = \langle e, r, s_1, \dots, s_n\rangle,$$
$$\rho_2((\langle e, s, r, s_1, \dots, s_n\rangle, (v, u))) = \langle e, u, v, s, r, s_1, \dots, s_n\rangle$$

and $\rho_1(s)$, $\rho_2(s)\uparrow$ otherwise. Therefore, for all θ the element $\Omega^{\sim}(\theta)$ is defined as follows.

0. If $t \in \tilde{\psi}_i(\langle s_1, \dots, s_{n_i}\rangle)$, $1 \le i \le l$, then $t \in \Omega^{\sim}(\theta)(\langle\langle 0, n_i + n, i\rangle, s_1, \dots, s_{n_i}, r_1, \dots, r_n\rangle)$.
1. $t \in \Omega^{\sim}(\theta)(\langle\langle 1, n, t\rangle, s_1, \dots, s_n\rangle)$.
2. $t \in \Omega^{\sim}(\theta)(\langle\langle 2, n+1\rangle, t, s_1, \dots, s_n\rangle)$.
3. $(t, r) \in \Omega^{\sim}(\theta)(\langle\langle 3, n+2\rangle, t, r, s_1, \dots, s_n\rangle)$.

4_0. $L^*(t) \in \Omega^{\sim}(\theta)(\langle\langle 4, n+1, 0\rangle, t, s_1, \dots, s_n\rangle)$.

4_1. $R^*(t) \in \Omega^{\sim}(\theta)(\langle\langle 4, n+1, 1\rangle, t, s_1, \dots, s_n\rangle)$.

5. If $t \in \theta(\langle e_1, r, s_1, \dots, s_n\rangle)$ for a certain $r \in \theta(\langle e_2, s_1, \dots, s_n\rangle)$, then

$$t \in \Omega^{\sim}(\theta)(\langle\langle 5, n, e_1, e_2\rangle, s_1, \dots, s_n\rangle).$$

6. If $t \in \theta(\langle e_1, r, s_1, \dots, s_n\rangle)$, $r \in M \cup \{0\}$, then $t \in \Omega^{\sim}(\theta)(\langle\langle 6, n+1, e_1, e_2\rangle, r, s_1, \dots, s_n\rangle)$.

If $t \in \theta(\langle e_2, u, v, s, r, s_1, \dots, s_n\rangle)$ for certain $u \in \theta(\langle\langle 6, n+1, e_1, e_2\rangle, s, s_1, \dots, s_n\rangle)$, $v \in \theta(\langle\langle 6, n+1, e_1, e_2\rangle, r, s_1, \dots, s_n\rangle)$, then

$$t \in \Omega^{\sim}(\theta)(\langle\langle 6, n+1, e_1, e_2\rangle, (s, r), s_1, \dots, s_n\rangle).$$

7. If $t \in \theta(\langle e, s_{j+1}, s_1, \dots, s_j, s_{j+2}, \dots, s_n\rangle)$, then

$$t \in \Omega^{\sim}(\theta)(\langle\langle 7, n, j, e\rangle, s_1, \dots, s_n\rangle).$$

8. If $t \in \theta(\langle e, s_1, \ldots, s_n \rangle)$, then

$$t \in \Omega^{\sim}(\theta)(\langle\langle 8, n+m+1, n\rangle, e, s_1, \ldots, s_n, r_1, \ldots, r_m\rangle).$$

It follows in particular that $\Omega^{\sim}(\tilde{\varphi}_0) = \Omega(\varphi_0)^{\sim}$ for all $\alpha_0 \in \mathscr{F}_0$; hence

$$\tilde{\sigma} = (\sup_n \Omega^n(O_0))^{\sim} = \sup_n \Omega^n(O_0)^{\sim} = \sup_n \Omega^{\sim n}(O_0^{\sim})$$

$$= \sup_n \Omega^{\sim n}(O) = \mu\theta.\Omega^{\sim}(\theta).$$

The elements ρ, ρ_1, ρ_2 are st-recursive (further details about this are left to the exercises), hence $\Omega^{\sim}$ is st-recursive in $\tilde{\psi}_1, \ldots, \tilde{\psi}_l$. Therefore, $\tilde{\sigma}$ is st-recursive in $\tilde{\psi}_1, \ldots, \tilde{\psi}_l$ by 10.8*, which completes the proof.

Proposition 24.2. If φ is a n-ary function and $\tilde{\varphi}$ is st-recursive in $N^{\sim} \cup \{\tilde{\psi}_1, \ldots, \tilde{\psi}_l\}$, then $\varphi \in PC(N, \psi_1, \ldots, \psi_l)$.

Proof. The function φ is prime computable in $\tilde{\varphi}$; hence it suffices to show that whenever $\varphi \in \mathscr{F}$ is st-recursive in $N^{\sim} \cup \{\tilde{\psi}_1, \ldots, \tilde{\psi}_l\}$, then $\varphi \in PC(N, \psi_1, \ldots, \psi_l)$.

The initial functions $L, R, K_0^* - K_3^*, \tilde{\psi}_1, \ldots, \tilde{\psi}_l$ and $\tilde{e}$, $e \in N$, are in $PC(N, \psi_1, \ldots, \psi_l)$, and it is immediate that the set $PC(N, \psi_1, \ldots, \psi_l)$ is closed under the operations $\circ, \Pi, St$. It remains to show that this set is closed under $\langle\ \ \rangle, [\ \ ]$ as well. The case of iteration follows.

Suppose that $\varphi \in PC(N, \psi_1, \ldots, \psi_l)$, i.e. $\varphi = \lambda s.\sigma(e_0, s)$ with a certain $e_0 \in N^*$. Then there is an index $e_1 \in N^*$ such that $\lambda s.\sigma(e_1, e, s) = (I, \varphi\lambda s.\sigma(e, s))$ for all e and $\lambda s.\sigma_{n+1}(e_1, e, s) \leq (I, \varphi\lambda s.\sigma_n(e, s))$ for all e, n. (We recall that $\sigma_n = \Omega^n(O_0)$.)

There is by lemma 21 of Moschovakis [1969] an index $e \in N^*$ such that $\sigma(e, s) = \sigma(e_1, e, s)$ for all s and $\sigma_n(e, s) \subseteq \sigma_n(e_1, e, s)$ for all s, n. Writing ψ for $\lambda s.\sigma(e, s)$, it follows that $\psi = (I, \varphi\psi)$, hence $[\varphi] \leq \psi$. On the other hand, $\lambda s.\sigma_0(e, s) \leq [\varphi]$ and whenever $\lambda s.\sigma_n(e, s) \leq [\varphi]$, then

$$\lambda s.\sigma_{n+1}(e, s) \leq \lambda s.\sigma_{n+1}(e_1, e, s) \leq (I, \varphi\lambda s.\sigma_n(e, s)) \leq (I, \varphi[\varphi]) = [\varphi],$$

hence

$$\psi = \lambda s.\sigma(e, s) = \sup_n \lambda s.\sigma_n(e, s) \leq [\varphi].$$

Therefore, $[\varphi] = \psi \in PC(N, \psi_1, \ldots, \psi_l)$.

The case of translation is treated similarly, though it may be omitted because the operation $\langle\ \ \rangle$ is prime st-recursive. (This will follow from the proof of 27.17; see also exercise 21.7.) This completes the proof.

The following Prime Computability Theorem characterizes the prime computability of Moschovakis in terms of st-recursiveness. A similar characterization may be found in Skordev [1980].

Proposition 24.3. Let φ be a n-ary function over M^*. Then the following are equivalent.

(1) $\varphi \in PC(N, \psi_1, \ldots, \psi_l)$.
(2) $\tilde{\varphi}$ is st-recursive in $N^{\sim} \cup \{\tilde{\psi}_1, \ldots, \tilde{\psi}_l\}$.
(3) $\tilde{\varphi}$ is prime st-recursive in $N^{\sim} \cup \{\tilde{\psi}_1, \ldots, \tilde{\psi}_l\}$.

In particular, if $\varphi, \psi_1, \ldots, \psi_l$ are unary, then $\varphi, \psi_1, \ldots, \psi_l$ can be substituted for $\tilde{\varphi}, \tilde{\psi}_1, \ldots, \tilde{\psi}_l$ in (2), (3).

This follows from 24.1, 24.2 and the fact that, as mentioned in the proof of 24.2, translation is prime st-recursive.

One may substitute $\check{N} = \{\check{s} = \lambda t.s / s \in N\}$ for $N^{\sim}$ since $\tilde{s} = K_0^* St(\check{s}) K_1^*$ and $\check{s} = \tilde{s} K_1^* K_4$. The set $N^{\sim}$ may also be replaced by $\mathscr{L}_0 = N^{*\sim}$ and obviously $\mathscr{L}_0^2 K_1 \subseteq \mathscr{L}_0$, hence the 'boldface' version results mentioned in the remark to exercise 10.9 hold for st-recursiveness in $N^{*\sim} \cup \{\tilde{\psi}_1, \ldots, \tilde{\psi}_l\}$.

If $\psi_1, \ldots, \psi_l$ are single-valued, then the above constructions can be carried out within the subspace of example 24.1 consisting of all the unary single-valued functions over M^*, which is a particular instance of example 22.2.

The characterization of the search computability of Moschovakis is now immediate since

$$SC(N, \psi_1, \ldots, \psi_l) = PC(N, U^*, \psi_1, \ldots, \psi_l),$$

where $U^* = \lambda s.M^*$ and $SC(N, \psi_1, \ldots, \psi_l)$ stands for the set of functions over M^* *search computable with constants derived from N in* $\psi_1, \ldots, \psi_l$. (Cf. Skordev [1980] about this equality.)

Proposition 24.4 (Search Computability Theorem). Proposition 24.3 holds with SC substituted for PC and U^* added to the initial elements in (2), (3).

As shown in Soskov [1983], the functions over M prime computable in $U = L \cup R$ are in essence those computable by the effective definitional schemes of Friedman [1971] or, more precisely, by the recursively enumerable definitional schemes of Shepherdson [1975], hence st-recursiveness also subsumes the latter notions. Notice that while st-recursiveness subsumes the prime and search computability of Moschovakis and Friedman's computability, by 23.1, 23.4 so does recursiveness as far as functions over M rather than M^* are concerned.

In order to express the notion of hyperprojective function of Moschovakis [1969], one should consider consecutive spaces $\mathscr{S}, \mathscr{S}'$ constructed from example 24.1 by 19.11. The corresponding IOS-notion is that of $\mathscr{B}'$-recursiveness with $\mathscr{B}' = \{\overline{K_0^*} - \overline{K_3^*}, St, Ex\}$, where Ex is the functional E of Moschovakis,

$$t \in Ex(\varphi)(s)$$
$$\text{iff } (t = 0 \,\&\, \forall r \varphi((r,s))\downarrow \,\&\, \exists r(0 \in \varphi((r,s)))) \vee (t = 1 \,\&\, \forall r \exists u \neq 0 (u \in \varphi((r,s)))).$$

We shall not go into further details since an analogous characterization via $\mathscr{B}'$-recursiveness will be obtained in chapter 29.

EXERCISES TO CHAPTER 24

Exercise 24.1. Show that the function Z is st-recursive, where $Z(0) = \bar{0}(0)$, $Z(s) = \bar{1}(s)$, if $s \in M$, and $Z(s) = R^2(s)$ otherwise.

Hint. $Z = K_0^* St(K_2^*) K_1^* (\bar{0}, K_0^* St(K_3^*) K_1^* R)$.

Exercise 24.2. Show that the function Nn is st-recursive, where $Nn(m) = 0$ for all m, and $Nn(s) = 1$ otherwise.

Hint. $Nn = R[Z(L, \breve{I}L, K_1^* St(K_2^*)K_1^*(R, \breve{I}L))]$.

Exercise 24.3. Show that the function Seq is st-recursive, where $Seq(s) = 0$, if $s = \langle s_1, \ldots, s_n \rangle$ for certain $n, s_1, \ldots, s_n$, and $Seq(s) = 1$ otherwise.

Hint. $Seq = R[\psi]$, where

$$\psi = Z(\breve{I}L, \breve{I}L, St(Z)K_6(K_1^* K_4 K_2^* L, K_6(\breve{I}L, St(K_4)K_1^* \circ St(Z)K_6(\breve{I}L, K_6(\breve{I}L, St(K_1^* St(K_2^*)K_1^*)K_6(K_1^* R, \breve{I}L)))))).$$

Exercise 24.4. Show that the function Eq is st-recursive, where $Eq(s) = 0$, if $s = (n, n)$ for a certain n, and $Eq(s) = 1$ otherwise.

Hint. $Eq = K_0^* St(Seq)K_1^*(K_4 Nn, \breve{I})$.

Remark. Nn, Seq, Eq make it easy to construct more sophisticated st-recursive elements such as ρ, ρ_1, ρ_2 in the proof of 24.1. For example,

$$\rho_1 = K_1^* St(K_0^* St(Seq)K_1^*)K_6(K_6 K_1^* St(L)K_1^*, O).$$

CHAPTER 25

Further examples

This chapter studies several 'less standard' first order examples of IOS involving intuitions connected with probabilistic nondeterminism and the reliable estimation of functions. They originate in examples 13, 18, 19 of Skordev [1980], chapter 3.

Proposition 25.1 (Example 25.1). Let M be an infinite set with a splitting scheme f_1, f_2. Take $\mathscr{F} = \{\varphi/\varphi: M^2 \to [0, \infty]\}$, $\varphi \le \psi$ iff $\forall st(\varphi(s,t) \le \psi(s,t))$,

$$\varphi\psi(s,t) = \sum_r \varphi(s,r)\psi(r,t) = \sup\left\{\sum_{r\in N} \varphi(s,r)\psi(r,t)/N \subseteq M \text{ is finite}\right\},$$

$$(\varphi, \psi)(f_1(s), t) = \varphi(s,t), (\varphi,\psi)(f_2(s),t) = \psi(s,t)$$

and $(\varphi,\psi)(s) = 0$ otherwise, $I(s,s) = 1$ and $I(s,t) = 0$ otherwise, $L = \lambda st.I(f_1(s), t)$ and $R = \lambda st.I(f_2(s), t)$. Then $\mathscr{S} = (\mathscr{F}, I, \Pi, L, R)$ is a $(**)_0$, $(***)_0$-complete OS.

Proof. The quadruple $(\mathscr{F}, \Pi, \lambda\theta.L\theta, \lambda\theta.R\theta)$ is both a CCPS and a SCPS by 16.4, 16.10 since all the subsets of $[0,\infty]$ have least upper bounds.

The equalities $\varphi I = I\varphi = \varphi$ are immediate. Let us verify that multiplication is associative and right distributive with respect to Π.

$$\begin{aligned}(\varphi\psi)\chi(s,t) &= \sum_r \varphi\psi(s,r)\chi(r,t) = \sum_r \left(\sum_u \varphi(s,u)\psi(u,r)\right)\chi(r,t)\\ &= \sum_r\sum_u \varphi(s,u)\psi(u,r)\chi(r,t) = \sum_u\sum_r \varphi(s,u)\psi(u,r)\chi(r,t)\\ &= \sum_u \varphi(s,u)\sum_r \psi(u,r)\chi(r,t) = \sum_u \varphi(s,u)\psi\chi(u,t) = \varphi(\psi\chi)(s,t)\end{aligned}$$

for all s, t, hence $(\varphi\psi)\chi = \varphi(\psi\chi)$. It follows that

$$\begin{aligned}(\varphi,\psi)\chi(f_1(s),t) &= \sum_r (\varphi,\psi)(f_1(s),r)\chi(r,t) = \sum_r \varphi(s,r)\chi(r,t)\\ &= \varphi\chi(s,t) = (\varphi\chi, \psi\chi)(f_1(s),t)\end{aligned}$$

and similarly $(\varphi,\psi)\chi(f_2(s),t) = \psi\chi(s,t) = (\varphi\chi,\psi\chi)(f_2(s),t)$, while $(\varphi,\psi)\chi(s,t) = (\varphi\chi,\psi\chi)(s,t) = 0$ otherwise, hence $(\varphi,\psi)\chi = (\varphi\chi,\psi\chi)$.

Let $\mathscr{H}$ be a well ordered subset of $\mathscr{F}$ and $\varphi = \lambda st \sup\{\theta(s,t)/\theta\in\mathscr{H}\}$. Following Skordev [1980], we show that $\varphi\psi = \sup(\mathscr{H}\psi)$ and $\psi\varphi = \sup\psi\mathscr{H}$ for all ψ. It is immediate that $\mathscr{H}\psi \le \varphi\psi$. Suppose that $\mathscr{H}\psi \le \tau$. If $r_1, \ldots, r_m$

are distinct members of M and $\theta_1,\ldots,\theta_m \in \mathscr{H}$, there is a $\theta \in \mathscr{H}$ such that $\theta_1,\ldots,\theta_m \leq \theta$; hence

$$\sum_{i=1}^{m} \theta_i(s, r_i)\psi(r_i, t) \leq \sum_{i=1}^{m} \theta(s, r_i)\psi(r_i, t) \leq \theta\psi(s, t) \leq \tau(s, t).$$

This is valid for all $m, \theta_1,\ldots,\theta_m, r_1,\ldots,r_m$; hence $\sum_{i=1}^{m} \varphi(s, r_i)\psi(r_i, t) \leq \tau(s, t)$ for all $m, r_1,\ldots,r_m$, which implies $\sum_r \varphi(s, r)\psi(r, t) \leq \tau(s, t)$. Therefore, $\varphi\psi(s, t) \leq \tau(s, t)$ for all s, t, hence $\varphi\psi \leq \tau$. Therefore, $\varphi\psi = \sup(\mathscr{H}\psi)$ and similarly $\psi\varphi = \sup \psi\mathscr{H}$. It follows in particular that $\circ$ is monotonic. Finally, 19.1–19.3 imply that $\mathscr{S}$ is a $(**)_0$, $(***)_0$-complete OS, which completes the proof.

It is worth mentioning however that the equality $(\sup \mathscr{H})\psi = \sup(\mathscr{H}\psi)$ does not hold for arbitrary $\mathscr{H}$. Indeed, the element

$$U = \sup\{L, R\} = \lambda st.(L(s, t) + R(s, t))$$

does not satisfy condition (2) of exercise 7.8 since $U(I, I) \nleq I$. The equality $U(\varphi, \psi) = \lambda st.(\varphi(s, t) + \psi(s, t))$ holds instead.

Proposition 25.2. Let $\mathscr{S}$ be the IOS of example 25.1 or a subspace of it. Then $\langle\ \rangle, \Delta, [\ \ ]$ are characterized as follows.

$$\langle\varphi\rangle(f_2^n(f_1(s)), f_2^n(f_1(t))) = \varphi(s, t),$$
$$\Delta(\varphi, \psi)(f_2^n(f_1(s)), t) = \sum_{r_1,\ldots,r_n} \varphi(s, r_1)\psi(r_1, r_2)\ldots\psi(r_n, t)$$

and $\langle\varphi\rangle(s, t)$, $\Delta(\varphi, \psi)(s, t) = 0$ otherwise,

$$[\varphi](s, t) = \sum_{n, r_0,\ldots,r_{n-1} \in f_2(M), r_n} I(s, r_0)\varphi(f_2^{-1}(r_0), r_1)\ldots\varphi(f_2^{-1}(r_{n-1}), r_n)I(r_n, f_1(t)).$$

Proof. The proof is based on the fact that $\langle\varphi\rangle = \Delta(\varphi L, R)$, $\Delta(\varphi, \psi) = \sup_n \Gamma^n(O)$, where $\Gamma = \lambda\theta.(\varphi, \theta\psi)$, while $[\varphi] = \sup_n \Gamma^n(O)$, $\Gamma = \lambda\theta.(I, \varphi\theta)$. We present the case of iteration only.

We have

$$\Gamma^1(O)(s, t) = (I, O)(s, t) = \sum_{r_0} I(s, r_0)I(r_0, f_1(t)).$$

Suppose that $n \geq 1$ and $\Gamma^n(O)(s, t) = \sum_{i<n, r_0,\ldots,r_{i-1} \in f_2(M), r_i} I(s, r_0)\varphi(f_2^{-1}(r_0), r_1)\cdots \varphi(f_2^{-1}(r_{i-1}), r_i)I(r_i, f_1(t))$. Then

$$\begin{aligned}
\Gamma^{n+1}(O)(s, t) &= (I, \varphi\Gamma^n(O))(s, t) = I(s, f_1(t)) + \sum_{r_0 \in f_2(M)} I(s, r_0)\varphi\Gamma^n(O)(f_2^{-1}(r_0), t)\\
&= I(s, f_1(t)) + \sum_{r_0 \in f_2(M)} I(s, r_0)\sum_r \varphi(f_2^{-1}(r_0), r)\Gamma^n(O)(r, t)\\
&= \sum_{r_0} I(s, r_0)I(r_0, f_1(t)) + \sum_{r_0 \in f_2(M)} I(s, r_0)\sum_r \varphi(f_2^{-1}(r_0), r)\\
&\quad \times \sum_{0<i<n+1, r_1,\ldots,r_{i-1} \in f_2(M), r_i} I(r, r_1)\varphi(f_2^{-1}(r_1), r_2)\ldots \varphi(f_2^{-1}(r_{i-1}), r_i)I(r_i, f_1(t))\\
&= \sum_{i<n+1, r_0,\ldots,r_{i-1} \in f_2(M), r_i} I(s, r_0)\varphi(f_2^{-1}(r_0), r_1)\ldots \varphi(f_2^{-1}(r_{i-1}), r_i)I(r_i, f_1(t)).
\end{aligned}$$

Using the equality $[\varphi](s, t) = \sup_n(\Gamma^n(O)(s, t))$, we get the desired characterization of $[\ \ ]$. This completes the proof.

Of course, the pairing space of example 25.1 can be augmented with multiplication in another way to become the IOS of example 21.1 with $E=[0,\infty]$. Propositions 21.2, 25.2 show that the corresponding probabilistic and fuzzy spaces have identical operations $\Pi, \langle\ \ \rangle$ but different $\circ, \Delta, [\ \ ]$.

The following space consists of *probabilistic functions.*

Proposition 25.3 (Example 25.2). Let $\mathscr{S}_0=(\mathscr{F}_0, I, \Pi_0, L, R)$ be the IOS of example 25.1. Take

$$\mathscr{F}=\left\{\varphi/\varphi: M^2\to[0,1]\,\&\,\forall s\left(\sum_t \varphi(s,t)\le 1\right)\right\}.$$

Then $\mathscr{S}=(\mathscr{F}, I, \Pi_0\restriction\mathscr{F}^2, L, R)$ is a $(**)_0$, $(***)_0$-complete OS.

Proof. It is immediate that $I, L, R\in\mathscr{F}$ and $\mathscr{F}$ is closed under Π_0. If $\varphi,\psi\in\mathscr{F}$, then

$$\sum_t \varphi\psi(s,t)=\sum_t\sum_r \varphi(s,r)\psi(r,t)=\sum_r \varphi(s,r)\sum_t \psi(r,t)\le\sum_r \varphi(s,r)\le 1$$

for all s; hence $\varphi\psi\in\mathscr{F}$. Therefore, $\mathscr{F}$ is closed under $\circ$.

As shown in Skordev [1980], $\mathscr{F}$ is closed under least upper bounds of well ordered subsets. In fact, let $\mathscr{H}$ be such a subset of $\mathscr{F}$ and $\varphi=\sup\mathscr{H}=\lambda st.\sup\{\theta(s,t)/\theta\in\mathscr{H}\}$. Let $t_1,\ldots,t_m$ be distinct members of M and $\theta_1,\ldots,\theta_m\in\mathscr{H}$. Then there is a $\theta\in\mathscr{H}$ such that $\theta_1,\ldots,\theta_m\le\theta$, hence

$$\sum_{i=1}^m \theta_i(s,t_i)\le\sum_{i=1}^m \theta(s,t_i)\le\sum_t \theta(s,t)\le 1.$$

This holds for all $m,\theta_1,\ldots,\theta_m$, $t_1,\ldots,t_m$, hence $\sum_{i=1}^m \varphi(s,t_i)\le 1$ for all m, $t_1,\ldots,t_m$, which implies $\sum_t\varphi(s,t)\le 1$, hence $\varphi\in\mathscr{F}$. Therefore, $\mathscr{S}$ is a $(**)_0$, $(***)_0$-complete subspace of $\mathscr{S}_0$ by 18.15, 18.16. The proof is complete.

A few words about the intuition behind example 25.2. The members of $\mathscr{F}$ may be regarded as semantical counterparts of programs processed by a computer as in chapter 21; in this case $\varphi(s,t)\in[0,1]$ is the probability of s being processed into t. This intuitive interpretation is extended to cover example 25.1 by regarding $\varphi(s,t)\in[0,\infty]$ as the average number of trajectories leading from s to t (from t to s, in another version); cf. Skordev [1980] for a further discussion.

Proposition 25.4 (Example 25.3). Let M be an infinite set with a splitting scheme f_1, f_2. Take $\mathscr{F}=\{\varphi/\varphi: M\to 2^M\}$, $\varphi\le\psi$ iff ψ is an extension of φ, i.e. $\operatorname{Dom}\varphi\subseteq\operatorname{Dom}\psi$ and $\psi(s)=\varphi(s)$ for all $s\in\operatorname{Dom}\varphi$, where $\operatorname{Dom}\varphi=\{s/\varphi(s)\ne\varnothing\}$. Take $\varphi\psi(s)=\psi(\varphi(s))$, if $s\in\operatorname{Dom}\varphi$, $\varphi(s)\subseteq\operatorname{Dom}\psi$, and $\varphi\psi(s)\uparrow$ otherwise, $(\varphi,\psi)(f_1(s))=\varphi(s)$, $(\varphi,\psi)(f_2(s))=\psi(s)$ and $(\varphi,\psi)(s)\uparrow$ otherwise, $I=\lambda s.s$, $L=f_1$ and $R=f_2$. Then $\mathscr{S}=(\mathscr{F}, I, \Pi, L, R)$ is a $(**)_0$-complete OS.

Proof. It easily follows that $\le$ is a partial order, the operations $\circ, \Pi$ are monotonic and I is a unit.

$$\begin{aligned}\operatorname{Dom}(\varphi\psi)\chi&=\{s\in\operatorname{Dom}\varphi\psi/\varphi\psi(s)\subseteq\operatorname{Dom}\chi\}\\&=\{s\in\operatorname{Dom}\varphi/\varphi(s)\subseteq\operatorname{Dom}\psi\,\&\,\psi(\varphi(s))\subseteq\operatorname{Dom}\chi\}\\&=\{s\in\operatorname{Dom}\varphi/\varphi(s)\subseteq\operatorname{Dom}\psi\chi\}=\operatorname{Dom}\varphi(\psi\chi)\end{aligned}$$

and if $s \in \mathrm{Dom}\,(\varphi\psi)\chi$, then

$$(\varphi\psi)\chi(s) = \chi(\psi(\varphi(s))) = \varphi(\psi\chi)(s);$$

hence $(\varphi\psi)\chi = \varphi(\psi\chi)$.

$$\begin{aligned}\mathrm{Dom}\,(\varphi, \psi)\chi &= \{s \in \mathrm{Dom}\,(\varphi, \psi)/(\varphi, \psi)(s) \subseteq \mathrm{Dom}\,\chi\} \\ &= \{f_1(s)/s \in \mathrm{Dom}\,\varphi \,\&\, \varphi(s) \subseteq \mathrm{Dom}\,\chi\} \\ &\quad \cup \{f_2(s)/s \in \mathrm{Dom}\,\psi \,\&\, \psi(s) \subseteq \mathrm{Dom}\,\chi\} \\ &= \{f_1(s)/s \in \mathrm{Dom}\,\varphi\chi\} \cup \{f_2(s)/s \in \mathrm{Dom}\,\psi\chi\} = \mathrm{Dom}\,(\varphi\chi, \psi\chi)\end{aligned}$$

and if $f_1(s) \in \mathrm{Dom}(\varphi, \psi)\chi$, then $(\varphi, \psi)\chi(f_1(s)) = \varphi\chi(s)$, respectively $(\varphi, \psi)\chi(f_2(s)) = \psi\chi(s)$, if $f_2(s) \in \mathrm{Dom}\,(\varphi, \psi)\chi$; hence $(\varphi, \psi)\chi = (\varphi\chi, \psi\chi)$.

$$\mathrm{Dom}\,L(\varphi, \psi) = \{s/f_1(s) \in \mathrm{Dom}\,(\varphi, \psi)\} = \mathrm{Dom}\,\varphi$$

and if $s \in \mathrm{Dom}\,\varphi$, then

$$L(\varphi, \psi)(s) = (\varphi, \psi)(f_1(s)) = \varphi(s),$$

hence $L(\varphi, \psi) = \varphi$. Similarly, $R(\varphi, \psi) = \psi$. Therefore, $\mathscr{S}$ is an OS.

Let $\mathscr{H}$ be a well ordered subset of $\mathscr{F}$ and $\psi \in \mathscr{F}$. Take $\varphi = \cup \mathscr{H}$. Then $\theta \leq \varphi$ for all $\theta \in \mathscr{H}$ since $\mathscr{H}$ is well ordered, hence $\mathscr{H}\psi \leq \varphi\psi$. Suppose that $\mathscr{H}\psi \leq \tau$. If $s \in \mathrm{Dom}\,\varphi\psi$, then $s \in \mathrm{Dom}\,\varphi$, hence there is a $\theta \in \mathscr{H}$ such that $s \in \mathrm{Dom}\,\theta$ and $\varphi(s) = \theta(s)$. It follows that $s \in \mathrm{Dom}\,\theta\psi$; hence $s \in \mathrm{Dom}\,\tau$ and $\tau(s) = \theta\psi(s) = \varphi\psi(s)$. This holds for all $s \in \mathrm{Dom}\,\varphi\psi$; hence $\varphi\psi \leq \tau$. Therefore, $\varphi\psi = \sup\,(\mathscr{H}\psi)$.

Let $\mathscr{H}$ and φ be the same as above. Then $L\mathscr{H} \leq L\varphi$. Suppose that $L\mathscr{H} \leq \tau$. If $s \in \mathrm{Dom}\,L\varphi$, then $f_1(s) \in \mathrm{Dom}\,\varphi$ and $L\varphi(s) = \varphi(f_1(s))$. There is a $\theta \in \mathscr{H}$ such that $f_1(s) \in \mathrm{Dom}\,\theta$ and $\theta(f_1(s)) = \varphi(f_1(s))$; hence $\tau(s) = L\theta(s) = L\varphi(s)$. Therefore, $L\,\varphi \leq \tau$; hence $L\,\varphi = \sup L\,\mathscr{H}$. Similarly, $R\varphi = \sup R\mathscr{H}$; hence $\mathscr{S}$ is $(**)_0$ complete. The proof is complete.

Proposition 25.5. The operations $\langle\ \ \rangle, \Delta, [\ \ ]$ of example 25.3 can be characterized as follows.

$$\langle\varphi\rangle(f_2^n(f_1(s))) = f_2^n(f_1(\varphi(s))),$$
$$\Delta(\varphi, \psi)(f_2^n(f_1(s))) = \varphi\psi^n(s),$$

and $\langle\varphi\rangle(s)$, $\Delta(\varphi, \psi)(s)\uparrow$ otherwise.

We say that t is a *φ-successor of $f_2(s)$* iff $t \in \varphi(s)$, while s is *φ-regular* iff $s \in f_1(M) \cup f_2(\mathrm{Dom}\,\varphi)$. Take the least subset D_φ of M such that whenever s is φ-regular and all its φ-successors are in D_φ, then $s \in D_\varphi$. Then $t \in [\varphi](s)$ iff $s \in D_\varphi$ and there are $n, r_0, \ldots, r_n$ such that $r_0 = s$, r_{i+1} is a φ-successor of r_i for all $i < n$, and $r_n = f_1(t)$.

Proof. We take $\Gamma = \lambda\theta.(\varphi, \theta\psi)$ and prove by induction on n that $\Gamma^n(O)(f_2^i(f_1(s))) = \varphi\psi^i(s)$, if $i < n$, and $\Gamma^n(O)(s)\uparrow$ otherwise. Exercise 18.4 gives $\Delta(\varphi, \psi) = \sup_n \Gamma^n(O)$, which implies the desired characterization of Δ and that of $\langle\ \ \rangle$ by $\langle\varphi\rangle = \Delta(\varphi L, R)$.

Let $t \in \sigma(s)$ iff $s \in D_\varphi$ and there are $n, r_0, \ldots, r_n$ such that $r_0 = s$, r_{i+1} is a φ-successor of r_i for all $i < n$, and $r_n = f_1(t)$. We show that $[\varphi] = \sigma$, following a similar argument from example 18, Skordev [1980], chapter 3.

Let us show first that $\text{Dom}\,\sigma = D_\varphi$. In order to get $D_\varphi \subseteq \text{Dom}\,\sigma$ it suffices to show that whenever s is φ-regular and all its φ-successors are in $\text{Dom}\,\sigma$, then $s \in \text{Dom}\,\sigma$. If $s = f_1(t)$, then $s \in D_\varphi$ and $\sigma(s) = t$, hence $s \in \text{Dom}\,\sigma$. Let $s = f_2(r), r \in \text{Dom}\,\varphi$. Take a $t \in \varphi(r)$. Then t is a φ-successor of s, hence $t \in \text{Dom}\,\sigma$. All φ-successors of s are in $\text{Dom}\,\sigma \subseteq D_\varphi$, hence $s \in D_\varphi$. It follows that $\sigma(t) \subseteq \sigma(s)$, hence $\sigma(s) \neq \varnothing$ and $s \in \text{Dom}\,\sigma$.

Our next objective is to prove that $(I, \varphi\sigma) \leq \sigma$. Suppose that $s \in \text{Dom}(I, \varphi\sigma)$ and write N for $(I, \varphi\sigma)(s)$. In other words, either $s \in f_1(M)$ or $s \in f_2(\text{Dom}\,\varphi)$ and $\varphi(f_2^{-1}(s)) \subseteq D_\varphi$, while

$$N = \{t/s = f_1(t) \vee t \in \sigma(\varphi(f_2^{-1}(s)))\}.$$

Then s is φ-regular and all its φ-successors are in D_φ; hence $s \in D_\varphi = \text{Dom}\,\sigma$. It follows easily that $N \subseteq \sigma(s)$. Conversely, suppose that $t \in \sigma(s)$. Then there are $n, r_0, \ldots, r_n$ such that $r_0 = s$, r_{i+1} is a φ-successor of r_i for all $i < n$, and $r_n = f_1(t)$. If $n = 0$, then $s = f_1(t)$; hence $t \in N$. If $n > 0$, then $s \in f_2(M)$ and $t \in \sigma(\varphi(f_2^{-1}(s)))$, hence $t \in N$ again. We get $\sigma(s) \subseteq N$; hence $\sigma(s) = (I, \varphi\sigma)(s)$ for all $s \in \text{Dom}(I, \varphi\sigma)$, which implies $(I, \varphi\sigma) \leq \sigma$.

Suppose that $(I, \varphi\tau) \leq \tau$. Then for all s

(1) $s \in f_1(M) \vee s \in f_2(\text{Dom}\,\varphi) \,\&\, \varphi(f_2^{-1}(s)) \subseteq \text{Dom}\,\tau \Rightarrow s \in \text{Dom}\,\tau \,\&\, \tau(s)$
$= \{t/s = f_1(t) \vee t \in \tau(\varphi(f_2^{-1}(s)))\}$.

If s is φ-regular and all its φ-successors are in $\text{Dom}\,\tau$, then $s \in \text{Dom}\,\tau$ by (1); hence $D_\varphi \subseteq \text{Dom}\,\tau$.

Let $N = \{s \in D_\varphi / \tau(s) = \sigma(s)\}$. Whenever s is φ-regular and all its φ-successors are in N, then all its φ-successors are in D_φ; hence $s \in D_\varphi$. Moreover, all φ-successors of s are in $\text{Dom}\,\tau$ since $D_\varphi \subseteq \text{Dom}\,\tau$, so (1) gives

$$\tau(s) = \{t/s = f_1(t) \vee t \in \sigma(\varphi(f_2^{-1}(s)))\};$$

hence $\tau(s) = (I, \varphi\sigma)(s) = \sigma(s)$ and $s \in N$. The definition of D_φ implies $D_\varphi \subseteq N$, hence $\tau(s) = \sigma(s)$ for all $s \in D_\varphi$. Therefore, $\sigma \leq \tau$.

We conclude that $\sigma = \mu\theta.(I, \varphi\theta)$, hence $[\varphi] = \sigma$. The proof is complete.

The spaces of examples 22.1, 25.3 have the same carriers $\mathscr{F}$ and operations $\Pi, \langle\ \ \rangle$, but different $\leq, \circ, \Delta, [\ \ ]$.

Example 25.3 is related to the idea of reliable estimation of functions. Following Skordev [1980], an element $\varphi \in \mathscr{F}$ is a *reliable estimate* of a function $f: M \dashrightarrow M$ iff $\text{Dom}\,\varphi \subseteq \text{Dom}\,f$ and $f(s) \in \varphi(s)$ for all $s \in \text{Dom}\,\varphi$. Given two reliable estimates φ, ψ of f, ψ is *better than* φ iff $\text{Dom}\,\varphi \subseteq \text{Dom}\,\psi$ and $\psi(s) \subseteq \varphi(s)$ for all $s \in \text{Dom}\,\varphi$.

The IOS of example 22.2 is based on $\mathscr{F}_0 = \{f/f: M \dashrightarrow M\}$. Thus the members of $\mathscr{F}$ are reliable estimates of members of $\mathscr{F}_0$ and, as observed in the cited work, if φ, ψ are reliable estimates of f, g, then $\varphi\psi$ is a reliable estimate of fg. It will also be shown in the exercises that $(\varphi, \psi), \langle\varphi\rangle, [\varphi]$ are reliable estimates of $(f, g), \langle f\rangle$ and $[f]$.

As far as reliable estimates are concerned, the relation 'is better than' seems more natural than 'is an extension of'. Modifying $\mathscr{S}$ in this way, one gets the following example.

Proposition 25.6 (Example 25.4). Let $\mathcal{S} = (\mathcal{F}, I, \Pi, L, R)$ be the OS of example 25.3 and $\mathcal{F}_1$ be obtained from $\mathcal{F}$ by substituting $\leq_1$ for $\leq$, where $\varphi \leq_1 \psi$ iff ψ is better than φ. Then $\mathcal{S}_1 = (\mathcal{F}_1, I, \Pi, L, R)$ is a μA_3-iterative OS and $\mathcal{S}, \mathcal{S}_1$ have identical operations $\langle \quad \rangle, [\quad]$.

Proof. It follows easily that $\leq_1$ is a partial ordering of $\mathcal{F}$ and $\circ, \Pi$ are monotonic with respect to it; hence $\mathcal{S}_1$ is an OS. Suppose for instance that $\varphi \leq_1 \varphi_1$, $\psi \leq_1 \psi_1$. Then

$$\operatorname{Dom} \varphi\psi = \operatorname{Dom} \varphi \cap \varphi^{-1}(\operatorname{Dom} \psi) \subseteq \operatorname{Dom} \varphi_1 \cap \varphi_1^{-1}(\operatorname{Dom} \psi_1) = \operatorname{Dom} \varphi_1\psi_1$$

and if $s \in \operatorname{Dom} \varphi\psi$, then

$$\varphi_1\psi_1(s) = \psi_1(\varphi_1(s)) \subseteq \psi_1(\varphi(s)) \subseteq \psi(\varphi(s)) = \varphi\psi(s);$$

hence $\varphi\psi \leq_1 \varphi_1\psi_1$.

Let $\mathcal{H}$ be a subset of $\mathcal{F}$ well ordered with respect to $\leq$ and let $\varphi = \sup \mathcal{H}$. Then $\varphi = \cup \mathcal{H}$ by the proof of 25.4. It is immediate that $\mathcal{H} \leq_1 \varphi$. Suppose that $\mathcal{H} \leq_1 \tau$. If $s \in \operatorname{Dom} \varphi$, then there is a $\theta \in \mathcal{H}$ such that $s \in \operatorname{Dom} \theta$ and $\theta(s) = \varphi(s)$. The inequality $\theta \leq_1 \tau$ implies $s \in \operatorname{Dom} \tau$ and $\tau(s) \subseteq \theta(s)$; hence $\tau(s) \subseteq \varphi(s)$. This holds for all $s \in \operatorname{Dom} \varphi$; hence $\varphi \leq_1 \tau$. Therefore, $\varphi = \sup_1 \mathcal{H}$. It follows from 18.18 that $\mathcal{S}_1$ is μA_3-iterative and has the same initial operations as $\mathcal{S}$. The proof is complete.

Other interesting first order IOS-examples can be constructed corresponding to related examples from Skordev's book, bearing in mind the general connection between operative and combinatory spaces to be established in chapter 27. Among them are probabilistic spaces using countably additive measures on σ-algebras of sets rather than discrete probabilities, spaces connected with the notion of $\forall$-definiteness and topological versions of examples 22.1, 22.2, 25.3, 25.4. Some of those examples can be modified to accommodate the complexity measure of data processing.

EXERCISES TO CHAPTER 25

Exercise 25.1. Let $\mathcal{S}$ be the IOS of example 25.1 and $\varphi^{-1} = \lambda st.\varphi(t,s)$. Show that $\langle\!\langle \quad \rangle\!\rangle = \lambda\varphi.(\varphi, \varphi^{-1})$ is a t-operation with a corresponding set $\mathcal{B}_0 = \{U\}$, and $\mathcal{S}, \langle\!\langle \quad \rangle\!\rangle$ satisfy the axiom $t\mu A_3$.

Hint. Adapt the proof of 21.12 with $+$ substituted for sup, where $\varphi + \psi = \lambda st.(\varphi(s,t) + \psi(s,t))$.

Exercise 25.2. Let $\mathcal{S}$ be the IOS of example 25.1 or 25.2 and let J be the same as in 21.13. Take $St(\varphi)(J(s,t), J(s,r)) = \varphi(t,r)$ and $St(\varphi)(s,t) = 0$ otherwise. Show that St is a t-operation satisfying $t\mu A_3$.

Hint. Follow the proof of 21.13.

Analogue to 21.14 and the relevant exercises to chapter 21 may also be established.

Exercise 25.3. Let $\mathcal{S}$ be the IOS of example 25.3 or 25.4 and let St be introduced as in 21.13. Show that St is a t-operation satisfying $t\mu A_3$.

Hint. Follow the proof of 21.13. Use 18.21, respectively a t-analogue to 18.18.

Exercise 25.4. Show that, unlike translation, iteration in examples 25.3, 25.4 is not always reached at level ω as a least fixed point.

Hint. Modifying a counterexample of Skordev [1980], take $s_0 \in f_1(M)$, $s_{n+1} = f_1(s_n)$, if $\exists m(n+1 = 2^{m+1})$, and $s_{n+1} = f_2(s_n)$ otherwise, $\varphi(s_0) = \{s_2, s_3, \ldots\}$ and $\varphi(s_{n+1}) = s_{n+3}$, then take $\Gamma = \lambda\theta.(I, \varphi\theta)$, $\sigma = \sup_n \Gamma^n(O)$. Show that $s_1 \notin \mathrm{Dom}\,\sigma$ and $s_{n+2} \in \mathrm{Dom}\,\sigma$ for all n; hence $(I, \varphi\sigma) \neq \sigma$.

The following exercise shows however that bounded multiple-valuedness ensures ω-accessibility.

Exercise 25.5 (Examples 25.5, 25.6). Prove that the functions φ such that $\varphi(s)$ is finite for all s form iterative subspaces of examples 25.3 and 25.4 in which the least fixed points of all inductive mappings are reached at level ω.

Exercise 25.6. Let $\mathscr{S}_0, \mathscr{S}, \mathscr{S}_1$ be the IOS of examples 22.2, 25.3, 25.4 based on the same M, f_1, f_2. Show that whenever $\varphi, \psi \in \mathscr{F} = \mathscr{F}_1$ are reliable estimates respectively of $f, g \in \mathscr{F}_0$, then (φ, ψ), $\langle \varphi \rangle$, $[\varphi]$ are reliable estimates of (f, g), $\langle f \rangle$, $[f]$.

Hint. Use 18.15 to show that $\mathscr{S}_0$ is a subspace of $\mathscr{S}$, which implies that $\mathscr{S}_0$ is a subspace of $\mathscr{S}_1$. Use the fact that φ is a reliable estimate of f iff $\varphi \leq_1 f$.

Remark. It follows that whenever all the members of $\mathscr{B}_0 \subseteq \mathscr{F}_0$ have reliable estimates in $\mathscr{B} \subseteq \mathscr{F} = \mathscr{F}_1$ and f is recursive in $\mathscr{B}_0$, then f has a reliable estimate φ recursive in $\mathscr{B}$.

Exercise 25.7 (Example 25.7). Let M be an infinite set with a splitting scheme f_1, f_2 and $u \notin M$. Take $\mathscr{F} = \{\varphi / \varphi : M \cup \{u\} \to 2^{M \cup \{u\}} \,\&\, \varphi(u) = u\}$, $\varphi \leq \psi$ iff $\forall s(\varphi(s) \subseteq \psi(s))$, $\varphi\psi = \lambda s.\psi(\varphi(s))$, $(\varphi, \psi)(f_1(s)) = \varphi(s)$, $(\varphi, \psi)(f_2(s)) = \psi(s)$ and $(\varphi, \psi)(s) = u$ otherwise, $I = \lambda s.s$, and extend f_1, f_2 to $L, R \in \mathscr{F}$. Show that $\mathscr{S} = (\mathscr{F}, I, \Pi, L, R)$ is a μA_3-iterative OS. Give explicit characterizations of $\langle\ \rangle$, $[\ ]$.

Remark. This modification of example 22.1 is an analogue to example 2 of Skordev [1980a]. Example 22.2 can be modified to form a subspace again. Intuitively, u indicates that an unproductive termination has occurred. Notice that $\varphi O = O$ fails, and if $f_1(M) \cup f_2(M) \subset M$, then $(O, O) \neq O$.

Exercise 25.8. Let $\mathscr{S}$ be the space of example 25.7, let J be as in 21.13 and define St as follows. If $r \in \varphi(t)$, $r \neq u$, take $J(s, r) \in St(\varphi)(J(s, t))$. If $u \in \varphi(t)$, take $u \in St(\varphi)(J(s, t))$. Take $St(\varphi)(s) = u$, provided $s \notin J(M_0, M)$. Show that St is a t-operation satisfying $t\mu A_3$.

Exercise 25.9. Show that the element $V = \lambda s.\{f_1(s), f_2(s)\}$ in example 25.4 satisfies condition (1) of exercise 7.14, while in example 25.7 $U = L \cup R$ satisfies condition (1) of exercise 7.10.

PART F

Connection with other theories

CHAPTER 26

Mathematical theory of programs

This chapter clarifies the connection between IOS and certain related algebraic systems of Computer Science, including the programming spaces of Skordev [1982] and the system of de Bakker and Scott introduced in the unpublished work de Bakker and Scott [1969]. Results of the general IOS-theory are transferred to appropriate extensions of the latter system.

The programming spaces of Skordev are easy to describe. Ignoring some minor technical differences, they are nothing else but OS augmented with iteration satisfying (££), i.e. IOS without translation. A version of this system is studied by Georgieva [1980].

The system of de Bakker and Scott is a formal one, hence a comparable formal system for IOS is needed. In fact we are going to introduce two such systems, $\mathscr{S}_0$ and $\mathscr{S}_1$.

System $\mathscr{S}_0$. Its language $\mathscr{L}_0$ has constants I, L, R and variables denoted by $\theta, \theta_1, \theta_2, \ldots$. Terms are constructed by the operations $\varphi\psi$, (φ, ψ), $\langle \varphi \rangle$ and $[\varphi]$. Formulas are equalities $\varphi = \psi$.

The axioms and rules of $\mathscr{S}_0$ are those of equality and substitution plus the following nonlogical axioms.

$$(\theta_1\theta_2)\theta_3 = \theta_1(\theta_2\theta_3), \quad \theta I = \theta, \quad I\theta = \theta,$$
$$(\theta_1, \theta_2)\theta = (\theta_1\theta, \theta_2\theta),$$
$$L(\theta_1, \theta_2) = \theta_1, \; R(\theta_1, \theta_2) = \theta_2,$$
$$(\theta L, \langle \theta \rangle R) = \langle \theta \rangle, \quad (I, \theta[\theta]) = [\theta].$$

What is interesting about this simple formal system is that it provides for the weak representability of all the partial recursive functions, which implies in turn that $\mathscr{S}_0$ is essentially undecidable, That is, all consistent extensions of $\mathscr{S}_0$ are undecidable. Consistency here means that $\bar{0} = O$ is not deducible, hence $\bar{0} = \bar{1}$ is not deducible. (As usual, $\bar{n}$, O stand for LR^n, $R[R]$.) The system $\mathscr{S}_0$ itself is consistent since every IOS is a model of it.

Proposition 26.1 (Undecidability Theorem). $\mathscr{S}_0$ is essentially undecidable.

Proof. Following an argument of Skordev [1980], take a partial recursive function $f:\omega \dashrightarrow \{0,1\}$ which has no general recursive extension. (For instance, if σ is a unary partial recursive function universal for all such functions and g is a supposed general recursive extension of $f = \lambda s.1 \dot{-} \bar{s}\sigma(s)$, then $g = \bar{n}\sigma$ for a certain n, hence $g(n) = 1 \dot{-} g(n)$, which is not the case.) The proof of 8.2 implies that there is an $\mathscr{L}_0$-term φ such that whenever $f(s) = t$, then $\bar{s}\varphi = \bar{t}$ is deducible in $\mathscr{S}_0$, written $\vdash_{\mathscr{S}_0} \bar{s}\varphi = \bar{t}$.

Suppose that $\mathscr{T}$ is a decidable consistent extension of $\mathscr{S}_0$. Then the function

$$g(s) = \begin{cases} 0, & \text{if } \vdash_{\mathscr{T}} \bar{s}\varphi = \bar{0}, \\ 1, & \text{if } \nvdash_{\mathscr{T}} \bar{s}\varphi = \bar{0} \end{cases}$$

is general recursive. If $f(s) = 0$, then $\vdash_{\mathscr{S}_0} \bar{s}\varphi = \bar{0}$; hence $\vdash_{\mathscr{T}} \bar{s}\varphi = \bar{0}$ and $g(s) = 0$. If $f(s) = 1$, then similarly $\vdash_{\mathscr{T}} \bar{s}\varphi = \bar{1}$; hence $\nvdash_{\mathscr{T}} \bar{s}\varphi = \bar{0}$ and $g(s) = 1$. Therefore g is a general recursive extension of f, which is not the case. The proof is complete.

The system $\mathscr{S}_1, \mathscr{S}_3, \mathscr{S}_4$ studied in this chapter are consistent extensions of $\mathscr{S}_0$, hence undecidable. One may also want to know whether they are incomplete with respect to a certain standard model or class of models, say the class of models consisting of partial single-valued (or multiple-valued) functions. The answer is positive; all consistent recursively axiomatizable extensions of $\mathscr{S}_0$ are incomplete with respect to a wide range of natural classes of models.

A structure $\mathscr{S}$ the language of which extends $\mathscr{L}_0$ is said to *have the representation property* iff whenever f is a partial recursive function, φ is a variable-free $\mathscr{L}_0$-term weakly representing f by the proof of 8.2 and $f(s)\uparrow$, then $\models_{\mathscr{S}} \bar{s}\varphi = O$, i.e. $\bar{s}\varphi = O$ is valid in $\mathscr{S}$. For instance, all μA_3-iterative OS $\mathscr{S}$ have this property.

The following statement is called the Incompleteness Theorem.

Proposition 26.2. Let $\mathscr{T}$ be a recursively axiomatizable extension of $\mathscr{S}_0$ and let $\mathscr{K}$ be a class of models of $\mathscr{T}$ which have the representation property, assuming $\not\models_{\mathscr{K}} \bar{0} = O$. Then there are $\mathscr{L}_0$-terms ρ, σ such that $\models_{\mathscr{K}} \rho = \sigma$ but $\nvdash_{\mathscr{T}} \rho = \sigma$.

Proof. The idea is again due to Skordev. Take f as in the previous proof, $g(s) = 0$, if $f(s) = 0$, and $g(s)\uparrow$ otherwise; then take a $\mathscr{L}_0$-term ψ weakly representing g by the proof of 8.2.

The function $h(s) = 1$, if $\vdash_{\mathscr{T}} \bar{s}\psi = O$, and $h(s)\uparrow$ otherwise, is partial recursive by the recursive axiomatizability of $\mathscr{T}$. By way of contradiction, suppose that $\models_{\mathscr{K}} \bar{s}\psi = O$ implies $h(s) = 1$ for all s.

If $f(s) = 0$, then $g(s) = 0$; hence $\vdash_{\mathscr{S}_0} \bar{s}\psi = \bar{0}$, which implies $\models_{\mathscr{K}} \bar{s}\psi = \bar{0}$ and $h(s)\uparrow$. Otherwise $g(s)\uparrow$ gives $\models_{\mathscr{K}} \bar{s}\psi = O$ by the representation property, hence $h(s) = 1$. We get $f \leq g \cup h$, which is not the case since $U(g, h)$ is a general recursive function by 22.4. Therefore, there is an s such that $\models_{\mathscr{K}} \bar{s}\psi = O$ but $\nvdash_{\mathscr{T}} \bar{s}\psi = O$. The proof is complete.

In particular, the role of $\mathscr{T}$ can be played by $\mathscr{S}_0$ itself. The latter system is however complete with respect to the canonical model consisting of all $\mathscr{L}_0$-terms factorized by assuming $\varphi = \psi$ if $\vdash_{\mathscr{S}_0} \varphi = \psi$.

A modification of the Incompleteness Theorem ensures that all consistent extensions of $\mathscr{S}_0$ which provide representability of the partial recursive functions are not recursively axiomatizable.

Proposition 26.3. Let $\mathscr{T}_1$ be a consistent extension of $\mathscr{S}_0$ such that whenever f is a partial recursive function, φ is a $\mathscr{L}_0$-term corresponding to f by the proof of 8.2 and $f(s)\uparrow$, then $\vdash_{\mathscr{T}_1} \bar{s}\varphi = O$. Then $\mathscr{T}_1$ is not recursively axiomatizable.

Proof. Suppose that $\mathscr{T}$ is a recursively axiomatized extension of $\mathscr{S}_0$ and $\mathscr{T}_1$ is an extension of $\mathscr{T}$. Repeating the proof of 26.2 with $\vdash_{\mathscr{T}_1}$ substituted for $\models_{\mathscr{K}}$, one gets $\mathscr{L}_0$-terms ρ, σ such that $\vdash_{\mathscr{T}_1} \rho = \sigma$ but $\nvdash_{\mathscr{T}} \rho = \sigma$, hence $\mathscr{T}$, $\mathscr{T}_1$ are not equivalent. This completes the proof.

It follows from 26.3 that the theory of μA_3-iterative OS is equivalent to no recursively axiomatized first order theory.

System $\mathscr{S}_1$. The language $\mathscr{L}_0$ is extended to a wider one $\mathscr{L}_1$ as follows. Terms are now constructed by the operations $\varphi\psi$, (φ, ψ) and the variable binding *μ-operation* $(\mu\theta.\varphi)$, while $\langle \varphi \rangle$, $[\varphi]$ stand respectively for $\mu\theta.(\varphi L, \theta R)$, $\mu\theta.(I, \varphi\theta)$ for a certain θ not free in φ, say the first such θ. Atomic formulas ϕ, Ψ are equalities $\varphi = \psi$ and inequalities $\varphi \leq \psi$. Formulas are implications $\vec{\phi} \to \vec{\Psi}$, where $\vec{\phi}$, $\vec{\Psi}$ are conjunctions of atomic formulas written as finite lists. If $\vec{\phi}$ is an empty list, then we write simply $\vec{\Psi}$ for $\to \vec{\Psi}$.

The logical axioms and rules of $\mathscr{S}_1$ are those of equality and substitution plus

$$\phi_1, \ldots, \phi_n \to \phi_i, \quad \text{all } 1 \leq i \leq n,$$

$$\frac{\vec{\phi} \to \Psi_i,\ 1 \leq i \leq n}{\vec{\phi} \to \Psi_1, \ldots, \Psi_n},$$

$$\frac{\vec{\phi} \to \vec{\Psi},\ \vec{\Psi} \to \vec{X}}{\vec{\phi} \to \vec{X}}.$$

The nonlogical axioms are as follows.

$$\begin{gathered}
\theta_1 \leq \theta_2, \theta_3 \leq \theta_4 \to \theta_1\theta_3 \leq \theta_2\theta_4, \\
(\theta_1\theta_2)\theta_3 = \theta_1(\theta_2\theta_3), \theta I = \theta, I\theta = \theta, \\
\theta_1 \leq \theta_2, \theta_3 \leq \theta_4 \to (\theta_1, \theta_3) \leq (\theta_2, \theta_4), \\
(\theta_1, \theta_2)\theta = (\theta_1\theta, \theta_2\theta), \\
L(\theta_1, \theta_2) = \theta_1, \quad R(\theta_1, \theta_2) = \theta_2, \\
\varphi(\mu\theta.\varphi/\theta) \leq \mu\theta.\varphi,
\end{gathered}$$

where $\varphi(\psi/\theta)$ is φ with ψ substituted for the free occurrences of θ. Finally, $\mathscr{S}_1$ has a nonlogical rule of inference,

$$(\mu) \qquad \frac{\vec{\phi}, \Psi \to \Psi(\varphi/\theta)}{\vec{\phi} \to \Psi(\mu\theta.\varphi/\theta)},$$

where Ψ is either $\theta\chi \leq \tau$ or $\langle\theta\rangle \leq \langle I\rangle\tau$, and θ is not free in $\vec{\phi}, \chi, \tau$.

All the theorems of the formal system $\mathscr{S}_1$ are obviously theorems of the theory of μA_1-iterative OS. Conversely, a number of the statements proved in chapters 4–9 are theorems or metatheorems of $\mathscr{S}_1$. Since $\mathscr{S}_1$ has no negation, the role of $\vec{\phi} \to \neg\vec{\Psi}$ is played by $\vec{\phi}, \vec{\Psi} \to L = R$.

System $\mathscr{S}_2$**.** The system of de Bakker and Scott. Its language $\mathscr{L}_2$ has constants I, O, function variables $\theta, \theta_1, \theta_2, \ldots$ and predicate variables P, P_1, $P_2, \ldots$. Terms are constructed by the operations $\varphi\psi$, $(P \to \varphi, \psi)$ and $\mu\theta.\varphi$. Formulas are introduced as in $\mathscr{L}_1$ and the logical axioms and rules of $\mathscr{S}_2$ are those of $\mathscr{S}_1$.

The nonlogical axioms include the axioms of a partially ordered semigroup with unit I and zero O, and the following axioms for conditionals due to McCarthy [1963].

$$(P \to I, I) \leq I,$$
$$\theta_1 \leq \theta_2, \theta_3 \leq \theta_4 \to (P \to \theta_1, \theta_3) \leq (P \to \theta_2, \theta_4),$$
$$(P \to (P \to \theta_1, \theta_2), \theta_3) = (P \to \theta_1, \theta_3),$$
$$(P \to \theta_1, (P \to \theta_2, \theta_3)) = (P \to \theta_1, \theta_3),$$
$$(P_1 \to (P_2 \to \theta_1, \theta_2), (P_2 \to \theta_3, \theta_4)) = (P_2 \to (P_1 \to \theta_1, \theta_3), (P_1 \to \theta_2, \theta_4)),$$
$$(P \to \theta_1, \theta_2)\theta = (P \to \theta_1\theta, \theta_2\theta).$$

The μ-operation satisfies the axiom

$$\varphi(\mu\theta.\varphi/\theta) \leq \mu\theta.\varphi$$

and Scott's μ-induction rule

$$\frac{\vec{\phi} \to \vec{\Psi}(O/\theta),\ \vec{\phi}, \vec{\Psi} \to \vec{\Psi}(\varphi/\theta)}{\vec{\phi} \to \vec{\Psi}(\mu\theta.\varphi/\theta)},$$

where θ is not free in $\vec{\phi}$.

This formal system has been extensively studied by de Bakker [1971], its relevance to Computer Science residing in the fact that $\mathscr{L}_2$-terms provide yet another presentation of unary recursive program schemes; more details on this connection will be given in the exercises. The unary (recursion-free) schemes correspond to the so called regular terms introduced as follows, $\mathrm{Fr}(\varphi)$ standing for the set of all free variables of φ.

Each variable θ is a term *regular in* θ. If $\theta \notin \mathrm{Fr}(\varphi)$ and ψ is regular in θ, then both φ and $\varphi\psi$ are regular in θ. If φ, ψ are regular in θ, then so is $(P \to \varphi, \psi)$. If φ is regular in θ, θ_1, then $\mu\theta.\varphi$ is regular in θ_1.

All function constants and variables are *regular terms*. If φ, ψ are regular, then so are $\varphi\psi$ and $(P \to \varphi, \psi)$. If φ is regular in θ and regular, then $\mu\theta.\varphi$ is regular.

In order to apply the IOS-theory to the theory of program schemes, we design a formal system to extend both $\mathscr{S}_1$ and $\mathscr{S}_2$.

System $\mathscr{S}_3$**.** Its language $\mathscr{L}_3$ is obtained from $\mathscr{L}_2$ by adding new function constants L, R, K and a predicate constant Ev. (One may also add a predicate constant Ze as in chapter 23.) The axioms and rules of $\mathscr{I}_3$ are those of $\mathscr{I}_2$ plus

$$L(Ev \rightarrow K\theta_1, K\theta_2) = \theta_1, \quad R(Ev \rightarrow K\theta_1, K\theta_2) = \theta_2.$$

The introduction of L, R, K, Ev corresponds to the additional assumptions of Böhm and Jacopini [1966] and de Bakker [1971], p. 46. In other words, we have introduced a counter, hence the $\mathscr{L}_3$-terms are unary recursive schemes with one counter, while the regular $\mathscr{L}_3$-terms are unary schemes with one counter.

Proposition 26.4. $\mathscr{I}_3$ is an extension of $\mathscr{I}_1$.

Proof. Writing (φ, ψ) for $(Ev \rightarrow K\varphi, K\psi)$, the language $\mathscr{L}_3$ is an extension of $\mathscr{L}_1$ and the axioms of $\mathscr{I}_1$ are deducible in $\mathscr{I}_3$, while the inference rules of $\mathscr{I}_1$ other than (μ) are rules of $\mathscr{I}_3$ as well. The rule (μ) however turns out to be a particular instance of Scott's rule. In order to prove this it suffices to show that $\underset{\mathscr{I}_3}{\vdash} \Psi(O/\theta)$ for all Ψ allowed in (μ). The assertion $\underset{\mathscr{I}_3}{\vdash} O\chi \leq \tau$ is immediate. If Ψ is $\langle \theta \rangle \leq \langle I \rangle \tau$, then we get $\underset{\mathscr{I}_3}{\vdash} \langle O \rangle \leq \langle I \rangle O$ by Scott's rule, hence $\underset{\mathscr{I}_3}{\vdash} \langle O \rangle \leq \langle I \rangle \tau$. This completes the proof.

The following technical statement is easily proved by using $\underset{\mathscr{I}_3}{\vdash} (P \rightarrow \varphi, \psi) = \tilde{P}(\varphi, \psi)$, where $\tilde{P}$ stands for $(P \rightarrow L, R)$.

Proposition 26.5. For every $\mathscr{L}_3$-term φ with predicate variables $P_1, \ldots, P_m$ there is an $\mathscr{L}_1$-term ψ and function variables $\theta_1, \ldots, \theta_{m+2}$ such that $\underset{\mathscr{I}_3}{\vdash} \varphi = \psi(K, \widetilde{Ev}, \tilde{P}_1, \ldots, \tilde{P}_m/\theta_1, \ldots, \theta_{m+2})$.

We introduce certain varieties of $\mathscr{L}_3$-terms in accordance with the notions of chapter 7, 'recursive' replaced by 'canonical' to avoid confusion. A term is *primitive* iff I does not occur in it and the μ-operation occurs as $\langle \quad \rangle$ only. A term is *canonical* (respectively, *prime canonical*) iff the μ-operation occurs in it only as $\langle \quad \rangle, [\quad]$ (as $[\quad]$).

Notice that all prime canonical terms are regular, which is not the case for the primitive or canonical terms. The proof of 23.1 shows that the prime canonical terms are structured schemes with one counter, while the canonical ones are structured schemes with one counter and a very special form of recursion compensating for the lack of a second counter.

Propositions 26.4, 26.5 make it possible to restate for $\mathscr{I}_3$ some basic IOS-results without having to repeat their proofs once again. To begin with, the following Recursion Elimination and Structurization Theorem corresponds to 9.15*.

Proposition 26.6. For every term φ there is a canonical term ψ such that $\underset{\mathscr{I}_3}{\vdash} \varphi = \psi$.

By 26.6, the following Normal Form Theorem and Enumeration Theorem correspond respectively to 9.4, 9.19.

Proposition 26.7. For every term φ and all $P_1, \ldots, P_m$, $\theta_1, \ldots, \theta_n$ there is a primitive term ψ such that $P_1,\ldots,P_m,\theta_1,\ldots,\theta_n \notin \mathrm{Fr}(\psi)$ and

$$\vdash_{\mathscr{I}_3} \varphi = \bar{1}[\psi(I,\langle \tilde{P}_1\rangle,\ldots,\langle \tilde{P}_m\rangle,\langle\theta_1\rangle,\ldots,\langle\theta_n\rangle)].$$

In particular, $\vdash_{\mathscr{I}_3} \varphi = \bar{1}[\psi]$ whenever $m = n = 0$.

Proposition 26.8. For all $P_1,\ldots,P_m,\theta_1,\ldots,\theta_n$ there is a canonical term σ such that whenever $\mathrm{Fr}(\varphi) \subseteq \{P_1,\ldots,P_m,\theta_1,\ldots,\theta_n\}$, then $\vdash_{\mathscr{I}_3} \varphi = \bar{k}\sigma$ for a certain $\bar{k}$.

The following statement shows that all unary schemes with one counter can be structurized—a well known result of Böhm and Jacopini [1966] reaffirmed by de Bakker [1971], Skordev [1982] and Georgieva [1980].

Proposition 26.9. Let φ be regular in $\theta_1,\ldots,\theta_n$ and regular. Then there is a prime canonical ψ such that $\theta_1,\ldots,\theta_n \notin \mathrm{Fr}(\psi)$ and $\vdash_{\mathscr{I}_3} \varphi = \psi(I,\theta_1,\ldots,\theta_n)$. In particular, if φ is regular, then $\vdash_{\mathscr{I}_3} \varphi = \psi$ for a certain prime canonical ψ.

Proof. By induction on the construction of φ; the induction step for the μ-operation follows.

Let $\vdash_{\mathscr{I}_3} \varphi_1 = \psi(I,\theta_1,\ldots,\theta_n)$ and $\varphi = \mu\theta_n.\varphi_1$. Writing ρ for $\psi(\bar{0}L,\ldots,\overline{n-2}L, R^{n-1}L, R)$, it follows that $\vdash_{\mathscr{I}_3} \varphi_1 = \rho((I,\theta_1,\ldots,\theta_{n-1}),\theta_n)$; hence $\vdash_{\mathscr{I}_3} \varphi = R[\rho](I,\theta_1,\ldots,\theta_{n-1})$ by 6.11. The proof is complete.

Combining 26.9 and 9.8, one gets a Normal Form Theorem for regular terms.

Proposition 26.10. For every regular term φ, every θ and n not less than the number of the free occurrences of θ in φ there is a term ψ free of I and μ-operation such that $\theta \notin \mathrm{Fr}(\psi)$ and

$$\vdash_{\mathscr{I}_3} \varphi = \bar{1}[\psi(I,\theta\bar{4},\ldots,\theta\overline{n+3})].$$

The proof of 23.1 suggests that [ ] is equivalent to looping and indeed this can be formally expressed by the equalities $\mu\theta.(P \to I,\varphi\theta) = \tilde{P}[\varphi\tilde{P}]$, $[\varphi] = (\mu\theta.(Ev \to I, K\varphi\theta))K$, θ not free in φ, which are easily deducible in $\mathscr{I}_3$.

As far as the construct $\langle\ \rangle$ is concerned, we know that its implementation requires an extra counter. For this purpose we introduce the following extension of the formal system $\mathscr{I}_3$.

System $\mathscr{I}_4$. Its language $\mathscr{L}_4$ extends $\mathscr{L}_3$ by adding new function constants L_2, R_2, K_2, a new predicate constant Ev_2 and, eventually, another predicate constants Ze_2. The new axioms are

$$LL_2 = L_2L,\ LR_2 = R_2L,\ LK_2 = K_2L,$$
$$RL_2 = L_2R,\ RR_2 = R_2R,\ RK_2 = K_2R,$$
$$KL_2 = L_2K,\ KR_2 = R_2K,$$
$$\widetilde{Ev}L_2 = L_2\widetilde{Ev},\ \widetilde{Ev}R_2 = R_2\widetilde{Ev},$$
$$L_2\widetilde{Ev}_2K_2 = L,\ R_2\widetilde{Ev}_2K_2 = R,$$
$$L\widetilde{Ev}_2 = (Ev_2 \to L^2, LR),\ R\widetilde{Ev}_2 = (Ev_2 \to RL, R^2).$$

$\mathscr{I}_4$ extends $\mathscr{I}_3$ in such a way that 26.6–26.10 hold for $\mathscr{L}_4$-terms and $\mathscr{I}_4$-deducibility as well. Thus we come to the following Recursion Elimination and Structurization Theorem.

Proposition 26.11. Let $\mathrm{Fr}(\varphi) = \{P_1, \ldots, P_m, \theta_1, \ldots, \theta_n\}$. Then there is a prime canonical term ψ such that

$$\vdash_{\mathscr{I}_3} \tilde{P}_i L_2 = L_2 \tilde{P}_i, \tilde{P}_i R_2 = R_2 \tilde{P}_i, 1 \leq i \leq m, \theta_j L_2 = L_2 \theta_j,$$
$$\theta_j R_2 = R_2 \theta_j, 1 \leq j \leq n \rightarrow \varphi = \psi.$$

In view of 26.6 φ may be assumed canonical, so the proof formalizes that of the implication (3)$\Rightarrow$(2) of 23.1. The new axioms of $\mathscr{I}_4$ ensure that $W = \widetilde{Ev_2 K_2}$, $W_1 = L_2$, $W_2 = R_2$ satisfy the assumptions of 21.10. In particular, $\langle I \rangle L_2 = L_2 \langle I \rangle$ and $\langle I \rangle R_2 = R_2 \langle I \rangle$ follow from Scott's rule.

In terminology of chapter 23 prime canonical $\mathscr{L}_4$-terms are structured unary C-schemes, while in general $\mathscr{L}_4$-terms are unary recursive C-schemes. Therefore, every unary recursive C-scheme is equivalent by 26.11 to a structured unary C-scheme, provided its free variables are interpreted by predicates and functions which do not use or affect the second counter. The proofs of 9.15*, 21.10 actually give an algorithm which transforms recursive program schemes into recursion-free ones. While this result has already been mentioned in chapter 23, we point out below that it also holds for other classes of program schemes.

So far $\mathscr{L}_3$, $\mathscr{L}_4$-terms have been regarded as single-valued unary recursive schemes, which assumes that terms are interpreted by functions $\varphi : \omega \times M \longrightarrow \omega \times M$, respectively $\varphi : \omega^2 \times M \longrightarrow \omega^2 \times M$, interpreting predicate symbols by predicates $P : \omega \times M \longrightarrow \{0, 1\}$, respectively $P : \omega^2 \times M \longrightarrow \{0, 1\}$. However, examples 22.1, 21.1, 25.1, 25.2, 25.7 provide other interesting interpretations for $\mathscr{I}_1$ which can be extended to interpretations for $\mathscr{I}_3$, $\mathscr{I}_4$ by introducing appropriate kinds of predicates. (The third and fourth axioms for conditionals will have to be modified or dropped in some cases. These axioms as well as the first and fifth axioms for conditionals and $\theta O = O$ were used in no proof in this chapter.) Consequently, $\mathscr{L}_3$, $\mathscr{L}_4$-terms could also be regarded as multiple-valued, fuzzy, probabilistic etc. unary recursive schemes. Further details are left for the exercises.

Unary recursive schemes can be turned into n-ary ones by adding to the languages $\mathscr{L}_3$, $\mathscr{L}_4$ constants $I_i^{(j)}$, $1 \leq i \neq j \leq n$, as in chapter 23. Similarly, a stack can be introduced by adding to $\mathscr{L}_3$, $\mathscr{L}_4$ new constants Si_i, So_i, $1 \leq i \leq n$. While we are not able to suggest satisfactory axioms for $I_i^{(j)}$ or Si_i, So_i, no additional axioms are needed to reaffirm 26.6–26.11 except that the new constants should commute with L_2, R_2.

It is noteworthy that the interpretations discussed above and, perhaps, all those interesting from the standpoint of Computer Science validate the infinitary (£££)-rule

$$\frac{\alpha_n \varphi = \alpha_{n+1} R, n \geq 0}{\alpha_0 R[\varphi] = O}.$$

where the terms α_n are constructed from I, L, R by multiplication. Therefore, the structures used for those interpretations have the representation property so that all recursively axiomatizable extensions of $\mathscr{A}_0$ (including $\mathscr{A}_1, \mathscr{A}_3, \mathscr{A}_4$) are incomplete with respect to them by 26.2.

Writing a, b, c for formulas of a certain language and introducing the formula construct $\{a\}\varphi\{b\}$, one may use the following Hoare-style inference rules for proving partial correctness of structured $\mathscr{L}_1$-programs.

$$\frac{\{a\}\varphi\{c\},\ \{c\}\psi\{b\}}{\{a\}\varphi\psi\{b\}},$$

$$\frac{\{a\}L_1\varphi\{b\},\ \{a\}R_1\psi\{b\}}{\{a\}(\varphi,\psi)\{b\}},$$

$$\frac{\{a\}R_1\{a\},\ \{a\}L_1\varphi L\{b\},\ \{b\}R\{b\}}{\{a\}\langle\varphi\rangle\{b\}},$$

$$\frac{\{a\}L_1\{b\},\ \{a\}R_1\varphi\{a\}}{\{a\}[\varphi]\{b\}},$$

where $L_1 = (I, O)$, $R_1 = (O, I)$. (Compare with Hoare [1969], Apt [1981].) If $\mathscr{L}_1$-terms are interpreted by single-valued functions, then $\{a\}\varphi\{b\}$ says that whenever $a(s)$ is true and $\varphi(s) = t$, then $b(t)$ is true. Inversely interpreted rules are helpful too, with $\{a\}\varphi\{b\}$ meaning that whenever $\varphi(s) = t$ and $b(t)$, then $a(s)$. The correctness of these two interpretations is shown in the exercises.

More generally, the systems considered in this chapter can also be related to certain varieties of Dynamic, Algorithmic, Process etc. Logics, thereby bringing these logics closer to 'real life'.

EXERCISES TO CHAPTER 26

Exercise 26.1. Given a system of inequalities $\varphi_i \le \theta_i$, $1 \le i \le m$, where φ_i are (regular in $\theta_1, \ldots, \theta_m$ and regular) $\mathscr{L}_2$-terms, show that there are (regular) $\mathscr{L}_2$-terms $\sigma_1, \ldots, \sigma_m$ such that $\theta_1, \ldots, \theta_m$ are not free in them, $\vdash_{\mathscr{A}_2} \varphi_i(\sigma_1, \ldots, \sigma_m/\theta_1, \ldots, \theta_m) \le \sigma_i$, $1 \le i \le m$, and $\vdash_{\mathscr{A}_2} \varphi_i \le \theta_i$, $1 \le i \le m \to \sigma_i \le \theta_i$, $1 \le i \le m$.

Hint. Induction on m. If $m = 1$, then take $\sigma_1 = \mu\theta_1.\varphi_1$. Given a system $\varphi_i \le \theta_i$, $1 \le i \le m+1$, take $\sigma = \mu\theta_{m+1}.\varphi_{m+1}$, then take $\sigma_1, \ldots, \sigma_m$ to correspond to the system $\varphi_i(\sigma/\theta_{m+1}) \le \theta_i$, $1 \le i \le m$ by the induction clause, and $\sigma_{m+1} = \sigma(\sigma_1, \ldots, \sigma_m/\theta_1, \ldots, \theta_m)$. (Respectively, use the fact that whenever φ, ψ are regular in $\theta_1, \ldots, \theta_m$ and regular, and φ is regular in θ_{m+1}, then $\varphi(\psi/\theta_{m+1})$ is regular in $\theta_1, \ldots, \theta_m$ and regular.)

Exercise 26.2. Show that for every (regular) $\mathscr{L}_2$-term σ there is a system $\varphi_i \le \theta_i$, $1 \le i \le m$ such that φ_i are of the form $I, O, \theta_k\theta_l$ (respectively, $k \ne 1, \ldots, m$) or $(P \to \theta_k, \theta_l)$ and σ is the first component of the least solution of that system in the sense of exercise 26.1.

Hint. Use induction on the construction of σ.

Exercise 26.3. Given an interpretation of $\mathscr{L}_3$ such that the axioms and rules of $\mathscr{J}_3$ apart from Scott's rule are valid and condition (***) is satisfied, show that Scott's rule is also valid.

Hint. Follow the proof of 18.14.

Exercise 26.4. Let $\mathscr{S}$ be a (***)-complete OS such that $(L, R) \leq I$ and $\varphi O = O$ for all φ. Extend it to an interpretation of $\mathscr{J}_3$.

Hint. Take $(P \to \varphi, \psi) = (L\varphi, R\psi)$.

Remark. This construction is not as unsatisfactory as it may seem. Indeed, the equality $(P \to \varphi, \psi) = \tilde{P}(\varphi, \psi)$ suggests that the expressive power of $\mathscr{J}_3$ is hardly increased by the presence of predicate variables.

Exercise 26.5. Add a new function constant F to $\mathscr{L}_3$ and find an interpretation for which the axioms of $\mathscr{J}_3$ and (μ) are valid, while Scott's rule is not.

Hint. Take the IOS $\mathscr{S}'$ of exercise 19.5 and interpret terms by members of the subspace consisting of all φ' such that $\varphi'(O) = O$. Define $(P \to \varphi', \psi')$ as $(\bar{L}\varphi', \bar{R}\psi')$, where $\bar{L} = \lambda\varphi.\lambda s.\varphi(3s)$, $\bar{R} = \lambda\varphi.\lambda s.\varphi(3s + 1)$. Take $L\theta \leq I$ for Ψ and interpret F as the element ρ' of the hint to exercise 19.5. Then $\Psi(O/\theta)$ and $\Psi \to \Psi(F(I, \theta)/\theta)$ are valid but $\Psi(\mu\theta.F(I, \theta)/\theta)$ fails.

Exercise 26.6. Let $\mathscr{S}$ be the IOS of example 21.1. Consider fuzzy predicates $P: M \times \{0, 1\} \to E$. Interpreting conditionals by $(P \to \varphi, \psi)(s, t) = \sup\{\inf\{P(s, 0), \varphi(s, t)\}, \inf\{P(s, 1), \psi(s, t)\}\}$ show that the axioms and rules of $\mathscr{J}_3$ are valid, provided $\geq$ is substituted for $=$ in the third and fourth axioms for conditionals.

Taking $E = \{\bot, \top\}$ in particular, one gets interpretations by multiple-valued functions and predicates.

Exercise 26.7. Let $\mathscr{J}$ be the IOS of example 25.1 (example 25.2). Consider probabilistic predicates $P: M \times \{0, 1\} \to [0, \infty]$ (respectively, $P: M \times \{0, 1\} \to [0, 1]$ such that $\forall s(P(s, 0) + P(s, 1) \leq 1)$). Interpreting conditionals by $(P \to \varphi, \psi)(s, t) = P(s, 0)\varphi(s, t) + P(s, 1)\psi(s, t)$, show that the axioms and rules of $\mathscr{J}_3$ apart from the third and fourth axioms for conditionals are valid.

Exercise 26.8. Prove that the Hoare rules suggested at the end of this chapter are valid, provided $\mathscr{L}_1$-terms are interpreted by members of example 22.2.

Hint. The rules for $\circ, \Pi$ are verified directly. Supposing $\{a\}R_1\{a\}$, $\{a\}L_1\varphi L\{b\}$ and $\{b\}R\{b\}$ true, show that the set $\mathscr{E} = \{\theta/\{a\}\theta\{b\}\}$ is closed under the mapping $\lambda\theta.(\varphi L, \theta R)$. Use 18.2 to get $\langle\varphi\rangle \in \mathscr{E}$. Similarly for [ ].

Exercise 26.9. Prove that the Hoare rules are valid, provided $\mathscr{L}_1$-terms are interpreted by members of example 22.1, where $\{a\}\varphi\{b\}$ is true iff whenever $t \in \varphi(s)$ and $a(s)$ is true, then so is $b(t)$ (or whenever $t \in \varphi(s)$ and $b(t)$, then $a(s)$).

Systems $\mathscr{J}_1^{st}, \mathscr{J}_3^{st}$. The systems $\mathscr{J}_1, \mathscr{J}_3$ extended by adding a new crop of function variables $x, x_1, x_2, \ldots,$ function constants $K_0 - K_4$, a term construct $St(\varphi)$, axioms $St(I)K_0 = K_0$, $\chi K_0 = (\chi L, \chi R)$, $St(St(I))K_1 = K_1$, $\chi_1\chi_2K_1K_2 = \chi_1\chi_2$, $K_3K_4 = I$, $\chi St(\theta) = \theta\chi$ where χ, χ_1, χ_2 range over $\mathscr{L}$-terms ($K_3, x, x_1, \ldots$ are $\mathscr{L}$-terms and whenever χ_1, χ_2 are $\mathscr{L}$-terms, then so is $\chi_1\chi_2K_1$), and the rule

$$\frac{\vec{\phi} \to \varphi x \psi \leq \rho x \sigma}{\vec{\phi} \to St(\varphi)\psi \leq St(\rho)\sigma},$$

where x does not occur in $\vec{\phi}, \varphi, \psi, \rho, \sigma$. Finally Ψ is allowed to stand for $St(\theta) \leq St(I)\tau$ in (μ).

Exercise 26.10. Prove that 26.6, 26.8 hold for $\mathscr{I}_3^{st}$. Establish an analogue to 26.7.

Hint. Show that St is a t-operation in $\mathscr{I}_1^{st}$.

CHAPTER 27

Skordev combinatory spaces

The present approach belongs to the algebraic-axiomatic trend in Recursion Theory started by Skordev [1976]. In a series of papers Skordev develops a general theory of recursion on the specially designed algebraic system of combinatory space and studies a wide variety of interesting spaces to represent certain particular concepts of effective computability, as well as to introduce such concepts in new areas. Most of the work on the subject has been carried out by Skordev himself, and his book Skordev [1980] offers a comprehensive presentation.

This chapter examines Skordev combinatory spaces from the point of view of operative spaces. Roughly speaking, the former are OS with a storing operation, with the translation operation eliminated in the case of iterative spaces.

We present the notion of combinatory space in a recent version proposed by Skordev [1980a] with somewhat different notation. A *combinatory space* is a 9-tuple

$$\mathcal{S}^* = (\mathcal{F}, I, \mathcal{C}, \Pi^*, L^*, R^*, \Sigma, T, F),$$

where $\mathcal{F}$ is a partially ordered semigroup with unit I, $\Pi^*:\mathcal{F}^2 \to \mathcal{F}, \Sigma:\mathcal{F}^3 \to \mathcal{F}$ is monotonic, $\mathcal{C} \subseteq \mathcal{F}$, $L^*, R^*, T, F \in \mathcal{F}$, $T \neq F$, $\mathcal{C}T \subseteq \mathcal{C}$, $\mathcal{C}F \subseteq \mathcal{C}$, $\Pi^*(\mathcal{C},\mathcal{C}) \subseteq \mathcal{C}$ and

$$\begin{gathered}
\forall x(x\varphi \leq x\psi) \Rightarrow \varphi \leq \psi,\\
(x, y)^*L^* = x,\ (x, y)^*R^* = y,\\
x(\varphi, \psi)^* = (x\varphi, x\psi)^*,\\
\varphi(I, x\psi)^* = (\varphi, x\psi)^*, \quad \psi(x, I)^* = (x, \psi)^*,\\
x(\chi \to \varphi, \psi) = (x\chi \to x\varphi, x\psi),\\
\chi(I \to x\varphi, x\psi) = (\chi \to x\varphi, x\psi),\\
(\chi \to \varphi, \psi)\rho = (\chi \to \varphi\rho, \psi\rho),\\
(T \to \varphi, \psi) = \varphi, \quad (F \to \varphi, \psi) = \psi,
\end{gathered}$$

where $(\varphi, \psi)^*$, $(\chi \to \varphi, \psi)$ stand for $\Pi^*(\varphi, \psi)$, $\Sigma(\chi, \varphi, \psi)$ and x, y range over $\mathcal{C}$.

Take for example $\mathcal{F} = \{\varphi/\varphi:\omega - \to \omega\}$, $\mathcal{C} = \{\lambda t.s/s \in \omega\}$, the operations $\circ, \Pi^*$ and branching of chapter 2 for $\circ, \Pi^*, \Sigma$ and the elements $\psi_0 - \psi_4$ considered there for L^*, R^*, I, T, F. Then $\mathcal{S}^* = (\mathcal{F}, I, \mathcal{C}, \Pi^*, L^*,$

$R^*, \Sigma, T, F)$ is a combinatory space. Some higher order examples will be given in the exercises.

An element φ is *normal* iff $\mathscr{C}\varphi \subseteq \mathscr{C}$. For example, I, T, F are normal. It follows easily that all normal elements φ are left distributive with respect to Π^*, Σ, i.e. $\varphi(\rho, \sigma)^* = (\varphi\rho, \varphi\sigma)^*$, $\varphi(\chi \to \rho, \sigma) = (\varphi\chi \to \varphi\rho, \varphi\sigma)$ for all χ, ρ, σ.

The following assertions are due to Skordev.

Proposition 27.1. If φ, ψ are normal, then $(\varphi, \psi)^* L^* = \varphi$. If φ is normal, then $(\varphi, \psi)^* R^* = \psi$.

Proof. If φ, ψ are normal, then

$$x(\varphi, \psi)^* L^* = (x\varphi, x\psi)^* L^* = x\varphi$$

for all x; hence $(\varphi, \psi)^* L^* = \varphi$.

It follows that $y(x, I)^* R^* = (x, y)^* R^* = y$ for all y, hence $(x, I)^* R^* = I$. Therefore, if φ is normal, then

$$x(\varphi, \psi)^* R^* = (x\varphi, x\psi)^* R^* = x\psi(x\varphi, I)^* R^* = x\psi I = x\psi$$

for all x; hence $(\varphi, \psi)^* R^* = \psi$.

Proposition 27.2. $xy = y$.

Proof. $xy = x(I, y)^* R^* = (x, y)^* R^* = y$.

In particular, 27.2 implies that all the elements in $\mathscr{C}$ are normal.

Proposition 27.3. $(\rho, I)^*(L^*\chi \to R^*\varphi, R^*\psi) = (\rho\chi \to \varphi, \psi)$.

Proof. We have

$$\begin{aligned} x(\rho, I)^*(L^*\chi \to R^*\varphi, R^*\psi) &= x\rho(I, x)^*(L^*\chi \to R^*\varphi, R^*\psi) = x\rho(\chi \to x\varphi, x\psi) \\ &= (x\rho\chi \to x\varphi, x\psi) = x(\rho\chi \to \varphi, \psi) \end{aligned}$$

for all x, which completes the proof.

Proposition 27.4. $(xT \to \varphi, \psi) = \varphi$, $(xF \to \varphi, \psi) = \psi$.

Proof. Using 27.2, we get

$$y(xT \to \varphi, \psi) = (xT \to y\varphi, y\psi) = x(T \to y\varphi, y\psi) = y\varphi$$

for all y, hence $(xT \to \varphi, \psi) = \varphi$ and similarly $(xF \to \varphi, \psi) = \psi$. The proof is complete.

The components of $\mathscr{S}^*$ enable us to construct a companion OS $\mathscr{S}$.

Proposition 27.5. Take $\Pi(\varphi, \psi) = (L^* \to R^*\varphi, R^*\psi)$, $L = (T, I)^*$ and $R = (F, I)^*$. Then $\mathscr{S} = (\mathscr{F}, I, \Pi, L, R)$ is an OS.

Proof. Π is obviously monotonic, while

$$\begin{aligned} (\varphi, \psi)\chi &= (L^* \to R^*\varphi, R^*\psi)\chi = (L^* \to R^*\varphi\chi, R^*\psi\chi) = (\varphi\chi, \psi\chi), \\ L(\varphi, \psi) &= (T, I)^*(L^* \to R^*\varphi, R^*\psi) = (T \to \varphi, \psi) = \varphi \end{aligned}$$

by 27.3 and similarly $R(\varphi, \psi) = \psi$, which completes the proof.

Proposition 27.6. $(\chi \to \varphi, \psi) = (\chi, I)^*(\varphi, \psi)$.

This follows from 27.3.

If the spaces $\mathscr{S}^*, \mathscr{S}$ are to be in some sense equivalent, then their initial operations should be expressible in terms of each other. Since there is nothing in $\mathscr{S}$ to match Π^*, we introduce the additional operation $St = \lambda\varphi.(L^*, R^*\varphi)^*$. Some elementary properties follow to show that St is a storing operation with a corresponding set $\mathscr{L} = \mathscr{C}^{\sim}$, where $\tilde{x} = (x, I)^*$. Notice that all $\tilde{x}$ are normal.

Proposition 27.7. $\tilde{x}St(\varphi) = \varphi\tilde{x}$.

Proof. $\tilde{x}(L^*, R^*\varphi)^* = (\tilde{x}L^*, \tilde{x}R^*\varphi)^* = (x, \varphi)^* = \varphi\tilde{x}$.

Proposition 27.8. If $\varphi\tilde{x}\psi \leq \rho\tilde{x}\sigma$ for all $\tilde{x}$, then $St(\varphi)\psi \leq St(\rho)\sigma$.

Proof. We have $(x, \varphi)^*\psi \leq (x, \rho)^*\sigma$ for all x. Multiplying on the left by yR^*, we get $(x, yR^*\varphi)^*\psi \leq (x, yR^*\rho)^*\sigma$ for all x, y; hence $(I, yR^*\varphi)^*\psi \leq (I, yR^*\rho)^*\sigma$. Multiplying on the left by yL^*, we get $(yL^*, yR^*\varphi)^*\psi \leq (yL^*, yR^*\rho)^*\sigma$ for all y, which implies $St(\varphi)\psi \leq St(\rho)\sigma$.

Proposition 27.9. $(\varphi, \psi)^* = K_0^* St(\varphi) K_1^* St(\psi)$, where $K_0^* = (I, I)^*$, $K_1^* = (R^*, L^*)^*$.

Proof. We have

$$\begin{aligned} xK_0^* St(\varphi) K_1^* St(\psi) &= x\tilde{x}St(\varphi)K_1^*St(\psi) = x\varphi\tilde{x}K_1^*St(\psi) = x\varphi(I, x)^*St(\psi) \\ &= x\varphi(I, x\psi)^* = (x\varphi, x\psi)^* = x(\varphi, \psi)^* \end{aligned}$$

for all x. The proof is complete.

Proposition 27.10. $\tilde{x}K_0 = (\tilde{x}L, \tilde{x}R)$, where $K_0 = St(I)(R^*L^* \to St(R^*)L, St(R^*)R)$.

Proof.

$$\begin{aligned} \tilde{x}K_0 &= (\tilde{x}R^*L^* \to \tilde{x}St(R^*)L,\ \tilde{x}St(R^*)R) \\ &= (L^* \to R^*\tilde{x}L,\ R^*\tilde{x}R) = (\tilde{x}L, \tilde{x}R). \end{aligned}$$

Proposition 27.11. $\tilde{x}\tilde{y}K_1 = (y, x)^{*\sim}$ where $K_1 = St^2(I)St(K_1^*)(R^*L^*, St(R^*))^*K_1^*$.

Proof.

$$\tilde{x}\tilde{y}K_1 = (y, I, x)^*(R^*L^*, St(R^*))^*K_1^* = (I, y, x)^*K_1^* = ((y, x)^*, I)^* = (y, x)^{*\sim}.$$

Proposition 27.12. $(y, x)^{*\sim}K_2 = \tilde{x}\tilde{y}$, where $K_2 = K_1^*(R^*L^*, St(R^*))^*St(K_1^*)$.

Proof.

$$\begin{aligned} (y, x)^{*\sim}K_2 &= (I, y, x)^*(R^*L^*, St(R^*))^*St(K_1^*) = (y, I, x)^*St(K_1^*) \\ &= (y, x, I)^* = \tilde{x}\tilde{y}. \end{aligned}$$

Proposition 27.13. St is a storing operation.

This follows from 27.7, 27.8, 27.10–27.12.

An element φ is *polynomial* in* $\mathscr{B} \subseteq \mathscr{F}$ iff $\varphi \in cl(\{I, L^*, R^*, T, F\} \cup \mathscr{B}/\circ, \Pi^*, \Sigma)$.

Proposition 27.14. The following are equivalent.

(1) φ is polynomial* in $\mathscr{B}$.

(2) φ is polynomial in $\mathscr{B}_0 \cup St(\mathscr{B})$, where

$$\mathscr{B}_0 = \{R^*, St(T), St(F), St(K_0^*), St(K_1^*), St^2(L^*), St^2(R^*)\}.$$

Proof. The implication (2)$\Rightarrow$(1) is immediate. Assuming (1), an easy induction on the construction of φ using 27.9, 27.6, 10.12, 10.13, 10.16 shows that $St(\varphi)$ is polynomial in $\mathscr{B}_0 \cup St(\mathscr{B})$; hence so is φ since $\varphi = LSt(\varphi)R^*$. The proof is complete.

A subset $\mathscr{E}$ of $\mathscr{F}$ is a *regular* segment* iff $\mathscr{E} = \{\theta / \chi\theta\rho \leq \tau \text{ for all } \langle \chi, \rho, \tau \rangle \in \mathscr{A}\}$ for a certain $\mathscr{A} \subseteq \{\varphi / \mathscr{C}\varphi \subseteq \mathscr{C}\} \times \mathscr{F}^2$. Assume from now on that the space $\mathscr{S}^*$ is *iterative* in the sense of Skordev [1980]. That is, an *iteration operation* $[\ ,\ ]: \mathscr{F}^2 \to \mathscr{F}$ exists for which $(\psi \to I, \varphi[\varphi, \psi]) = [\varphi, \psi]$ and $[\varphi, \psi]$ is a member of all regular* segments closed under $\lambda\theta.(\psi \to I, \varphi\theta)$.

Proposition 27.15. The companion OS $\mathscr{S}$ of $\mathscr{S}^*$ is iterative. Moreover, the axiom μA_3 holds for mappings $\lambda\theta.(I, \varphi\theta)$ and the assertion of 5.7** holds.

Proof. Taking $[\varphi] = [R^*\varphi, L^*]R^*$, we have

$$(I, \varphi[\varphi]) = (L^* \to R^*, R^*\varphi[R^*\varphi, L^*]R^*) = (L^* \to I, R^*\varphi[R^*\varphi, L^*])R^*$$
$$= [R^*\varphi, L^*]R^* = [\varphi].$$

Let us show that $[\varphi]$ is a member of any regular segment $\mathscr{E}$ closed under $\lambda\theta.(I, \varphi\theta)$. Consider the regular* segment $\mathscr{E}^* = \{\theta / \theta R^* \in \mathscr{E}\}$. If $\theta \in \mathscr{E}^*$, then $\theta R^* \in \mathscr{E}$; hence $(L^* \to I, R^*\varphi\theta)R^* = (L^* \to R^*, R^*\varphi\theta R^*) = (I, \varphi\theta R^*) \in \mathscr{E}$, which implies $(L^* \to I, R^*\varphi\theta) \in \mathscr{E}^*$. Therefore, $[R^*\varphi, L^*] \in \mathscr{E}^*$; hence $[\varphi] \in \mathscr{E}$.

In order to apply 5.13, we construct an operation $\langle\ \ \rangle_1$ such that $\bar{n}\langle\varphi\rangle_1 = \varphi\bar{n}$ for all n, φ.

Let $\sigma = (L^{*2} \to (L^*R^*, R^*R)^*L, (L^*R^*, R^*R)^*R)$. Then

$$x(\bar{0}, \varphi\bar{k})^*\sigma = (x\bar{0}, x\varphi\bar{k})^*\sigma = x\varphi\bar{k}(x\bar{0}, I)^*\sigma = x\varphi\bar{k}(xT \to (x, R)^*L, (x, R)^*R)$$
$$= x\varphi\bar{k}(x, R)^*L = (x, x\varphi\overline{k+1})^*L = x(I, \varphi\overline{k+1})^*L$$

for all x; hence $(\bar{0}, \varphi\bar{k})^*\sigma = (I, \varphi\overline{k+1})^*L$, while

$$x(\overline{n+1}, \varphi\bar{k})^*\sigma = x\varphi\bar{k}(x\overline{n+1}, I)^*\sigma = x\varphi\bar{k}(x\bar{n}F \to (x\bar{n}, R)^*L, (x\bar{n}, R)^*R)$$
$$= x\varphi\bar{k}(x\bar{n}, R)^*R = (x\bar{n}, x\varphi\overline{k+1})^*R$$

for all x; hence $(\overline{n+1}, \varphi\bar{k})^*\sigma = (\bar{n}, \varphi\overline{k+1})^*R$.

Take $\langle\varphi\rangle_1 = (L^*, R^*, [I]\varphi\bar{0})^*[\sigma]R^*$. Using the easy equalities $(T, \rho)^* \circ [\sigma] = \rho$, $(\bar{n}F, \rho)^*[\sigma] = \rho\sigma[\sigma]$, we get

$$\bar{0}\langle\varphi\rangle_1 = (T, I, \varphi\bar{0})^*[\sigma]R^* = (I, \varphi\bar{0})^*R^* = \varphi\bar{0},$$
$$\overline{n+1}\langle\varphi\rangle_1 = (\bar{n}F, \bar{n}, \varphi\bar{0})^*[\sigma]R^*$$
$$= (\bar{n}, \varphi\bar{0})^*\sigma[\sigma]R^* = (\overline{n-1}, \varphi\bar{1})^*\sigma[\sigma]R^* = \cdots = (\bar{0}, \varphi\bar{n})^*\sigma[\sigma]R^*$$

$$=(I,\varphi\overline{n+1})^*L[\sigma]R^*=(I,\varphi\overline{n+1})^*R^*=\varphi\overline{n+1};$$

hence $\bar{n}\langle\varphi\rangle_1=\varphi\bar{n}$ for all n. By 5.13, the proof is complete.

Proposition 27.16. $[\varphi,\psi]=(\psi,I)^*[\varphi(\psi,I)^*]$.

Proof. Using 27.6 and 6.10, we get

$$[\varphi,\psi]=\mu\theta.(\psi\to I,\varphi\theta)=\mu\theta.(\psi,I)^*(I,\varphi\theta)=(\psi,I)^*[\varphi(\psi,I)^*].$$

Proposition 27.17. *St* is a t-operation whose corresponding set of functional elements is the set $\mathscr{B}_0$ of 27.14. The additional assumptions of exercises 10.2–10.5 are also satisfied.

Proof. The mappings $\Sigma_0=\lambda\theta_1\theta_2.\theta_1\theta_2$, $\Sigma_1=\lambda\theta_1\theta_2.K_0(\theta_1,\theta_2)$, $\Sigma_3=\lambda\theta.K_0[\theta K_0]St(I)$, $\Sigma_4=\lambda\theta.K_1\theta K_2$ and $\Sigma_5=\lambda\theta.L\theta R^*$ are price recursive in $\mathscr{B}_0$ and satisfy the equalities (0), (1), (3)–(5) of chapter 10 by 27.15, 10.18. By 27.14, the proof of 27.15 gives a mapping Σ_7 prime recursive in $\mathscr{B}_0$ such that $c(\Sigma_7)=1$ and $\langle\varphi\rangle=\Sigma_7(St(\varphi))$ for all φ. The proof is complete.

Now we are able to show that recursiveness of $\mathscr{S}^*$ and st-recursiveness of $\mathscr{S}$ are equivalent. The former will be marked by an asterisk to distinguish it from IOS-recursiveness of $\mathscr{S}$. Of course, an element φ is *recursive* in* $\mathscr{B}$ iff

$$\varphi\in cl(\{I,L^*,R^*,T,F\}\cup\mathscr{B}/\circ,\Pi^*,\Sigma,[\ ,\ ]).$$

The notion of mapping recursive* in $\mathscr{B}$ is introduced via parametrization as usual.

Proposition 27.18. Let $\Gamma:\mathscr{F}^n\to\mathscr{F}$, $n\geq 1$, and $\mathscr{B}\subseteq\mathscr{F}$. Then the following are equivalent.

(1) Γ is recursive* in $\mathscr{B}$.

(2) Γ is st-recursive in $\mathscr{B}$.

(3) Γ is prime st-recursive in $\mathscr{B}$.

The same equivalences hold for elements.

Proof. Propositions 27.9, 27.6, 27.16 ensure (1)⇒(2), while 27.17 gives (2)⇒(3). The implication (3)⇒(1) is obvious. This completes the proof.

The last statement makes it possible to derive some basic theorems of the theory of Skordev combinatory spaces as particular instances of results established in chapter 10. These include a Normal Form Theorem, an Enumeration Theorem, a Secton Recursion Theorem etc. The space $\mathscr{S}$ is (£££)-iterative, which gives by 8.3*** a Representation Theorem for $\mathscr{S}^*$ and yields the implications of 26.1–26.3.

Assume that $\mathscr{S}^*$ satisfies a slightly stronger μ-axiom, namely that every unary mapping Γ recursive* in $\mathscr{F}$ has a fixed point which is a member of all regular* segments closed under Γ. Then a First Recursion Theorem for $\mathscr{S}^*$ is also furnished by 10.8* (or exercise 10.5) too.

It follows that in this case the companion space $\mathscr{S}$ satisfies the axiom tμA_3 and the stronger tμ-axiom of exercise 10.9. The other assumptions of exercise 10.9 are also satisfied, provided one takes

$$K_4=R^*,\quad K_5=(L^*,L^*,R^*)^*,\quad K_6=(R^*L^*,L^*,R^{*2})^*.$$

While 27.5, 27.15 guarantee that all (iterative) combinatory spaces have companion (iterative) OS based on the same semigroups, the converse fails. Therefore, there are more OS than combinatory spaces.

Proposition 27.19. There is an IOS $\mathscr{S}$ whose semigroup $\mathscr{F}$ is a semigroup of no combinatory space.

Proof. Let $\mathscr{S}_0 = (\mathscr{F}_0, I, \Pi_0, L, R)$ be the IOS of example 4.3 and $\mathscr{F} = \{\varphi \in \mathscr{F}_0 / \varphi$ is recursive$\}$. Then $\mathscr{S} = (\mathscr{F}, I, \Pi_0 \restriction \mathscr{F}^2, L, R)$ is a μA_3-iterative subspace of $\mathscr{S}_0$ by exercise 18.6.

Suppose that $\mathscr{F}$ is a semigroup of a certain combinatory space $\mathscr{S}^* = (\mathscr{F}, I, \mathscr{C}, \Pi^*, L^*, R^*, \Sigma, T, F)$. Then $I \notin \mathscr{C}$ since $I \in \mathscr{C}$ would give $\mathscr{C} = \{I\}$ by 27.2, hence $T = F$, which is not the case. We shall use the unwinding method to show that whenever $\varphi \neq I$ and $\varphi\varphi = \varphi$, then $\varphi(L, R) = \varphi$. In particular, $x(L, R) = x$ for all $x \in \mathscr{C}$, hence $(L, R) = I$. That will complete the proof since the last equality is false.

Let φ be a recursive member of $\mathscr{F}_0$, $\varphi \neq I$ and $\varphi\varphi = \varphi$. Then $\varphi = \bar{I}[\sigma]$ with a certain primitive σ by 9.3. It follows easily that for all $\alpha \in \mathscr{D}$ either $\alpha\sigma \in \mathscr{D}$ or $\alpha\sigma \in \Pi(\mathscr{F}, \mathscr{F})$. Start constructing a sequence $\{\alpha_n\}$ such that $\alpha_0 = \bar{0}$, $\alpha_n\sigma = \alpha_{n+1}R$ and $\varphi = \alpha_n\sigma[\sigma]$. Then either $\alpha_n\sigma \in \mathscr{D}$ or $\alpha_n\sigma \in \Pi(\mathscr{F}, \mathscr{F})$ at the n-th step. If $\alpha_n\sigma = \beta R$ for a certain β, then take $\alpha_{n+1} = \beta$ and continue. Otherwise stop.

Suppose that $\alpha_n\sigma = \beta L$ for a certain β. Then $\varphi = \beta L[\sigma] = \beta$; hence $\varphi(L, R) = \varphi$ by $\varphi \neq I$.

Suppose that $\alpha_n\sigma = I$ or $\alpha_n\sigma \in \Pi(\mathscr{F}, \mathscr{F})$. Then $\varphi \in \Pi(\mathscr{F}, \mathscr{F})$; hence $\varphi(3k+2)\uparrow$ for all k. If $\varphi(s)$ is not of the form $3k+2$, then $\varphi(L, R)(s) = \varphi(s)$. Supposing $\varphi(s) = 3k+2$, one gets $\varphi(\varphi(s))\uparrow$ contrary to the equality $\varphi(\varphi(s)) = \varphi(s)$. Therefore, $\varphi(L, R)(s) = \varphi(s)$ for all s; hence $\varphi(L, R) = \varphi$.

If an infinite sequence $\{\alpha_n\}$ is obtained, then $\varphi = O$ by (£££); hence $\varphi(L, R) = \varphi$ again. This completes the proof.

To further elucidate the connections between OS and Skordev combinatory spaces it is desirable to characterize those OS augmented by the storing operation which have companion combinatory spaces.

EXERCISES TO CHAPTER 27

The first five exercises are due to D. Skordev.

Exercise 27.1. Let $\mathscr{S}^*$ be a combinatory space. Prove the following equalities.

a. $((\chi \to \varphi, \psi), \rho)^* = (\chi \to (\varphi, \rho)^*, (\psi, \rho)^*)$.

b. $((\chi \to \varphi, \psi) \to \rho, \sigma) = (\chi \to (\varphi \to \rho, \sigma), (\psi \to \rho, \sigma))$.

c. $(\rho \to (\sigma \to \varphi, \psi), (\sigma \to \chi, \tau)) = (\sigma \to (\rho \to \varphi, \chi), (\rho \to \psi, \tau))$.

d. $(I, \varphi)^* K_1^* = (\varphi, I)^*, (\varphi, I)^* K_1^* = (I, \varphi)^*$.

e. $(\varphi, \psi)^* St(\chi) = (\varphi, \psi\chi)^*$.

Exercise 27.2. Assuming that $\mathscr{S}^*$ meets the stronger axiom $\psi(x\varphi, I)^* = (x\varphi, \psi)^*$, prove the following equalities.

a. $\rho(x\chi \to \varphi, \psi) = (x\chi \to \rho\varphi, \rho\psi)$.
b. $(\varphi, \psi)^* K_1^* = (\psi, \varphi)^*$.
c. $(\varphi, \psi)^*(L^*\rho, R^*\sigma)^* = (\varphi\rho, \psi\sigma)^*$.
Hint for a. Use exercise 27.1a to get

$$\rho(x\chi \to \varphi, \psi) = \rho(x(\chi \to T, F), I)^*(\varphi, \psi) = (x(\chi \to T, F), \rho)^*(\varphi, \psi)$$
$$= (x\chi \to \rho\varphi, \rho\psi).$$

Exercise 27.3. Let $\mathscr{S}^*$ be iterative and let $\mathscr{S}$ be its companion IOS. Prove that $(\chi \to \varphi, \psi) = (\chi, R)^*[R^*\psi L][\varphi L]$.

Remark. It follows in particular that $(\varphi, \psi) = (L^*, R)^*[R^{*2}\psi L][R^*\varphi L]$; hence the operations Σ, Π can be eliminated respectively in $\mathscr{S}^*, \mathscr{S}$.

In the following two exercises $\mathscr{S}^*$ is assumed to satisfy the axiom of exercise 27.2 and to have an element U^* such that $x, L^*, R^* \leq U^*$, $U^*x \leq x$ for all x. Such an element exists in the spaces corresponding to examples 21.1–21.3 but not examples 21.4, 22.2, 22.4, 25.1–25.4. In the sequel $\tilde{U}$ stands for $(U^*, I)^*$.

Exercise 27.4. Prove the following assertions.
a. $U^*x = x$, $\tilde{U}R^* = I$.
b. $\forall x(\tilde{x}\varphi \leq \tau) \Rightarrow \tilde{U}\varphi \leq \tau$.
c. $xU^* = U^*$.
d. $\tilde{U}St(\varphi) = \varphi\tilde{U}$.
e. $\tilde{U}K_0 = (\tilde{U}L, \tilde{U}R)$.
f. $\tilde{U}\tilde{U} = \tilde{U}K_2$.
g. $St(\tilde{U}) = \tilde{U}St^2(I)(R^*L^*, St(R^*))^*$.
Hint for b. If $\forall x(\tilde{x}\varphi \leq \tau)$, then $St(\varphi) \leq St(I)R^*\tau$ by ($). Multiply on the left by $\tilde{U}$.
Hint for c. $\forall x(\varphi \leq x\psi)$ implies $\forall x(x\varphi \leq \psi)$.

Exercise 27.5. Show that φ is recursive* in $\{U^*\} \cup \mathscr{B}$ iff there is a ψ recursive* in $\mathscr{B}$ such that $\varphi = \tilde{U}\psi$.

Hint. Using exercise 27.4, show that $\varphi = \tilde{U}R^*\varphi$, $\tilde{U}\varphi\tilde{U}\psi = \tilde{U}K_2St(\varphi)\psi$, $(\tilde{U}\varphi, \tilde{U}\psi) = \tilde{U}K_0(\varphi, \psi)$, $St(\tilde{U}\varphi) = \tilde{U}St^2(I)(R^*L^*, St(R^*))^*St(\varphi)$ and $[\tilde{U}\varphi] = \tilde{U}K_0[K_2St(\varphi)K_0]R^*$. For the last equality observe that $[\sigma]$ is a member of all regular* segments closed under $\lambda\theta.(I, \sigma\theta)$ by the proof of 27.15.

Remark. This exercise establishes a normal form for search computability by 24.4, 27.18.

Exercise 27.6. Let (M, J, L', R') be a pairing space, let E be a partially ordered set with at least two distinct members, let $\Omega: M \times E^2 \to E$ and let $T, F: M \to M$ be monotonic such that $\Omega(T(s), e, d) = e$, $\Omega(F(s), e, d) = d$ for all $s \in M, e, d \in E$. Take

$$M' = \{\varphi / \varphi: M \to E \,\&\, \varphi \text{ is monotonic}\},$$
$$\mathscr{F}_1 = \{\phi / \phi: M \times M' \to E \,\&\, \phi \text{ is monotonic}\},$$

where ϕ is monotonic iff whenever $s \leq t$, $\varphi \leq \psi$, then $\phi(s,\varphi) \leq \phi(t,\psi)$. Also take

$$\mathscr{C}_1 = \{c^* = \lambda s\varphi.\varphi(c)/c \in M\},\ \phi \leq \Psi \text{ iff } \forall s\varphi(\phi(s,\varphi) \leq \Psi(s,\varphi)),$$
$$\phi\Psi = \lambda s\varphi.\phi(s, \lambda t.\Psi(t,\varphi)),$$
$$\Pi^*(\phi,\Psi) = \lambda s\varphi.\phi(s, \lambda t.\Psi(s, \lambda r.\varphi(J(t,r)))),$$
$$\Sigma(X,\phi,\Psi) = \lambda s\varphi.X(s, \lambda t.\Omega(t, \phi(s,\varphi), \Psi(s,\varphi))),$$
$$I_1 = \lambda s\varphi.\varphi(s),\ L^* = \lambda s\varphi.\varphi(L'(s)),\ R^* = \lambda s\varphi.\varphi(R'(s)),$$

$T_1 = \lambda s\varphi.\varphi(T(s))$ and $F_1 = \lambda s\varphi.\varphi(F(s))$. Show that $\mathscr{S}^* = (\mathscr{F}_1, I_1, \mathscr{C}_1, \Pi^*, L^*, R^*, \Sigma, T_1, F_1)$ is a combinatory space.

Remark. Smaller M' and $\mathscr{F}_1$ may also be taken provided the above definitions still make sense. This generalizes a construction suggested in Skordev [1980a], examples 4, 5, which can be obtained by taking the pairing space of 16.1 for (M, J, L', R').

Exercise 27.7. Let $\mathscr{S}^*, E$ be as above and suppose that all well ordered subsets of E have least upper bounds. Show that $\mathscr{S}^*$ is iterative.

Hint. Following the proof of 18.13, show that every monotonic mapping $\Gamma: \mathscr{F}_1 \to \mathscr{F}_1$ has a fixed point which is a member of all regular* segments closed under Γ.

Exercise 27.8. Let (M, J, L', R'), E, Ω, T, F be the same as in exercise 27.6 and N be a nonempty set. Take

$$M' = \{\varphi/\varphi: M \times N \to E \,\&\, \varphi \text{ is monotonic}\},$$
$$\mathscr{F}_1 = \{\phi/\phi: M \times N \times M' \to E \,\&\, \phi \text{ is monotonic}\}.$$

Modify the construction of exercise 27.6 to get a combinatory space $\mathscr{S}^*$ based on $\mathscr{F}_1$ and prove that it is iterative, provided all the well ordered subsets of E have least upper bounds.

Hint. Take $c^* = \lambda sx\varphi.\varphi(c,x)$,

$$\phi\Psi = \lambda sx\varphi.\phi(s,x,\lambda ty.\Psi(t,y,\varphi)),$$
$$(\phi,\Psi)^* = \lambda sx\varphi.\phi(s,x,\lambda ty.\Psi(s,y,\lambda rz.\varphi(J(t,r),z))),$$
$$(X \to \phi,\Psi) = \lambda sx\varphi.X(s,x,\lambda ty.\Omega(t,\phi(s,x,\varphi),\Psi(s,x,\varphi))) \quad \text{etc.}$$

The following exercise shows that hierarchy OS are rich enough to have companion combinatory spaces.

Exercise 27.9. Let $\mathscr{S}, \mathscr{S}', \mathscr{S}''$ be consecutive OS and suppose that $\mathscr{S}''$ admits a transfer operation Tf. Using exercise 27.6, construct a combinatory space $\mathscr{S}^*$ based on $\mathscr{F}''$ such that whenever $\mathscr{S}^*$ is iterative, then so is $\mathscr{S}''$ and relative recursiveness* is equivalent to (prime) tf-recursiveness.

Hint. Take $E = M = \mathscr{F}$, $J = \Pi$, $\Omega(\chi,\varphi,\psi) = \chi(\varphi,\psi)$, $T = \check{L}$, $F = \check{R}$, $M' = \mathscr{F}'$ and $\mathscr{F}_1 = \{\phi: \mathscr{F} \times \mathscr{F}' \to \mathscr{F} / \lambda\theta'.\lambda\theta.\phi(\theta,\theta') \in \mathscr{F}''\}$; one need not distinguish between $\mathscr{F}_1$ and $\mathscr{F}''$. The operation Π^* is correctly introduced by exercise 14.2. Assuming $\mathscr{S}^*$ iterative, construct a companion IOS $\mathscr{S}_1$ by 27.5, 27.15. Use exercise 14.1 to show that $Tf = St$. Apply 27.13, then pass from $\mathscr{S}_1$ to $\mathscr{S}''$ by exercise 7.2.

CHAPTER 28

Recursive functionals

This chapter deals with some initial concepts and results of a recursion theory on monotonic functionals comparable with the theory developed in the first part of Kechris and Moschovakis [1977], thus paving the way for an adequate approach to Kleene-recursiveness in higher types presented in the next chapter. There is also a direct connection with the topics of chapter 30.

Given two arbitrary sets M, N, we recall that $\varphi: M \dashrightarrow N$ is a single-valued *function* and write $\mathscr{F}$ for the set of all such functions. A monotonic single-valued *functional* is a $\phi: M \times \mathscr{F} \dashrightarrow N$ such that whenever $\phi(s, \varphi) = u$ and $\varphi \leq \psi$, then $\phi(s, \psi) = u$. We are interested both in functions and functionals recursive in functionals, where the relevant notions of recursiveness are to be provided by an appropriate IOS.

Proposition 28.1 (Example 28.1). Let M, N be nonempty sets and f_1, f_2 be a splitting scheme for M. Take

$$\mathscr{F}_1 = \{\phi/\phi: M \times \mathscr{F} \dashrightarrow N \,\&\, \phi \text{ is monotonic}\},$$
$$\phi \leq \Psi \text{ iff } \Psi \text{ is an extension of } \phi,$$
$$\phi\Psi = \lambda s\varphi.\phi(s, \lambda t.\Psi(t, \varphi)),$$
$$\Pi_1(\phi, \Psi)(f_1(s), \varphi) = \phi(s, \varphi),$$
$$\Pi_1(\phi, \Psi)(f_2(s), \varphi) = \Psi(s, \varphi)$$

and $\Pi_1(\phi, \Psi)(s, \varphi)\uparrow$ otherwise, $I_1 = \lambda s\varphi.\varphi(s)$, $L_1 = \lambda s\varphi.\varphi(f_1(s))$ and $R_1 = \lambda s\varphi.\varphi(f_2(s))$. Then $\mathscr{S}_1 = (\mathscr{F}_1, I_1, \Pi_1, L_1, R_1)$ is a $(**)_0$-complete OS.

Proof. Taking $\Pi(\varphi, \psi)(f_1(s)) = \varphi(s)$, $\Pi(\varphi, \psi)(f_2(s)) = \psi(s)$ and $\Pi(\varphi, \psi)(s)\uparrow$ otherwise, $L' = \lambda\varphi.\lambda s.L_1(s, \varphi)$ and $R' = \lambda\varphi.\lambda s.R_1(s, \varphi)$, the quadruple $\mathscr{S} = (\mathscr{F}, \Pi, L', R')$ is a subspace of the pairing space of 16.4 with $E = \{\bot, \top\}$, $e = \bot$, restricting oneself to single-valued φ. Proposition 16.10 implies that $\mathscr{S}$ is strongly complete.

Let $\mathscr{S}'$ be obtained from $\mathscr{S}$ by 19.9. Then $\mathscr{S}'$ is a $(**)_0$-complete OS; hence so is its isomorphic copy $\mathscr{S}_1$, assigning to each mapping $\varphi': \mathscr{F} \to \mathscr{F}$ the functional $\lambda s\varphi.\varphi'(\varphi)(s)$. The proof is complete.

While the operations translation and iteration of $\mathscr{S}_1$ are introduced by $\langle\phi\rangle = \mu\Theta.(\phi L_1, \Theta R_1)$, $[\phi] = \mu\Theta.(I_1, \phi\Theta)$, the following statement characterizes them explicitly in the spirit of 3.4. Indeed, example 3.2 is a particular instance of example 28.1.

Proposition 28.2. Let $\mathscr{S}_1$ be the IOS of example 28.1. Then

$$\langle\phi\rangle(f_2^n(f_1(s)), \varphi) = \phi(s, \lambda t.\varphi(f_2^n(f_1(t)))),$$
$$\Delta(\phi, \Psi)(f_2^n(f_1(s)), \varphi) = \phi\Psi^n(s, \varphi)$$

and $\langle\phi\rangle(s, \varphi)$, $\Delta(\phi, \Psi)(s, \varphi)\uparrow$ otherwise. The function $\sigma = \lambda s.[\phi](s, \varphi)$ is the least satisfying the equalities

$$\sigma(f_1(s)) = \varphi(s), \quad \sigma(f_2(s)) = \phi(s, \sigma).$$

In other words, $\sigma = \mu\theta.(\varphi, \lambda s.\phi(s, \theta))$.

Proof. The operation Π_1 is continuous by exercise 18.1 since $(L_1, R_1) \leq I_1$. Therefore, $\Delta(\phi, \Psi) = \sup_m \Theta_m$ by exercise 18.4, where $\Theta_0 = O_1 = \lambda s\varphi.\uparrow$ and $\Theta_{m+1} = (\phi, \Theta_m\Psi)$. An easy induction on m gives that $\Theta_m(f_2^n(f_1(s)), \varphi) = \phi\Psi^n(s, \varphi)$, if $n < m$, and $\Theta_m(s, \varphi)\uparrow$ otherwise, which implies the desired characterization of Δ and that of $\langle\ \rangle$ by 6.32.

The characterization of [ ] follows from the proof of 12.25. This completes the proof.

Example 28.2. Let M, N_1, N be nonempty sets and let f_1, f_2 be a splitting scheme for M. Take the IOS $\mathscr{S}_1 = (\mathscr{F}_1, I_1, \Pi_1, L_1, R_1)$ of example 28.1 with $M \times N_1$, $\lambda sx.(f_1(s), x)$, $\lambda sx.(f_2(s), x)$ playing the role of M, f_1, f_2. Then

$$\mathscr{F} = \{\varphi/\varphi: M \times N_1 \dashrightarrow N\},$$
$$\mathscr{F}_1 = \{\phi/\phi: M \times N_1 \times \mathscr{F} \dashrightarrow N \,\&\, \phi \text{ is monotonic}\},$$
$$\phi\Psi = \lambda sx\varphi.\phi(s, x, \lambda ty.\Psi(t, y, \varphi)),$$
$$(\phi, \Psi)(f_1(s), x, \varphi) = \phi(s, x, \varphi),$$
$$(\phi, \Psi)(f_2(s), x, \varphi) = \Psi(s, x, \varphi),$$

and $(\phi, \Psi)(s, x, \varphi)\uparrow$ otherwise, $I_1 = \lambda sx\varphi.\varphi(s, x)$, $L_1 = \lambda sx\varphi.\varphi(f_1(s), x)$ and $R_1 = \lambda sx\varphi.\varphi(f_2(s), x)$.

Proposition 28.3. The initial operations of example 28.2 may be characterized as follows.

$$\langle\phi\rangle(f_2^n(f_1(s)), x, \varphi) = \phi(s, x, \lambda ty.\varphi(f_2^n(f_1(t)), y)),$$
$$\Delta(\phi, \Psi)(f_2^n(f_1(s)), x, \varphi) = \phi\Psi^n(s, x, \varphi)$$

and $\langle\phi\rangle(s, x, \varphi)$, $\Delta(\phi, \Psi)(s, x, \varphi)\uparrow$ otherwise,

$$\lambda sx.[\phi](s, x, \varphi) = \mu\theta.(\varphi, \lambda sx.\phi(s, x, \theta)).$$

This follows from 28.2.

While example 28.2 is a particular instance of example 28.1, an isomorphic copy of the latter can be obtained from the former by taking a singleton for N_1. In the sequel we shall speak about example 28.1 in some instances and example 28.2 in others, bearing this connection in mind.

A general theory of recursive functionals could have its starting point in the theory developed in part B of this book. The central notion is that of a functional ϕ recursive in a collection of functionals $\mathscr{B}_1 \subseteq \mathscr{F}_1$, for which the following three characteristic theorems hold in paticular.

Every functional ϕ when recursive in $\mathscr{B}_1$ has a normal form $\phi = \bar{1}[\Sigma]$ with a certain Σ primitive in $\mathscr{B}_1$ by 9.3. A similar normal form is assumed in the very definition of the corresponding notion of $\mathscr{I}$-recursiveness introduced by Kechris and Moschovakis [1977].

Whenever $\mathscr{B}_1$ is finite, then there is by the Enumeration Theorem 9.18 a functional Σ recursive in $\mathscr{B}_1$ such that every functional ϕ recursive in $\mathscr{B}_1$ equals $\bar{n}\Sigma$ for a certain n.

Whenever a mapping $\Gamma:\mathscr{F}_1 \to \mathscr{F}_1$ is recursive in $\mathscr{B}_1$, then its least fixed point $\mu\Theta.\Gamma(\Theta)$ is recursive in $\mathscr{B}_1$ by 9.13*. This assertion seems stronger than the First Recursion Theorem (also called Induction Completeness Theorem) 3.3 of the cited work, which is rather an analogue to 6.16. Take for example $\phi = \mu\Theta.(I_1, \Psi\Theta^2)$, the least functional satisfying the equalities $\Theta(f_1(s), \varphi) = \varphi(s)$, $\Theta(f_2(s), \varphi) = \Psi(s, \lambda t.\Theta(t, \lambda r.\Theta(r, \varphi)))$. Proposition 9.13* implies that ϕ is recursive in Ψ, while the First Recursion Theorem of Moschovakis says nothing about ϕ.

Among the interesting functionals to be taken in $\mathscr{B}_1$ are those embodying quantification. Given a monotonic quantifier Q over M, we define a corresponding functional Q_M,

$$Q_M(s, \varphi) = \begin{cases} L_1R_1(s, \varphi), & \text{if } Qt(L_1(t, \varphi) = 0), \\ R_1^2(s, \varphi), & \text{if } Q^{\cup}t \neg(L_1(t, \varphi) = 0), \\ \uparrow, & \text{otherwise,} \end{cases}$$

where 0 is a fixed number of N. Notice that

$$Q_M(X, \phi, \Psi)(s, \varphi) = \begin{cases} \phi(s, \varphi), & \text{if } Qt(X(t, \varphi) = 0), \\ \Psi(s, \varphi), & \text{if } Q^{\cup}t \neg(X(t, \varphi) = 0), \\ \uparrow, & \text{otherwise.} \end{cases}$$

In fact, Q_M is the mapping Q_σ of exercise 13.1 slightly modified to overcome the obstacle that $\mathscr{S}$ is a pairing space rather than an OS. Similarly, if $N_1 = N' \times N''$ and Q is a monotonic quantifier over N', then the functional $Q_{N'}$ such that

$$Q_{N'}(s, x', x'', \varphi) = L_1R_1(s, x', x'', \varphi), \text{ if } Qy'(L_1(s, y', x'', \varphi) = 0)$$
$$Q_{N'}(s, x', x'', \varphi) = R_1^2(s, x', x'', \varphi), \text{ if } Q^{\cup}y' \neg(L_1(s, y', x'', \varphi) = 0)$$

and $Q_{N'}(s, x', x'', \varphi)\uparrow$ otherwise, embodies Q-quantification.

As far as functions recursive in functionals are concerned, the theory of chapter 13 applies in part since $\mathscr{S}$ is now only a pairing space. While for all functions $\psi \in \mathscr{F}$ the functionals $\check{\psi} = \lambda s\varphi.\psi(s)$ are in $\mathscr{F}_1$, there is no multiplication in $\mathscr{S}$, hence no functionals $\check{\psi}, Id, Ml$. (Multiplication can sometimes be introduced. Whenever $N = M$, then one may define $\varphi\psi$ as $\lambda sx.\psi(\varphi(s, x), x)$, thus transforming $\mathscr{S}$ into the IOS of example 22.4.)

The central notion of $\mathscr{B}_1$-recursiveness corresponds to $\mathscr{B}'$-recursiveness of chapter 13. Given a subset $\mathscr{B}_1$ of $\mathscr{F}_1$ and a nonempty subset $\mathscr{B}$ of $\mathscr{F}$, a function φ is *$\mathscr{B}_1$-recursive in $\mathscr{B}$* iff the functional $\check{\varphi}$ is recursive in $\check{\mathscr{B}} \cup \mathscr{B}_1$, while a mapping $\Gamma:\mathscr{F} \to \mathscr{F}$ is *$\mathscr{B}_1$-recursive in $\mathscr{B}$* iff the functional

$\Gamma^{\cup} = \lambda s\varphi.\Gamma(\varphi)(s)$ is recursive in $\check{\mathscr{B}} \cup \mathscr{B}_1$. Notions of prime $\mathscr{B}_1$-recursiveness, $\mathscr{B}_1$-primitiveness etc. are also introduced.

Ordinary properties as 13.1–13.7 hold in the present context. Others such as 13.8, 13.9* cannot now be claimed or even formulated. Sparse as the present theory may be, it includes the following key Normal Form Theorem, First Recursion Theorem and Enumeration Theorem. (Given are nonparametrized versions.)

Proposition 28.4. Let $\mathscr{B}_0 \subseteq \mathscr{F}_1$ be as in chapter 7 and let φ be $\mathscr{B}_1$-recursive in $\mathscr{B}$. Then $\varphi = L'(\mu\theta.\Gamma(\theta))$ for a certain Γ such that $\Gamma^{\cup}$ is a member of $\mathscr{F}_1$ strictly polynomial in $\check{\mathscr{B}} \cup \mathscr{B}_0 \cup \langle \mathscr{B}_1 \rangle$. (Therefore, Γ is $\mathscr{B}_1$-primitive in $\mathscr{B}$.)

Proof. Recall that $\mathscr{B}_0$ is a finite set of primitive functionals such that $cl(\mathscr{B}_0/^{\circ}, [\ \ ])$ contains all the recursive functionals. It follows from 7.11, 12.32 that $\check{\varphi}$ is prime recursive in $\check{\mathscr{B}} \cup \mathscr{B}_0 \cup \langle \mathscr{B}_1 \rangle$. Repeating the proof of 13.11 with *Id* replaced by $\check{\psi}$ for a certain $\psi \in \mathscr{B}$, we get the desired normal form.

A modified version of 13.15 asserts that whenever φ is $\mathscr{B}_1$-recursive in $\mathscr{B}$, then

$$\varphi = L'(R'(\mu\theta.\Gamma((\psi_0, \dots, \psi_m, \theta, \langle \Gamma_1 \rangle(\theta), \dots, \langle \Gamma_n \rangle(\theta))))),$$

where $\psi_0, \dots, \psi_m \in \mathscr{B}$, $\Gamma_1^{\cup}, \dots, \Gamma_n^{\cup} \in \mathscr{B}_1$ and $\Gamma^{\cup}$ is a strictly primitive member of $\mathscr{F}_1$.

Proposition 28.5. Whenever a mapping $\Gamma: \mathscr{F} \to \mathscr{F}$ is $\mathscr{B}_1$-recursive in $\mathscr{B}$, the function $\mu\theta.\Gamma(\theta)$ exists and is $\mathscr{B}_1$-recursive in $\mathscr{B}$.

Proof. 12.27 gives $(\mu\theta.\Gamma(\theta))\check{} = R_1[\Gamma^{\cup} R_1]$.

Systems of equalities are easily solved, allowing for simultaneous recursion.

Proposition 28.6. Let $\mathscr{B}, \mathscr{B}_1$ be finite and let $\mathscr{U}$ stand for the set of all functions $\mathscr{B}_1$-recursive in $\mathscr{B}$. Then there is a $\sigma \in \mathscr{U}$ which is univeral for $\mathscr{U}$, i.e. $\mathscr{U} = \{L'(R'^n(\sigma))/n \in \omega\}$.

Proof. Take a $\psi \in \mathscr{B}$ and a functional Σ recursive in $\check{\mathscr{B}} \cup \mathscr{B}_1$ and universal for all such functionals by 9.18, then take $\check{\sigma} = \Sigma\check{\psi}$. The function σ will do.

As the remarks to 13.18 indicate, the essential assumption above is that $\mathscr{B}_1$ be finite.

The results of chapter 9 concerning universal elements and mappings remain valid with minor modifications due to the fact that L', R' are no longer of the form $\bar{L}, \bar{R}$. For instance, the Second Recursion Theorem asserts that whenever $\sigma \in \mathscr{U}$ is principal universal for $\mathscr{U}$ and $\varphi \in \mathscr{U}$, then there is an n such that $L'(R'^n(\varphi)) = L'(R'^n(\sigma))$. (A function σ is principal universal for $\mathscr{U}$ iff for all $\varphi \in \mathscr{U}$ there is a natural in the sense of chapter 9 primitive recursive functional Ψ such that $\forall n(\bar{n}\check{\varphi} = \bar{n}\Psi\check{\sigma})$.)

The following statement introduces a t-operation *St* in the space $\mathscr{S}_1$. As a result we have notions of st-*recursiveness* for functionals and $\mathscr{B}_1$, st-*recursiveness* for functions. The general theory of chapter 10 applies to the former, while the remarks following 13.20* concern the latter.

Proposition 28.7. Let $\mathscr{S}_1$ be the IOS of example 28.1 and let M_0, J be the same as in 21.13. Take

$$St(\phi)(J(s,t),\varphi) = \phi(t, \lambda r.\varphi(J(s,r)))$$

and $St(\phi)(s,\varphi)\uparrow$ otherwise. Then St is a t-operation over $\mathscr{F}_1$ satisfying axiom $t\mu A_3$.

Proof. Take $\mathscr{L} = M\tilde{}_0$, where $\tilde{s} = \lambda t\varphi.\varphi(J(s,t))$,

$$K_0(J(s,f_1(t)),\varphi) = \varphi(f_1(J(s,t))),$$
$$K_0(J(s,f_2(t)),\varphi) = \varphi(f_2(J(s,t))),$$
$$K_1(J(s,J(t,r)),\varphi) = \varphi(J(J(s,t),r)),$$
$$K_2(J(J(s,t),r),\varphi) = \varphi(J(s,J(t,r)))$$

and $K_0(s,\varphi)$, $K_1(s,\varphi)$, $K_2(s,\varphi)\uparrow$ otherwise.

It follows easily that $\tilde{s}K_0 = (\tilde{s}L_1, \tilde{s}R_1)$, $\widetilde{st}K_1 = J(t,s)\tilde{}$ and $J(t,s)\tilde{}K_2 = \widetilde{st}$ for all $s, t \in M_0$. The equality $\tilde{s}St(\phi) = \phi\tilde{s}$ and the implication

$$\forall\tilde{s}(\tilde{s}\phi \leq \tilde{s}\Psi) \Rightarrow St(I_1)\phi \leq St(I_1)\Psi$$

follow by the definition of St; hence the axiom ($) is valid by exercise 10.7. Therefore, St is a storing operation, hence it is a t-operation and satisfies $t\mu A_3$ by 10.18, 18.21. The proof is complete.

Now we can briefly describe the approach of Kechris and Moschovakis [1977].

We consider monotonic functionals

$$\phi: M^n \times \mathscr{PI}_{k_1} \times \cdots \times \mathscr{PI}_{k_m} \longrightarrow \omega,$$

where $\mathscr{PI}_k = \{f / f: M^k \longrightarrow \omega\}$, $\omega \subseteq M$. Given a class of functionals $\mathscr{I}$, a functional ϕ is *$\mathscr{I}$-recursive* iff

$$\phi(s_1,\ldots,s_n,f_1,\ldots,f_m) = \Psi^\infty(n_1,\ldots,n_k,s_1,\ldots,s_n,f_1,\ldots,f_m)$$

for certain $\Psi \in \mathscr{I}$, $n_1,\ldots,n_k \in \omega$, where Ψ^∞ is the least functional satisfying the equality

$$\Theta(t_1,\ldots,t_{n+k},f_1,\ldots,f_m)$$
$$= \Psi(t_1,\ldots,t_{n+k},\lambda r_1\ldots r_{n+k}.\Theta(r_1,\ldots,r_{n+k},f_1,\ldots,f_m),f_1,\ldots,f_m).$$

In order to get a nontrivial theory the class $\mathscr{I}$ is assumed to be suitable, i.e. to contain certain specific functionals and to be closed under several operations on functionals. A functional ϕ is said to be recursive in a list of functionals $\Psi_1,\ldots,\Psi_n$ iff ϕ is $\mathscr{I}_0[\Psi_1,\ldots,\Psi_n]$-recursive, where $\mathscr{I}_0[\Psi_1,\ldots,\Psi_n]$ is the smallest suitable class containing $\Psi_1,\ldots,\Psi_n$. Recursion in so called normal lists in studied, including recursion in functionals embodying quantifiers.

While $\mathscr{I}$-recursiveness has its origins in Inductive Definability Theory, it is worth mentioning that the basic notion of inductive relation is in turn described via $\mathscr{I}$-recursiveness in the referred work of Kechris and Moschovakis (the 'boldface' version) and Kolaitis [1978] (the 'lightface' one).

We shall establish no precise characterization of $\mathscr{I}$-recursiveness, noticing only that it is st-recursiveness in the following particular instance of example 28.1.

Proposition 28.8 (Example 28.3). Let $\omega \subseteq M$ and $M^* = \bigcup_n M^n$. Take

$$\mathscr{F} = \{\varphi/\varphi:\omega \times M^* \dashrightarrow \omega\},$$
$$\mathscr{F}_1 = \{\phi/\phi:\omega \times M^* \times \mathscr{F} \dashrightarrow \omega \,\&\, \phi \text{ is monotonic}\},$$
$$\phi\Psi = \lambda sx\varphi.\phi(s, x, \lambda ty.\Psi(t, y, \varphi)),$$
$$(\phi, \Psi)(2s, x, \varphi) = \phi(s, x, \varphi),$$
$$(\phi, \Psi)(2s+1, x, \varphi) = \Psi(s, x, \varphi),$$

$I_1 = \lambda sx\varphi.\varphi(s, x)$, $L_1 = \lambda sx\varphi.\varphi(2s, x)$ and $R_1 = \lambda sx\varphi.\varphi(2s+1, x)$. Then $\mathscr{S}_1 = (\mathscr{F}_1, I_1, \Pi_1, L_1, R_1)$ is a $(**)_0$-complete OS. Introduce multiplication $\varphi\psi = \lambda sx.\psi(\varphi(s, x), x)$ in $\mathscr{F}$ and take $(\varphi, \psi)(2s, x) = \varphi(s, x)$, $(\varphi, \psi)(2s+1, x) = \psi(s, x)$, $I = \lambda sx.s$, $L = \lambda sx.2s$, $R = \lambda sx.2s+1$. Then $\mathscr{S} = (\mathscr{F}, I, \Pi, L, R)$ and $\mathscr{S}_1$ are consecutive IOS. (That is, so are $\mathscr{S}$ and the isomorphic copy $\mathscr{S}'$ of $\mathscr{S}_1$ consisting of mappings over $\mathscr{F}$.)

Proof. The IOS $\mathscr{S}, \mathscr{S}_1$ are particular instances respectively of examples 22.4, 28.2 and they are consecutive by 19.11. This completes the proof.

Explicit characterizations of the initial operations of $\mathscr{S}_1$ are obtained by 28.3, observing that $\omega = \bigcup_n f_2^n(f_1(\omega))$. The theory of chapter 13 applies completely since $\mathscr{S}$ is now an IOS.

The following specific functions and functionals may be assumed initial.

$$Id = \lambda sx\varphi.s, \quad \overline{Ml} = \lambda sx\varphi.L\varphi R\varphi(s, x), \quad \psi_0 = \lambda sx.R(s)\mathrm{sg}s,$$
$$\psi_1 = \lambda sx.\overline{lh(x)}(s), \text{ where } lh(t_1, \ldots, t_n) = n,$$
$$\psi_2(s, t, x) = L(s), \text{ if } t \in \omega, \text{ otherwise } \psi_2(s, t, x) = R(s),$$
$$\Psi_0(s, \varphi) = \varphi(s, s), \; \Psi_0(s, t, x, \varphi) = \varphi(s, s, t, t, x),$$
$$\Psi_1(\bar{n}(s), x, \varphi) = \varphi(s, x_1), \text{ if } lh(x) > n$$

and x_1 is x with its first component moved to the $n+1$-th place,

$$\Psi_2(s, t, x, \varphi) = \varphi(t, s, x), \text{ if } t \in \omega,$$
$$\Psi_3(s, t, x, \varphi) = \varphi(s, x)$$

and $\Psi_1(s, x, \varphi) - \Psi_3(s, x, \varphi)\uparrow$ otherwise.

The operation St is introduced by 28.7, using the pairing function $J((s, x), (t, y)) = (\overline{lh(x)}\bar{s}(t), x, y)$.

EXERCISES TO CHAPTER 28

Exercise 28.1. Let $\mathscr{S}$ be the IOS of example 28.2 and let M_0, J be as in 21.13. Take $St(\phi)(J(s, t), x, \varphi) = \phi(t, x, \lambda ry.\varphi(J(s, r), y))$ and $St(\phi)(s, x, \varphi)\uparrow$ otherwise. Show that St is a t-operation satisfying axiom $\mathrm{t}\mu\mathrm{A}_3$.

Hint. Take a fixed $x_0 \in N_1$ and $J_1((s, x_0), (t, x)) = (J(s, t), x)$ for all $s \in M_0$, $t \in M$, $x \in N_1$. Then the storing operation corresponding to J_1 by 28.7 is exactly St.

Exercise 28.2. Show that the operations St introduced by 28.7 and exercise 28.1 satisfy the assumptions of exercises 10.2–10.5, 10.9 for appropriate functionals K_4–K_6.

Hint. Cf. exercise 21.7.

Remarks. In view of the comments to 10.9, a 'boldface' version of the theory may also be pursued.

Exercise 28.3. Let $\mathscr{S}_1, St$ be as in 28.7 with $N = M$. Using the same pairing function J, introduce an operation St_0 over $\mathscr{F}$ by exercise 21.2. Show that $St(\bar{\varphi}) = \overline{St_0(\varphi)}$ for all φ, where $\bar{\varphi} = \lambda s\psi . \varphi\psi(s)$.

Remark. Therefore, the isomorphism between the IOS $\mathscr{S}$, $\bar{\mathscr{F}}$ agrees with the operation St. A similar assertion holds for the operation St of exercise 28.1.

Exercise 28.4. Let $\mathscr{S}_1, St$ be as in 28.7 (exercise 28.1) and $M_0 = M$. Construct an iterative combinatory space $\mathscr{S}^*$ based on $\mathscr{F}_1$ such that St is exactly the operation of 27.13.

Hint. Use exercise 27.6 (exercise 27.8) with $E = N \cup \{\perp\}$ and $s \leq t$ iff $s = t$ in M.

CHAPTER 29

Higher recursion theory

The notion of Kleene-recursive function with finite type arguments was introduced by Kleene [1959], enabling the recursion theory of such functions to grow considerably in the subsequent period. Here we suggest an alternative approach based on the notion of $\mathscr{B}_1$-recursiveness of chapter 28, hence quite similar to the approach of Kechris and Moschovakis [1977]. The latter work provides a comprehensive introduction to the subject, while this chapter simply marks the place of Higher Recursion Theory from the Algebraic Recursion Theory viewpoint.

Take $T^{(0)} = \omega$, $T^{(j+1)} = \{\alpha^{j+1}/\alpha^{j+1}: T^{(j)} \to \omega\}$ (the members of $T^{(j)}$ are called *j-objects*), $T = \bigcup_j T^{(j)}$ and

$$T^* = \bigcup_n T^n = \{x = (\alpha_1, \ldots, \alpha_n)/n \in \omega \,\&\, \alpha_1, \ldots, \alpha_n \in T\},$$

assuming that the i-objects in x precede the j-objects whenever $i < j$. Let

$$T^*_m = \{x \in T^*/ar(x) = m\},$$

where $ar(x) = \langle n_0, \ldots, n_{j_k} \rangle, n_i$ is the number of all i-objects in $x = (\alpha^{j_1}, \ldots, \alpha^{j_k})$ for all $i \leq j_k$. Writing p_i for the i-th prime number, the coding function used is $\langle k_0, \ldots, k_n \rangle = p_0^{k_0+1} \ldots p_n^{k_n+1}$ with inverses $(\langle k_0, \ldots, k_n \rangle)_i = k_i$ for $i \leq n$ and $(k)_i \uparrow$ otherwise.

A function $f: T^* \dashrightarrow \omega$ is called *m-ary* if $\mathrm{Dom}\, f \subseteq T^*_m$. Such functions with finite type arguments are called functionals in Kleene [1959], but here the terms 'function' and 'functional' are used in the sense of chapter 28.

Example 29.1. The IOS $\mathscr{S}_1$ of example 28.3 with $M = T$.

Here, one has consecutive IOS $\mathscr{S} = (\mathscr{F}, I, \Pi, L, R)$ and $\mathscr{S}_1 = (\mathscr{F}_1, I_1, \Pi_1, L_1, R_1)$, where

$$\begin{aligned}
&\mathscr{F} = \{\varphi/\varphi: \omega \times T^* \dashrightarrow \omega\},\\
&\varphi\psi = \lambda sx.\psi(\varphi(s, x), x),\\
&(\varphi, \psi)(2s, x) = \varphi(s, x), (\varphi, \psi)(2s + 1, x) = \psi(s, x),\\
&I = \lambda sx.s, L = \lambda sx.2s, R = \lambda sx.2s + 1,\\
&\mathscr{F}_1 = \{\phi/\phi: \omega \times T^* \times \mathscr{F} \dashrightarrow \omega \,\&\, \phi \text{ is monotonous}\},\\
&\phi\Psi = \lambda sx\varphi.\phi(s, x, \lambda ty.\Psi(t, y, \varphi)),\\
&(\phi, \Psi)(2s, x, \varphi) = \phi(s, x, \varphi), (\phi, \Psi)(2s + 1, x, \varphi) = \Psi(s, x, \varphi),\\
&I_1 = \bar{I}, L_1 = \bar{L}, R_1 = \bar{R},
\end{aligned}$$

$\breve{\psi}$ standing for $\lambda sx\varphi.\psi\varphi(s,x)$. Other familiar notations are $\bar{n}$ for LR^n or $L_1R_1^n, \breve{\psi} = \lambda sx\varphi.\psi(s,x)$. We also recall that

$$\langle\phi\rangle(\bar{n}(s), x, \varphi) = \phi(s, x, \lambda ty.\varphi(\bar{n}(t), y)),$$
$$\lambda sx.[\phi](s, x, \varphi) = \mu\theta.(\varphi, \lambda sx.\phi(s, x, \theta)).$$

The following functions and functionals are to be assumed initial.

$$Id = \breve{I},\ Ml = \lambda sx\varphi.L\varphi R\varphi(s,x),$$
$$\psi_0 = \lambda sx.R(s)\mathrm{sg}s,$$
$$\psi_1 = \lambda sx.ar(x)(s),$$
$$\psi_2(s, t, \alpha^1, x) = \alpha^1(t)$$

(and $\psi_2(s,x)\uparrow$ otherwise; this will often be omitted below),

$$\Psi_0 = \lambda sx\varphi.\varphi(s,s,x),\ \Psi_1(\overline{kj}(s), x, \varphi) = \varphi(x_1),$$

where x_1 is (s,x) with its first j-object moved to the $k+1$-th place,

$$\Psi_2(\bar{j}(s), \alpha^j, x, \varphi) = \alpha^j(\lambda\alpha^{j-2}.\varphi(s, \alpha^{j-2}, \alpha^j, x)),\ \text{if } j \geq 2,$$

$\Psi_3(\bar{j}(s), x, \varphi) = \varphi(s, x_1)$, where x_1 is x with its first j-object dropped.

Instead of ψ_2, Ψ_2 one may take the 'more powerful' ψ_2^+, Ψ_2^+, where

$$\psi_2^+(\bar{j}(s), \alpha^j, \alpha^{j+1}, x) = \alpha^{j+1}(\alpha^j),$$
$$\Psi_2^+(\bar{j}(s), x, \varphi) = L\varphi(s, \lambda\alpha^j.R\varphi(s, \alpha^j, x), x).$$

The corresponding two basic sets $\mathscr{B}_1$,

$$\mathscr{K}_1 = \{\breve{\psi}_0, \breve{\psi}_1, \breve{\psi}_2, Id, Ml, \Psi_0, \Psi_1, \Psi_2, \Psi_3\}$$

and

$$\mathscr{K}_1^+ = \{\breve{\psi}_0, \breve{\psi}_1, \breve{\psi}_2^+, Id, Ml, \Psi_0, \Psi_1, \Psi_2^+, \Psi_3\}$$

give two notions of computability in higher types, $\mathscr{K}_1$*-recursiveness* and $\mathscr{K}_1^+$*-recursiveness*. We recall that φ is $\mathscr{K}_1$-recursive in $\mathscr{B} \subseteq \mathscr{F}$ iff $\breve{\varphi}$ is recursive in $\breve{\mathscr{B}} \cup \mathscr{K}_1$ in the sense of $\mathscr{S}_1$. Functions $f{:}T^* \longrightarrow \omega$ are represented by members $f^* = \lambda sx.f(x)$ of $\mathscr{F}$, hence f is $\mathscr{K}_1$-recursive in $g_0,\ldots,g_l$ iff f^* is $\mathscr{K}_1$-recursive in $g_0^*,\ldots,g_l^*$. Similarly for $\mathscr{K}_1^+$-recursiveness.

The original relative Kleene-recursiveness is proved by Kleene [1959] to be nontransitive, hence it can be equivalent to $\mathscr{B}_1$-recursiveness for no $\mathscr{B}_1$. The characterization established below shows that absolute Kleene-recursiveness is nevertheless equivalent to $\mathscr{K}_1$-recursiveness. (As proved by Kechris and Moschovakis [1977], it is also equivalent to $\mathscr{I}$-recursiveness.) The exercises to this chapter show that relative Kleene-recursiveness is wider than relative $\mathscr{K}_1$-recursiveness and narrower than relative $\mathscr{K}_1^+$-recursiveness.

It is now time to formulate the definition of m-ary function f Kleene-recursive in a list of functions $g_0,\ldots,g_l$, where g_i is m_i-ary, $i \leq l$. (The following definition is not exactly the original one but a slightly modified equivalent version from Kechris and Moschovakis [1977].) For the sake of simplicity assume that $l = 1$ and say, $g_0{:}\omega \times T^{(2)} \longrightarrow \omega$, $g_1{:}T^{(1)} \times T^{(3)} \longrightarrow \omega$, i.e. $m_0 = \langle 1,0,1\rangle$ and $m_1 = \langle 0,1,0,1\rangle$. A mapping $\Omega{:}\mathscr{F} \to \mathscr{F}$ is introduced via the following clauses for $\Omega(\varphi)$.

0_0. $\Omega(\varphi)(\langle 0, ar(n,x), 0, e_1\rangle, n, x) = g_0(n, \lambda\alpha^1.\varphi(e_1, \alpha^1, x))$.
0_1. $\Omega(\varphi)(\langle 0, ar(x), 1, e_1, e_2\rangle, x) = g_1(\lambda\alpha^0.\varphi(e_1, \alpha^0, x), \lambda\alpha^2.\varphi(e_2, \alpha^2, x))$.
1. $\Omega(\varphi)(\langle 1, ar(t,x)\rangle, t, x) = t+1$.
2. $\Omega(\varphi)(\langle 2, ar(x), t\rangle, x) = t$.
3. $\Omega(\varphi)(\langle 3, ar(t,x)\rangle, t, x) = t$.
4. $\Omega(\varphi)(\langle 4, ar(x), e_1, e_2\rangle, x) = \varphi(e_2, \varphi(e_1, x), x)$.
5. $\Omega(\varphi)(\langle 5, ar(r,t,l,m,n,x)\rangle, r, t, l, m, n, x) = \begin{cases} p^t, & \text{if } m = n, \\ lt, & m \neq n. \end{cases}$
6. $\Omega(\varphi)(\langle 6, ar(x), j, k, e_1\rangle, x) = \varphi(e_1, x_1)$, where x has at least $k+1$ j-objects and x_1 is obtained from x by moving the first j-object to the $k+1$-th place.
7. $\Omega(\varphi)(\langle 7, ar(t, \alpha^1, x)\rangle, t, \alpha^1, x) = \alpha^1(t)$.
8. $\Omega(\varphi)(\langle 8, ar(\alpha^j, x), j, e_1\rangle, \alpha^j, x) = \alpha^j(\lambda\alpha^{j-2}.\varphi(e_1, \alpha^j, \alpha^{j-2}, x))$, if $j \geq 2$.
9. $\Omega(\varphi)(\langle g, ar(t,x,y), ar(x)\rangle, t, x, y) = \varphi(t, x)$.

Otherwise $\Omega(\varphi)(s,x)\uparrow$.

The mapping Ω is monotonic; hence $\sigma = \mu\theta.\Omega(\theta)$ exists. Moreover, it follows from the proof of 18.1 that $\sigma = \sigma_\zeta$ for a certain ordinal ζ, where $\sigma_\xi = \sup_{\eta<\xi}\Omega(\sigma_\eta)$ for all ξ.

A function f is *Kleene-recursive in* $g_0, \ldots, g_l$ iff $f = \lambda x.\sigma(e,x)$ with a certain index (natural number) e. Traditionally, $\{e\}(x)$ is written for $\sigma(e,x)$. We shall also write $\{e\}_\xi(x)$, $\{e\}_{<\xi}(x)$ respectively for $\sigma_\xi(e,x)$ $(\sup_{\eta<\xi}\sigma_\eta)$ (e,x). If $\sigma(e,x)\downarrow$, then (e,x) has an *ordinal of computation*, $|e,x| = \min\{\xi/\sigma_\xi(e,x)\downarrow\}$.

The initial functions $g_0, \ldots, g_l$ may include $j+2$-objects ${}^{j+2}F_Q$, where Q is a monotonic quantifier over $T(^j)$ and

$${}^{j+2}F_Q(\alpha^{j+1}) = \begin{cases} 0, & \text{if } Q\alpha^j(\alpha^{j+1}(\alpha^j) = 0), \\ 1, & \text{if } Q^\smile\alpha^j(\alpha^{j+1}(\alpha^j) \neq 0). \end{cases}$$

Kleene-recursiveness in lists α^{j+2}, ${}^{j+2}E$ has been studied extensively, ${}^{j+2}E$ standing for ${}^{j+2}F_\exists$.

Let $f: T^* \longrightarrow \omega$ be m-ary, writing explicitly $f(z, \alpha^{j_1+1}, \ldots, \alpha^{j_k+1})$ with z ranging over $\omega^{(m)_0}$. Bearing clause 0 in mind, assign a functional $\tilde{f} \in \mathscr{F}_1$ to f such that

$$\tilde{f}(s, z, x, \varphi) = f(z, \lambda\alpha^{j_1}.\bar{0}\varphi(s, \alpha^{j_1}, x), \ldots, \lambda\alpha^{j_k}.\overline{k-1}\,\varphi(s, \alpha^{j_k}, x)).$$

The main result of this chapter states that a function f is Kleene-recursive in $g_0, \ldots, g_l$ iff f is $\mathscr{K}_1 \cup \{\tilde{g}_0, \ldots, \tilde{g}_l\}$-recursive, i.e. iff $f^{*\smile}$ is recursive in $\mathscr{K}_1 \cup \{\tilde{g}_0, \ldots, \tilde{g}_l\}$. In order to prove the 'only if'-part of the equivalence, some $\mathscr{K}_1$-recursive functions and functionals will be constructed by using the following auxiliary statement.

Proposition 29.1. For every partial recursive function $f: \omega^n \longrightarrow \omega$ there is a $\mathscr{K}_1$-recursive function $\varphi \in \mathscr{F}$ such that $\varphi(s_1, \ldots, s_n, x) = f(s_1, \ldots, s_n)$ for all x and $\varphi(s,x)\uparrow$ otherwise.

Proof. Take $I^n_k(s_1, \ldots, s_n, x) = s_k$, and $I^n_k(s,x)\uparrow$ whenever $(ar(s,x))_0 < n$, $1 \leq k \leq n$. It follows that $I^n_1 L(s,x) = 2s$, provided $(ar(s,x))_0 \geq n$, and similarly for $I^n_1 R$, $I^n_1\psi_0$. The characterization established by 22.7 shows that there is a function φ which is recursive in ψ_0, I^n_k, $1 \leq k \leq n$ in the sense of $\mathscr{S}$

and corresponds to f as required. The equalities $I_1^{n\vee} = (\overline{n-1\bar{0}\Psi_1})^n Id$ and $I_k^{n\vee} = (\overline{n-1\bar{0}\Psi_1})^{k-1} Id$ for $1 < k \leq n$ imply that the functions I_k^n are $\mathscr{K}_1$-recursive, hence so is φ. The proof is complete.

Proposition 29.2. If f is Kleene-recursive in $g_0, \ldots, g_l$, then f is $\mathscr{K}_1 \cup \{\widetilde{g_0}, \ldots, \widetilde{g_l}\}$-recursive.

Proof. The function f has a Kleene index e, hence $f^* = \chi\sigma$, where $\chi = \lambda sx.e$ is $\mathscr{K}_1$-recursive by 29.1. But is the universal function σ $\mathscr{K}_1 \cup \{\widetilde{g_0}, \ldots, \widetilde{g_l}\}$-recursive? Noting that $\sigma = \mu\theta.\Omega(\theta)$, it suffices by 13.16 to prove that Ω is a $\mathscr{K}_1 \cup \{\widetilde{g_0}, \ldots, \widetilde{g_l}\}$-recursive mapping, i.e. $\Omega^{\vee} = \lambda sx\varphi.\Omega(\varphi)(s, x)$ is a functional recursive in $\mathscr{K}_1 \cup \{\widetilde{g_0}, \ldots, \widetilde{g_l}\}$.

There are by 29.1 $\mathscr{K}_1$-recursive functions $\varphi_0, \ldots, \varphi_6$ such that

$$\varphi_0(s, n, x) = n + 1,\ \varphi_1 = I_2^2,\ \varphi_i = \lambda sx.(s)_i,\ i = 2, 3, 4,$$

$$\varphi_5(s, r, t, l, m, n, x) = \begin{cases} r^t, & \text{if } m = n, \\ lt, & \text{if } m \neq n, \end{cases}$$

$\varphi_6(s, x) = \overline{(s)_0}(s)$, if $(s)_0 = 1, \ldots, 9$ and $lh(s)$ equals respectively 2, 3, 2, 4, 2, 5, 2, 4, 3; $\varphi_6(s, x) = \overline{(s)_2}\,\overline{(s)_0}(s)$, if $(s)_0 = 0$ and either $(s)_2 = 0$, $lh(s) = 4$, or $(s)_2 = 1$, $lh(s) = 5$, where $lh(s) = n > 0$ means that $s = \langle (s)_0, \ldots, (s)_{n-1} \rangle$.

The function $\rho = \lambda sx.\overline{(s)_1}(s)$ is $\mathscr{K}_1$-recursive by 29.1, and there is by 8.1 a recursive χ such that $\bar{s}\bar{s}\chi = \bar{0}$ and $\bar{s}\bar{t}\chi = \bar{1}$ whenever $s \neq t$. Therefore, the function $\varphi_7 = \psi_1 \langle \rho \rangle \chi(I, O)$ is $\mathscr{K}_1$-recursive and $\varphi_7(s, x) = s$, provided $(s)_1 = ar(x)$.

Take $\phi_0 = \bar{1}\bar{0}\Psi_1\bar{0}\Psi_3$. Then $\phi_0(s, n, x, \varphi) = \varphi(n, x)$.

Take $\phi_1 = \bar{1}\bar{0}\Psi_1$. Then $\phi_1(s, n, x, \varphi) = \varphi(n, s, x)$ and

$$\Psi_0 Ml(\overline{\phi_0\varphi_2}, \overline{\phi_1\varphi_3})(s, x, \varphi) = \varphi((s)_3, \varphi((s)_2, x), x).$$

The function $\rho = \lambda sx.\overline{(s)_3(s)_2}((s)_4)$ is $\mathscr{K}_1$-recursive. Taking $\phi_2 = \bar{\rho}(\langle \phi_1 \rangle R_1 L_1, R_1)\Psi_1$, we get that $\phi_2(s, x, \varphi) = \varphi((s)_4, x_1)$, if $(s)_2 = j$, $(s)_3 = k$, $(ar(x))_j > k$ and x_1 is x with its first j-object moved to the $k+1$-th place.

The function $\rho = \lambda sx.\overline{(s)_2}((s)_3)$ is $\mathscr{K}_1$-recursive. Take $\phi_3 = \bar{\rho}\Psi_2$. Then

$$\phi_3(s, \alpha^j, x, \varphi) = \alpha^j(\lambda\alpha^{j-2}.\varphi((s)_3, \alpha^{j-2}, \alpha^j, x)),$$

provided $(s)_2 = j \geq 2$.

Using ψ_1, Ψ_1, Ψ_3, we construct finally a functional ϕ_4 recursive in $\mathscr{K}_1$ such that $\phi_4(s, n, x, y, \varphi) = \varphi(n, x)$, provided $(s)_2 = ar(x)$.

It is now evidently the case that

$$\begin{aligned} \Omega^{\cup} = \overline{\varphi_7\varphi_6}((\overline{\widetilde{g_0}}(\varphi_3, O_1), \overline{\widetilde{g_1}}(\overline{\varphi_3}, \overline{\varphi_4}, O_1)), \check{\phi}_0, \check{\phi}_2, \check{\phi}_1, \Psi_0 Ml(\overline{\phi_0\varphi_2}, \overline{\phi_1\varphi_3}), \\ \check{\varphi}_5, \phi_2, \check{\psi}_2, \phi_3, \phi_4, O_1), \end{aligned}$$

hence $\Omega^{\cup}$ is recursive in $\mathscr{K}_1 \cup \{\widetilde{g_0}, \ldots, \widetilde{g_l}\}$. The proof is complete.

Proposition 29.3. If an m-ary function f is $\mathscr{K}_1 \cup \{\widetilde{g_0}, \ldots, \widetilde{g_l}\}$-recursive, then f is Kleene-recursive in $g_0, \ldots, g_l$.

Proof. We are going to show that for every functional ϕ recursive in $\mathscr{K}_1 \cup \{\widetilde{g_0}, \ldots, \widetilde{g_l}\}$ there is an index e_1 such that $\lambda n.\{e_1\}(n)$ is a primitive recursive (hence total) function and

$$\phi(s, x, \lambda ty.\{\{e\}(ar(y))\}(t, y)) = \{\{e_1\}(ar(x))\}(\langle e, s\rangle, x)$$

for all e, s, x. This will easily imply the desired conclusion. Indeed, take e_1 to correspond to $f^{*\smile}$. Then

$$f(x) = f^{*\smile}(0, x, \lambda ty.\{\{0\}(ar(y))\}(t, y)) = \{\{e_1\}(m)\}(\langle 0, 0\rangle, x)$$

for all x; hence f is Kleene-recursive in $g_0, \ldots, g_l$.

Using induction on the construction of ϕ, we shall find a corresponding index e_1 and prove in addition that whenever $\{\{e_1\}(ar(x))\}_\xi(\langle e, s\rangle, x) = u$, then $\phi(s, x, \lambda ty.\{\{e\}(ar(y))\}_{<\xi}(t, y)) = u$ for all ξ.

1. Let $\phi = \widetilde{g_0}$. Then

$$\begin{aligned}
&\widetilde{g_0}(s, n, x, \lambda ty.\{\{e\}(ar(y))\}(t, y))\\
&\quad = g_0(n, \lambda\alpha^1.\{\{e\}(ar(\alpha^1, x))\}(2s, \alpha^1, x))\\
&\quad = \{\langle 0, 4ar(x), 0, \{e\}(3ar(x))\rangle\}(n, 2s, x)\\
&\quad = \{h_1(ar(x), e)\}(n, 2s, x)\\
&\quad = \{\langle 6, 4ar(x), 0, 1, h_1(ar(x), e)\rangle\}(2s, n, x)\\
&\quad = \{h_2(ar(x), e)\}(\{\langle 2, 2ar(x), 2s\rangle\}(n, x), n, x)\\
&\quad = \{\langle 4, 2ar(x), \langle 2, 2ar(x), 2s\rangle, h_2(ar(x), e)\rangle\}(n, x)\\
&\quad = \{h_3(ar(x), e, s)\}(n, x)\\
&\quad = \{\langle 9, 8ar(x), 2ar(x)\rangle\}(h_3(ar(x), e, s), n, x, \langle e, s\rangle)\\
&\quad = \{h_4(ar(x))\}(h_3(ar(x), e, s), \langle e, s\rangle, n, x)\\
&\quad = \{h_4(ar(x))\}(\{h_5(ar(x))\}(\langle e, s\rangle, n, x), \langle e, s\rangle, n, x)\\
&\quad = \{\langle 4, 4ar(x), h_5(ar(x)), h_4(ar(x))\rangle\}(\langle e, s\rangle, n, x)\\
&\quad = \{\{e_1\}(ar(x))\}(\langle e, s\rangle, n, x)
\end{aligned}$$

with functions h_1, h_2, h_3 Kleene-recursive in $g_0, \ldots, g_l$, primitive recursive functions h_4, h_5 and index e_1 chosen quite obviously, using the fact that all partial recursive functions are Kleene-recursive in $g_0, \ldots, g_l$ (Kleene [1959]). For instance,

$$h_4 = \lambda s.\langle 6, 8s, 0, 1, \langle 6, 8s, 0, (s)_0 + 2, \langle 9, 8s, 2s\rangle\rangle\rangle.$$

The above transformations increase the ordinal of computation. Supposing $\{\{e_1\}(ar(x))\}_\xi(\langle e, s\rangle, n, x) = u$, one gets

$$\{\langle 4, 4ar(x), h_5(ar(x)), h_4(ar(x))\rangle\}_\xi(\langle e, s\rangle, n, x) = u,$$

hence

$$\{h_4(ar(x))\}_{<\xi}(\{h_5(ar(x))\}_{<\xi}(\langle e, s\rangle, n, x), \langle e, s\rangle, n, x) = u$$

etc. Finally, $g_0(n, \lambda\alpha^1.\{\{e\}(ar(\alpha^1, x))\}_{<\xi}(2s, \alpha^1, x)) = u$ gives $\widetilde{g_0}(s, n, x, \lambda ty.\{\{e\}(ar(y))\}_{<\xi}(t, y)) = u$.

2. The functionals $\tilde{g}_1, L_1, A_1 = (R_1, R_1L_1)$, Id, Ml, $\psi_0, \psi_1, \psi_2, \Psi_0, \Psi_1, \Psi_2$

and Ψ_3 are treated similarly; R_1 is replaced by A_1 to eliminate the pairing operation by 6.9.

It remains to show that the operations $\circ, \langle\ \ \rangle, [\ \ ]$ of $\mathscr{S}_1$ preserve the property in question.

3. The case of multiplication. Assume that e_1, e_2 correspond respectively to ϕ, Ψ. There is a primitive recursive function h such that $\{\{e_2\}(ar(x))\}(\langle e, s\rangle, x) = \{\{h(e)\}(ar(x))\}(s, x)$ and h increases the ordinal of computation, i.e. $|\{e_2\}(ar(x)), \ \langle e, s\rangle, x| < |\{h(e)\}(ar(x)), s, x|$ whenever $\{\{e_2\}(ar(x))\}(\langle e, s\rangle, x)\downarrow$. Therefore,

$$\begin{aligned}
&\phi\Psi(s, x, \lambda ty.\{\{e\}(ar(y))\}(t, y)) \\
&\quad = \phi(s, x, \lambda rz.\Psi(r, z, \lambda ty.\{\{e\}(ar(y))\}(t, y))) \\
&\quad = \phi(s, x, \lambda rz.\{\{e_2\}(ar(z))\}(\langle e, r\rangle, z)) \\
&\quad = \phi(s, x, \lambda rz.\{\{h(e)\}(ar(z))\}(r, z)) \\
&\quad = \{\{e_1\}(ar(x))\}(\langle h(e), s\rangle, x) \\
&\quad = \{\{e_3\}(ar(x))\}(\langle e, s\rangle, x)
\end{aligned}$$

for an appropriate e_3.

Suppose that $\{\{e_3\}(ar(x))\}_\xi(\langle e, s\rangle, x) = u$. Then $\{\{e_1\}(ar(x))\}_{<\xi}(\langle h(e), s\rangle, x) = u$; hence

$$\phi(s, x, \lambda rz.\{\{h(e)\}(ar(z))\}_{<\xi}(r, z)) = u$$

by the inductive assumption for ϕ. It follows that

$$\phi(s, x, \lambda rz.\{\{e_2\}(ar(z))\}_{<\xi}(\langle e, r\rangle, z)) = u;$$

hence

$$\phi(s, x, \lambda rz.\Psi(r, z, \lambda ty.\{\{e\}(ar(y))\}_{<\xi}(t, y))) = u$$

by the induction assumption for Ψ and the monotonicity of ϕ. Therefore,

$$\phi\Psi(s, x, \lambda ty.\{\{e\}(ar(y))\}_{<\xi}(t, y)) = u,$$

hence e_3 corresponding to $\phi\Psi$.

4. The case of translation. Assume that e_1 corresponds to ϕ. Recalling that there are primitive recursive functions h_1, h_2 such that $s = \overline{h_1(s)}(h_2(s))$ for all s, it follows that

$$\begin{aligned}
&\langle\phi\rangle(s, x, \lambda ty.\{\{e\}(ar(y))\}(t, y) \\
&\quad = \langle\phi\rangle(\overline{h_1(s)}(h_2(s)), x, \lambda ty.\{\{e\}(ar(y))\}(t, y)) \\
&\quad = \phi(h_2(s), x, \lambda ty.\{\{e\}(ar(y))\}(\overline{h_1(s)}(t), y)) \\
&\quad = \phi(h_2(s), x, \lambda ty.\{\{h_3(e, s)\}(ar(y))\}(t, y)) \\
&\quad = \{\{e_1\}(ar(x))\}(\langle h_3(e, s), h_2(s)\rangle, x) \\
&\quad = \{\{e_2\}(ar(x))\}(\langle e, s\rangle, x)
\end{aligned}$$

for a certain e_2 and a primitive recursive function h_3 which increase the ordinal of computation. Proceeding as in the case of multiplication, one shows that e_2 corresponds to $\langle\phi\rangle$.

5. The case of iteration. Assume that e_1 corresponds to ϕ. In view of the fact that $\Sigma = [\phi]$ satisfies the equality

$$\Sigma(s, x, \varphi) = \begin{cases} \varphi(s/2, x), & \text{if } s \text{ is even,} \\ \phi((s-1)/2, x), \lambda ty.\Sigma(t, y, \varphi), & \text{if } s \text{ is odd,} \end{cases}$$

take primitive recursive functions h_1, h_2 such that

$$\{\{h_1(e_2, e)\}(ar(x))\}(s, x) = \{\{e_2\}(ar(x))\}(\langle e, s\rangle, x),$$
$$\{\{h_2(e_2)\}(ar(x))\}(s, x)$$
$$= \begin{cases} \{\{(s)_0\}(ar(x))\}((s)_1/2, x), & \text{if } (s)_1 \text{ is even.} \\ \{\{e_1\}(ar(x))\}(\langle h_1(e_2, (s)_0), ((s)_1 - 1)/2\rangle, x), & \text{if } (s)_1 \text{ is odd} \end{cases}$$

for all e_2, e, s, x and h_1, h_2 increase the ordinal of computation. There is by the (Second) Recursion Theorem for Kleene-recursiveness an index e_2 such that $\{e_2\}(n) = \{h_2(e_2)\}(n)$ for all n.

Writing ρ, τ for $\lambda sx.\{\{e\}(ar(x))\}(s, x)$, $\lambda sx.\{\{h_1(e_2, e)\}(ar(x))\}(s, x)$ respectively, one gets

$$(\check{\rho}, \phi\check{\tau})(s, x, \varphi) = \begin{cases} \{\{e\}(ar(x))\}(s/2, x), & \text{if } s \text{ is even,} \\ \{\{e_1\}(ar(x))\}(\langle h_1(e_2, e), (s-1)/2\rangle, x), & \text{if } s \text{ is odd} \end{cases}$$
$$= \{\{h_2(e_2)\}(ar(x))\}(\langle e, s\rangle, x)$$
$$= \{\{e_2\}(ar(x))\}(\langle e, s\rangle, x) = \check{\tau}(s, x, \varphi)$$

for all s, x, φ, hence $(\check{\rho}, \phi\check{\tau}) = \check{\tau}$, which implies $\Sigma\check{\rho} \leq \check{\tau}$ by (££). However, the last inequality says precisely that whenever $\Sigma(s, x, \lambda ty.\{\{e\}(ar(y))\}(t, y)) = u$, then $\{\{e_2\}(ar(x))\}(\langle e, s\rangle, x) = u$.

Conversely, a transfinite induction on ξ shows that whenever $\{\{e_2\}(ar(x))\}_\xi(\langle e, s\rangle, x) = u$, then $\Sigma(s, x, \lambda ty.\{\{e\}(ar(y))\}_{<\xi}(t, y)) = u$. Suppose this is true for all $\eta < \xi$ and $\{\{e_2\}(ar(x))\}_\xi(\langle e, s\rangle, x) = u$. Then

$$\{\{h_2(e_2)\}(ar(x))\}_\xi(\langle e, s\rangle, x) = u.$$

If s is even, then $\{\{e\}(ar(x))\}_{<\xi}(s/2, x) = u$, hence

$$\Sigma(s, x, \lambda ty.\{\{e\}(ar(y))\}_{<\xi}(t, y)) = u.$$

Suppose that s is odd. Then

$$\{\{e_1\}(ar(x))\}_{<\xi}\left(\left\langle h_1(e_2, e), \frac{s-1}{2}\right\rangle, x\right) = u$$

hence

$$\phi\left(\frac{s-1}{2}, x, \lambda rz.\{\{h_1(e_2, e)\}(ar(z))\}_{<\xi}(r, z)\right) = u.$$

Therefore

$$\phi\left(\frac{s-1}{2}, x, \lambda rz.\{\{e_2\}(ar(z))\}_{<\xi}(\langle e, r\rangle, z)\right) = u,$$

which implies by the inductive hypothesis for ξ and the monotonicity of ϕ that

$$\phi\left(\frac{s-1}{2}, x, \lambda rz.\Sigma(r, z, \lambda ty.\{\{e\}(ar(y))\}_{<\xi}(t, y))\right) = u;$$

hence $\Sigma(s, x, \lambda ty.\{\{e\}(ar(y))\}_{<\xi}(t, y)) = u$. Therefore, e_2 corresponds to Σ. The proof is complete.

Proposition 29.4 (Kleene-Recursiveness Theorem). Let f be a m-ary function. Then f is Kleene-recursive in $g_0, \ldots, g_l$ iff f is $\mathscr{K}_1 \cup \{\tilde{g}_0, \ldots, \tilde{g}_l\}$-recursive. In particular, f is Kleene-recursive iff f is $\mathscr{K}_1$-recursive.

Follows by 29.2, 29.3.

The notions of relative $\mathscr{K}_1$-recursiveness and relative $\mathscr{K}_1^+$-recursiveness can be introduced by duly modifying the original definition of Kleene. Namely, consider the following alternative clauses which also originate in Kleene [1959].

0_i^-. $\Omega(\varphi)(\langle 0, ar(x), i\rangle, x) = g_i(x)$.
7^+. $\Omega(\varphi)(\langle 7, ar(\alpha^j, \alpha^{j+1}, x), j\rangle, \alpha^j, \alpha^{j+1}, x) = \alpha^{j+1}(\alpha^j)$.
8^+. $\Omega(\varphi)(\langle 8, ar(x), j, e_1, e_2\rangle, x) = \varphi(e_2, \lambda\alpha^j.\varphi(e_1, \alpha^j, x), x)$.

Let us call Kleene-recursiveness and Kleene$^+$-recursiveness the notions obtained by replacing clause 0 by clause 0^-, respectively clauses 0, 7, 8 by clauses $0^-, 7^+, 8^+$.

Proposition 29.5. Let f be a m-ary function. Then f is Kleene$^-$-recursive in $g_0, \ldots, g_l$ iff it is $\mathscr{K}_1$ -recursive in $g_0, \ldots, g_l$.

Proposition 29.6. Let f be a m-ary function. Then f is Kleene$^+$-recursive in $g_0, \ldots, g_l$ iff it is $\mathscr{K}_1^+$-recursive in $g_0, \ldots, g_l$.

The proofs of 29.5, 29.6 closely follow that of 29.4.

The notion of Kleene-recursiveness can also be formulated via representability in hierarchies of IOS. This is one of the earlier results of the IOS-theory, analogous to a similar one of Platek [1966]. Such representability does not however provide an adequate setting for recursion in higher types, in which we agree with Feferman [1977].

EXERCISES TO CHAPTER 29

Exercise 29.1. Show that ψ_2 is recursive in ψ_2^+.

Exercise 29.2. Show that Ψ_2 is recursive in $\mathscr{K}_1^+$.

Hint. $\Psi_2 = (O_1, O_1, D_1\langle\Psi_2^+\rangle C_1(R_1\check{\psi}_2^+, [I_1]))$, where C_1, D_1 are the elements of 6.35, 6.41.

Exercise 29.3. Show that ψ_2^+ is $\mathscr{K}_1$-recursive.

Hint. $\psi_2^+ = R_1[(\check{\psi}_2, D_1\langle R_1\Psi_2\rangle R_1)]$.

Exercise 29.4. Show that f is $\mathscr{K}_1 \cup \{\tilde{f}\}$-recursive.

Hint. If f is m-ary, $f(z, \alpha^{j_1+1}, \ldots, \alpha^{j_k+1})$ and $j_1 < j_2 < \cdots < j_k$, then $f^{*\vee} = \bar{\psi}_1(O_1, I_1)^m(\tilde{f}(\bar{j}_1, \ldots, \bar{j}_k, O_1)\check{\psi}_2^+, O_1)$. If $j_n = j_{n+1}$ for some n, then Ψ_1 is to be used. Suppose for instance that $m = \langle (m)_0, 2\rangle$, i.e. $k = 2$ and $j_1 = j_2 = 0$. Then $f^{*\vee} = \overline{\psi_1}(O_1, I_1)^m(\tilde{f}(\bar{0}, \bar{1}\bar{0}\Psi_1\bar{0}, O_1)\check{\psi}_2^+, O_1)$.

Exercise 29.5. Show that $\breve{f}$ is recursive in $\{f^{*\check{}}\} \cup \mathscr{K}_1^+$.

Remark. It follows from 29.4 and exercises 29.1, 29.2, 29.4 and 29.5 that relative $\mathscr{K}_1$-recursiveness implies relative Kleene-recursiveness, while the latter implies relative $\mathscr{K}_1^+$-recursiveness.

Exercise 29.6. Show that Ψ_2^+ is not recursive in $\mathscr{K}_1$.

Hint. Take $\omega^* = \bigcup_n \omega^n$, $\rho = I \restriction \omega^*$, $\tau = I \restriction (T^* \setminus \omega^*)$ and prove that whenever ϕ is recursive in $\mathscr{K}_1$, then $\bar{\rho}\phi = \bar{\rho}\phi\bar{\rho}$. This equality fails for Ψ_2^+ since $\bar{\rho}\Psi_2^+\check{\tau} \neq \bar{\rho}\Psi_2^+\bar{\rho}\check{\tau}$.

Exercise 29.7. Prove that all $\mathscr{K}_1$-recursive number functions are partial recursive.

Hint. Assign to each $\varphi \in \mathscr{F}$ a $\varphi^*: \omega \longrightarrow \omega$ such that $\varphi^*(\overline{s_n} \ldots \overline{s_1}\bar{n}(s_0)) = \varphi(s_0, \ldots, s_n)$. Show that for every ϕ recursive in $\mathscr{K}_1$ there is a μ-recursive mapping Γ such that if $\breve{\psi} = \phi\breve{\varphi}$, then $\psi^* = \Gamma(\varphi^*)$.

Remark. A modification of 29.1 shows that conversely, all partial recursive functions are $\mathscr{K}_1$-recursive.

Exercise 29.8. Construct a function f such that $\breve{f}$ is not recursive in $\{f^{*\check{}}\} \cup \mathscr{K}_1$.

Hint. Adapting a counterexample of Kleene [1963], take a partial recursive function $f_0: \omega \longrightarrow \{0, 1\}$ which has no general recursive extension. There is (by 9.3) a binary primitive recursive function h such that $f_0(s) = 0$ iff $\exists t(h(s, t) = 0)$. Take $g(s, t, \alpha^2) = 0$, if $h(s, t) > 0$, and $g(s, t, \alpha^2)\uparrow$ otherwise, then take $f = \lambda s\alpha^2 . \alpha^2(\lambda t . g(s, t, \alpha^2))$. By way of contradiction suppose that $\breve{f}$ is recursive in $\{f^{*\check{}}\} \cup \mathscr{K}_1$. Then $\breve{f}$ is recursive in $\mathscr{K}_1$ since both g and f are $\mathscr{K}_1$-recursive. Let $\varphi = \lambda s\alpha^1 . 1$ and $\breve{\chi} = \Psi_0 \breve{f} \breve{\varphi}$. Then χ is $\mathscr{K}_1$-recursive, hence partial recursive by exercise 29.7. Show that $f_0 \cup \chi$ is a general recursive extension of f_0, a contradiction.

Exercise 29.9. Using the above counterexample, show that relative Kleene-recursiveness is not transitive.

Hint. χ is Kleene-recursive in f which is Kleene-recursive, while χ is not partial recursive, hence not Kleene-recursive.

Remark. Kleene-recursiveness also fails to support a First Recursion Theorem. (Cf. Platek [1966]. Kleene has recently revisited the subject, assuming such a theorem in his new definition.) Proposition 29.4 and exercises 29.4, 29.8 show that these peculiarities are due to the privileged status of the initial functions as compared with other functions Kleene-recursive in the initial ones.

Exercise 29.10. Construct a $\mathscr{K}_1^+$-recursive number function which is not partial recursive.

Hint. Consider the function χ constructed in the hint to exercise 29.8.

Exercise 29.11. Let $\rho(\bar{j}(s), \alpha^{j+1}, x) = 1$, if $\forall \alpha^j(\alpha^{j+1}(\alpha^j) \neq 0)$, and $\rho(s, x)\uparrow$ otherwise. Show that ρ is $\mathscr{K}_1^+$-recursive.

Hint. Take $\varphi = \lambda sx.1$ and $\psi = \psi_2^+ \psi_0(O, I)$. Then $\bar{j}\check{\rho} = \bar{j}\Psi_2^+(\check{\varphi}, \bar{j}\check{\psi})$ for all j, hence $\check{\rho} = D_1 \langle \Psi_2^+ \rangle C_1([I_1]\check{\varphi}, \check{\psi})$.

Remark. It follows that the universal quantification halves ${}^{j+2}E_u$ of all ${}^{j+2}E$ are $\mathscr{K}_1^+$-recursive, where

$$
{}^{j+2}E_u(\alpha^{j+1}) = \begin{cases} \uparrow, & \text{if } \exists \alpha^j(\alpha^{j+1}(\alpha^j) = 0), \\ 1, & \text{if } \forall \alpha^j(\alpha^{j+1}(\alpha^j) \neq 0). \end{cases}
$$

CHAPTER 30

Inductive definability theory

Moschovakis [1977] proposes the program of developing Recursion Theory within the framework of Inductive Definability Theory. His approach is well illustrated by the handling of Higher Recursion Theory in Kechris and Moschovakis [1977]. A similar conceptual approach is suggested by Feferman [1977]. We believe the previous chapters of this book support the view that Recursion Theory has its own independent foundations. Moreover, the present chapter shows some notions and results of Inductive Definability Theory to be particular instances of more general ones of Recursion Theory.

We begin with some introductory definitions from Moschovakis [1974].

Assume that an *abstract structure* $\mathscr{V} = (M, \mathscr{R}_1, \dots, \mathscr{R}_l)$ is given, where M is a set, $\omega \subseteq M$, $\mathscr{R}_i$ is a n_i-ary relation, $\mathscr{R}_i \subseteq M^{n_i}$, $1 \leq i \leq l$. The language $\mathscr{L}^{\mathscr{V}}$ of $\mathscr{V}$ has individual variables $x, y, z, \dots$, relation constants $\mathscr{R}_1, \dots, \mathscr{R}_l$ and n-ary relation variables $S, T, U, \dots$ for all $n \geq 1$. We write $\vec{x}, \vec{S}$ for tuples of individual and relation variables; if $\vec{S} = S_1, \dots, S_k$, then $\neg\vec{S}$ will stand for $\neg S_1, \dots, \neg S_k$.

Relation symbols $\mathscr{R}$ are relation constants and variables. The expressions $\mathscr{R}(x_1, \dots, x_n)$ are formulas and whenever φ, ψ are formulas, then so are $\varphi \& \psi$, $\varphi \vee \psi, \neg\varphi$, $\exists y\varphi$ and $\forall y\varphi$. We write explicitly $\varphi(\vec{x}, \vec{S})$ to fix a list $\vec{x}, \vec{S}$ which contains all relation and free individual variables in φ. Notice that $\mathscr{L}^{\mathscr{V}}$ is a '*lightface*' language since no individual constants are allowed, while its '*boldface*' version in Moschovakis [1974] has constants for the members of M.

The notion of a relation symbol $\mathscr{R}$ *positive in* φ is introduced as follows. $\mathscr{R}$ is positive in $\mathscr{R}(x_1, \dots, x_n)$ and all formulas φ which have no occurrences of $\mathscr{R}$; if $\mathscr{R}$ is positive in φ, ψ, then it is positive in $\varphi\&\psi$, $\varphi \vee \psi$, $\exists y\varphi$ and $\forall y\varphi$. A formula φ is *positive* iff it is free of negation. If $\varphi(\vec{x}, \vec{S})$ is positive, then all the relation variables in $\vec{S}$ are obviously positive in φ.

For all signatures $(n, m_1, \dots, m_k)$ we have *second order relations*

$$\mathscr{P} = \mathscr{P}(\vec{x}, \vec{S}) \subseteq M^n \times 2^{M^{m_1}} \times \cdots \times 2^{M^{m_k}}$$

First order relations are those with $k = 0$. A relation $\mathscr{P}(\vec{x}, \vec{S})$ is (*positive*) *elementary on* $\mathscr{V}$ iff there is a (positive) formula $\varphi(\vec{x}, \vec{S}) \in \mathscr{L}^{\mathscr{V}}$ such that $\mathscr{P}(\vec{x}, \vec{S})$ iff $\varphi(\vec{x}, \vec{S})$ is true. Given a formula $\varphi(\vec{x}, \vec{S}, S)$ of signature $(n, m_1, \dots, m_k, n)$ with S positive in φ, a second order relation $\mathscr{I}_\varphi(\vec{x}, \vec{S})$ is constructed by taking

$$\mathscr{I}^\xi_\varphi = \{(\vec{x}, \vec{S}) / \varphi(\vec{x}, \vec{S}, \{\vec{x}' / (\bigcup_{\eta < \xi} \mathscr{I}^\eta_\varphi)(\vec{x}', \vec{S})\})\}$$

and $\mathscr{I}_\varphi = \bigcup_\xi \mathscr{I}_\varphi^\xi$. A relation $\mathscr{P}(\vec{x}, \vec{S})$ is (*positive*) *inductive on* $\mathscr{V}$ iff there is a (positive) formula $\varphi(\vec{y}, \vec{x}, \vec{S}, S)$ and natural numbers $s_1, \ldots, s_m$ such that

$$\mathscr{P}(\vec{x}, \vec{S}) \Leftrightarrow \mathscr{I}_\varphi(s_1, \ldots, s_m, \vec{x}, \vec{S}).$$

It should be stressed that second order relations positive inductive on $\mathscr{V}$ are monotonic in their relation arguments.

While the main goal of this chapter is to establish that inductiveness is a particular instance of IOS-recursiveness, the following statement shows that it is sufficient to deal with the notion of positive inductiveness.

Proposition 30.1. A second order relation $\mathscr{P}(\vec{x}, \vec{S})$ is elementary (inductive) on $\mathscr{V}$ iff there is a relation $Q(\vec{x}, \vec{T}, \vec{S}, \vec{S}')$ positive elementary (respectively, positive inductive) on $\mathscr{V}$ such that

$$\mathscr{P}(\vec{x}, \vec{S}) \Leftrightarrow Q(\vec{x}, \neg \mathscr{R}_1, \ldots, \neg \mathscr{R}_l, \vec{S}, \neg \vec{S}).$$

Proof. This follows from the evident fact that whenever $\varphi(\vec{x}, \vec{S}, \vec{U}) \in \mathscr{L}^{\mathscr{V}}$ and $\vec{U}$ are positive in φ, then there is a positive formula $\psi(\vec{x}, \vec{T}, \vec{S}, \vec{S}', \vec{U})$ such that $\varphi(\vec{x}, \vec{S}, \vec{U})$ is true iff $\psi(\vec{x}, \neg \mathscr{R}_1, \ldots, \neg \mathscr{R}_l, \vec{S}, \neg \vec{S}, \vec{U})$ is true.

Let us design now an IOS suitable for treating second order relations.

Example 30.1. The space $\mathscr{S}_1$ is constructed as follows. Take the set $\mathscr{F}$ of all relations $X \subseteq \omega \times \bigcup_n M^n$ concentrated on M^n for a certain n, where X *concentrates on* M^n iff

$$X(s, x_1, \ldots, x_m) \Leftrightarrow \forall x_{m+1} \ldots \forall x_n X(s, x_1, \ldots, x_n),$$
$$X(s, x_1, \ldots, x_k) \Leftrightarrow X(s, x_1, \ldots, x_n)$$

for all $m \leq n \leq k$, then take $\mathscr{T} = \omega$,

$$t(X) = \min\{n / X \text{ concentrates on } M^n\},$$

$$X \leq Y \text{ iff } X \subseteq Y,$$

$$\Pi(X, Y) = \{(2s, \vec{x}) / X(s, \vec{x})\} \cup \{(2s+1, \vec{x}) / Y(s, \vec{x})\},$$

$$L'(X) = \{(s, \vec{x}) / X(2s, \vec{x})\}, R'(X) = \{(s, \vec{x}) / X(2s+1, \vec{x})\}.$$

Get a (**)-complete OS $\mathscr{S}'$ from the $\mathscr{T}$-SCPS $\mathscr{S} = (\mathscr{F}, \Pi, L', R')$ by 19.9. Finally, $\mathscr{S}_1 = (\mathscr{F}_1, I_1, \Pi, L_1, R_1)$ is the isomorphic copy of $\mathscr{S}'$ obtained by transforming mappings $\varphi' : \mathscr{F} \to \mathscr{F}$ into relations $\{(s, \vec{x}, Y) / \varphi'(Y)(s, \vec{x})\}$.

We therefore have $\mathscr{F}_1 = \{\phi / \phi \subseteq \omega \times (\bigcup_n M^n) \times \mathscr{F} \,\&\, \phi$ is monotonic and normal$\}$, where ϕ is normal iff there is an n such that whenever Y concentrates on M^m, $m \geq n$, then so does $\{(s, \vec{x}) / \phi(s, \vec{x}, Y)\}$,

$$\phi \leq \Psi \text{ iff } \phi \subseteq \Psi,$$

$$\phi\Psi(s, \vec{x}, Y) \Leftrightarrow \phi(s, \vec{x}, \lambda t\vec{y}. \Psi(t, \vec{y}, Y)),$$

writing $\lambda t\vec{y}. \Psi(t, \vec{y}, Y)$ for $\{(t, \vec{y}) / \Psi(t, \vec{y}, Y)\}$,

$(\phi, \Psi)(s, \vec{x}, Y) \Leftrightarrow s$ is even $\& \ \phi(s/2, \vec{x}, Y) \vee s$ is odd $\& \ \Psi((s-1)/2, \vec{x}, Y)$,

$I_1(s, \vec{x}, Y) \Leftrightarrow Y(s, \vec{x})$, $L_1(s, \vec{x}, Y) \Leftrightarrow Y(2s, \vec{x})$ and $R_1(s, \vec{x}, Y) \Leftrightarrow Y(2s+1, \vec{x})$.

The equality $\bar{n}\langle\phi\rangle = \phi\bar{n}$ implies

$$\langle\phi\rangle(\bar{n}(s), \vec{x}, Y) \Leftrightarrow \phi(s, \vec{x}, \lambda ty.\, Y(\bar{n}(t), \vec{y})),$$

where $\bar{n}(s)$ stands for $2^n(2s+1)-1$ as in example 22.5.

Iteration in higher order spaces is characterized as in 12.25; hence $\lambda s\vec{x}.[\phi](s, \vec{x}, Y)$ is the least X in $\mathscr{F}$ satisfying the equivalence

$$X(s, \vec{x}) \Leftrightarrow s \text{ is even } \& \ Y\left(\frac{s}{2}, \vec{x}\right) \vee s \text{ is odd } \& \ \phi\left(\frac{s-1}{2}\cdot \vec{x}, X\right).$$

Consider the members $\Psi_0, \Psi_{\&}, \Psi_{\vee}, \Psi_{\exists}, \Psi_{\forall}$ of $\mathscr{F}_1$ such that

$$\begin{aligned}
&\Psi_0(s, \vec{x}, Y) \Leftrightarrow Y(0, \vec{x}),\\
&\Psi_{\&}(s, \vec{x}, Y) \Leftrightarrow Y(2s, \vec{x}) \,\&\, Y(2s+1, \vec{x}),\\
&\Psi_{\vee}(s, \vec{x}, Y) \Leftrightarrow Y(2s, \vec{x}) \vee Y(2s+1, \vec{x}),\\
&\Psi_{\exists}(s, \vec{x}, Y) \Leftrightarrow \exists y\, Y(s, y, \vec{x}),\\
&\Psi_{\forall}(s, \vec{x}, Y) \Leftrightarrow \forall y\, Y(s, y, \vec{x})
\end{aligned}$$

for all $\vec{x} \in M^n$, $n \geq t(Y)$ in the case of $\Psi_0, \Psi_{\&}, \Psi_{\vee}$, respectively $n+1 \geq t(Y)$ in the case of $\Psi_{\exists}, \Psi_{\forall}$. (In fact, the equivalences for $\Psi_0, \Psi_{\&}, \Psi_{\forall}$ hold for all $\vec{x}$.)

For all $i_0, \ldots, i_m$, $m > 0$, consider the member $\Psi_{i_0,\ldots,i_m}$ of $\mathscr{F}_1$ such that for all Y the relation $\lambda s\vec{x}.\Psi_{i_0,\ldots,i_m}(s, \vec{x}, Y)$ concentrates on M^n, $n = \max\{i_0, \ldots, i_m\}$, and

$$\Psi_{i_0,\ldots,i_m}(x_0, \ldots, x_n, Y) \Leftrightarrow Y(x_{i_0}, \ldots, x_{i_m}).$$

Fix the set of initial elements

$$\mathscr{M}_1 = \{\Psi_0, \Psi_{\&}, \Psi_{\vee}, \Psi_{\exists}, \Psi_{\forall}, \Psi_{i_0,\ldots,i_m} / m > 0 \,\&\, i_0, \ldots, i_m \in \omega\}.$$

One last definition completes our preliminaries. To each second order relation $\mathscr{P}(\vec{x}, \vec{S})$ of signature $(n, m_1, \ldots, m_k)$ which is monotonic on $\vec{S}$ assign the unique $\mathscr{P}^* \in \mathscr{F}_1$ such that for all Y the relation $\lambda s\vec{x}.\mathscr{P}^*(s, \vec{x}, Y)$ concentrates on M^n and

$$\mathscr{P}^*(s, \vec{x}, Y) \Leftrightarrow \mathscr{P}(\vec{x}, \lambda_{m_1}\vec{y}.\bar{0}(0, \vec{y}, Y), \ldots, \lambda_{m_{k-1}}\vec{y}.\overline{k-2}(0, \vec{y}, Y), \lambda_{m_k}\vec{y}.R_1^{k-1}(0, \vec{y}, Y)$$

for all s, Y and $\vec{x} \in M^n$, writing $\lambda_m \vec{y}\cdot$———for $\{\vec{y} \in M^m /$———$\}$. In particular, $\mathscr{R}^*(s, \vec{x}, Y) \Leftrightarrow \mathscr{R}(\vec{x})$ for all first order relations $\mathscr{R}$. Notice that

$$\begin{gathered}
\mathscr{P}^*(\phi_1, \ldots, \phi_k)(s, \vec{x}, Y) \Leftrightarrow \mathscr{P}(\vec{x}, \lambda_{m_1}\vec{y}.\phi_1(0, \vec{y}, Y), \ldots, \lambda_{m_k}\vec{y}.\phi_k(0, \vec{y}, Y)),\\
\mathscr{P}(\vec{x}, \vec{S}) \Leftrightarrow \mathscr{P}^*(S_1^*, \ldots, S_k^*)(s, \vec{x}, Y)
\end{gathered}$$

for all s, Y and $\vec{x} \in M^n$.

Proposition 30.2. If $\mathscr{P}(\vec{x}, \vec{S})$ is positive elementary on $\mathscr{V}$, then $\mathscr{P}^*$ is polynomial in $\{\mathscr{R}_1^*, \ldots, \mathscr{R}_l^*\} \cup \mathscr{M}_1$.

Proof. We follow the construction of the positive formula φ which defines $\mathscr{P}(x_1, \ldots, x_n, S_1, \ldots, S_k)$.

1. Let $\underline{\varphi(\vec{x}, \vec{S})}$ be $S_i(x_{i_1}, \ldots, x_{i_m})$, where $1 \leq i \leq k$, $1 \leq i_1, \ldots, i_m \leq n$. Then $\mathscr{P}^* = \Psi_0 \overline{i-1}\, \Psi_{0,i_1,\ldots,i_m}$, if $i < k$, and $\mathscr{P}^* = \Psi_0 R_1^{k-1} \Psi_{0,i_1,\ldots,i_m}$, if $i = k$.

2. Let $\varphi(\vec{x},\vec{S})$ be $\mathscr{R}_i(x_{i_1},\ldots,x_{i_m})$, where $1 \le i \le l$, $1 \le i_1,\ldots,i_m \le n$. Then $\mathscr{P}^* = \Psi_{0,i_1,\ldots,i_m}\mathscr{R}_i^*$.

3. Let $\mathscr{P}_1(\vec{x},\vec{S})$, $\mathscr{P}_2(\vec{x},\vec{S})$ be defined respectively by φ,ψ. If $\mathscr{P}$ is defined by $\varphi \,\&\, \psi$, then

$$\mathscr{P}^*(s,\vec{x},Y) \Leftrightarrow \mathscr{P}_1^*(s,\vec{x},Y) \,\&\, \mathscr{P}_2^*(s,\vec{x},Y)$$

for all s, Y and $\vec{x} \in M^n$; hence $\mathscr{P}^* = \Psi_{\&}(\mathscr{P}_1^*,\mathscr{P}_2^*)$. Similarly, if $\mathscr{P}$ is defined by $\varphi \vee \psi$, $\exists y\varphi$ or $\forall y\varphi$, then $\mathscr{P}^* = \Psi_{\vee}(\mathscr{P}_1^*,\mathscr{P}_2^*)$, $\mathscr{P}^* = \Psi_{\exists}\mathscr{P}_1^*$, $\mathscr{P}^* = \Psi_{\forall}\mathscr{P}_1^*$ respectively. The proof is complete.

Proposition 30.3. If $\mathscr{P}(\vec{x},\vec{S})$ is positive inductive on $\mathscr{V}$, then $\mathscr{P}^*$ is recursive in $\{\mathscr{R}_1^*,\ldots,\mathscr{R}_l^*\} \cup \mathscr{M}_1$.

Proof. There is a positive formula $\varphi(\vec{y},\vec{x},\vec{S},S)$ and numbers $s_1,\ldots,s_m$ such that

$$\mathscr{P}(\vec{x},\vec{S}) \Leftrightarrow \mathscr{I}_\varphi(s_1,\ldots,s_m,\vec{x},\vec{S}).$$

Let $\mathscr{Q}(\vec{y},\vec{x},\vec{S},S)$ be the positive elementary on $\mathscr{V}$ relation of signature $(m+n, m_1,\ldots,m_k, m+n)$ defined by φ. Write $(\vec{z},Y)^\nabla$ for

$$\vec{z}, \lambda_{m_1}\vec{z}'.\bar{0}(0,\vec{z}'\,Y),\ldots,\lambda_{m_{k-1}}\vec{z}'.\overline{k-2}\,(0,\vec{z}',Y), \lambda_{m_k}\vec{z}'.R_1^{k-1}(0,\vec{z}',Y),$$

$\vec{z}$ ranging over M^{m+n}. Then

$$\begin{aligned}\mathscr{I}_\varphi^*(s,\vec{y},\vec{x},Y) &\Leftrightarrow \mathscr{I}_\varphi((\vec{y},\vec{x},Y)^\nabla) \Leftrightarrow \mathscr{Q}((\vec{y},\vec{x},Y)^\nabla, \lambda_{m+n}\vec{z}.\mathscr{I}_\varphi((\vec{z},Y)^\nabla))\\ &\Leftrightarrow \mathscr{Q}((\vec{y},\vec{x},Y)^\nabla, \lambda_{m+n}\vec{z}.\mathscr{I}_\varphi^*(0,\vec{z},Y))\\ &\Leftrightarrow \mathscr{Q}^*(\bar{0},\ldots,\overline{k-2},R_1^{k-1},\mathscr{I}_\varphi^*)(s,\vec{y},\vec{x},Y)\end{aligned}$$

for all s, Y and $\vec{y} \in M^m$, $\vec{x} \in M^n$. Both

$$\lambda s\vec{z}.\mathscr{I}_\varphi^*(s,\vec{z},Y),\ \lambda s\vec{z}.\mathscr{Q}^*(\bar{0},\ldots,\overline{k-2},R_1^{k-1},\mathscr{I}_\varphi^*)(s,\vec{z},Y)$$

concentrate on M^{m+n} for all Y, hence

$$\mathscr{I}_\varphi^* = \mathscr{Q}^*(\bar{0},\ldots,\overline{k-2},R_1^{k-1},\mathscr{I}_\varphi^*).$$

Suppose that $\Theta = \mathscr{Q}^*(\bar{0},\ldots,\overline{k-2},R_1^{k-1},\Theta)$. The relation $\lambda s\vec{z}.\mathscr{Q}^*(\bar{0},\ldots,\overline{k-2},R_1^{k-1},\Theta)(s,\vec{z},Y)$ concentrates on M^{m+n} for all Y, hence so does $\lambda s\vec{z}.\Theta(s,\vec{z},Y)$. It follows that

$$\begin{aligned}\Theta(s,\vec{y},\vec{x},Y) &\Leftrightarrow \mathscr{Q}^*(\bar{0},\ldots,\overline{k-2},R_1^{k-1},\Theta)(s,\vec{y},\vec{x},Y)\\ &\Leftrightarrow \mathscr{Q}((\vec{y},\vec{x},Y)^\nabla, \lambda_{m+n}\vec{z}.\Theta(0,\vec{z},Y)),\end{aligned}$$

hence $\Theta(s,\vec{y},\vec{x},Y) \Leftrightarrow \Theta(0,\vec{y},\vec{x},Y)$ and $\mathscr{I}_\varphi((\vec{y},\vec{x},Y)^\nabla) \Rightarrow \Theta(s,\vec{y},\vec{x},Y)$, i.e. $\mathscr{I}_\varphi^*(s,\vec{y},\vec{x},Y) \Rightarrow \Theta(s,\vec{y},\vec{x},Y)$ for all s, Y and $\vec{y} \in M^m$, $\vec{x} \in M^n$. Observing that both $\lambda s\vec{z}.\mathscr{I}_\varphi^*(s,\vec{z},Y)$ and $\lambda s\vec{z}.\Theta(s,\vec{z},Y)$ concentrate on M^{m+n}, we conclude that $\mathscr{I}_\varphi^* \le \Theta$. Therefore,

$$\mathscr{I}_\varphi^* = \mu\Theta.\mathscr{Q}^*(\bar{0},\ldots,\overline{k-2},R_1^{k-1},\Theta),$$

hence $\mathscr{I}_\varphi^*$ is recursive in $\{\mathscr{R}_1^*,\ldots,\mathscr{R}_l^*\} \cup \mathscr{M}_1$ by 6.11 since so is $\mathscr{Q}^*$ by 30.2.

The proof of 22.3 implies $\Delta(R_1,L_1)(s,\vec{x},Y)\Leftrightarrow Y(s+1,\vec{x})$. Multiplying $\Psi_0,\Delta(R_1,L_1)$ and $\Psi_{i_0,\dots,i_u}$ for appropriate $u,i_0,\dots,i_u$, one gets a ϕ such that $\lambda s\vec{x}.\phi(s,\vec{x},Y)$ concentrates on M^n and

$$\phi(s,x_1,\dots,x_n,Y)\Leftrightarrow Y(s,s_1,\dots,s_m,x_1,\dots,x_n)$$

for all Y. Therefore, $\mathscr{P}^*=\phi\mathscr{I}^*_\varphi$; hence $\mathscr{P}^*$ is recursive in $\{\mathscr{R}^*_1,\dots,\mathscr{R}^*_l\}\cup\mathscr{M}_1$, which completes the proof.

The structure $\mathscr{V}$ is said to be *ω-acceptable* iff the subset ω of M admits a *coding scheme* $\mathscr{C}=(\omega,\le,\langle\ \ \rangle)$ such that the relations ω, $\le$, Seq, *lh*, *q* and their negations are positive elementary on $\mathscr{V}$, where $\le$ is the ordinary total ordering of ω, $\langle\ \ \rangle:\bigcup_n\omega^n\to\omega$ is injective,

$$\begin{aligned}\mathrm{Seq}(x)&\Leftrightarrow\exists n\exists x_1\dots\exists x_n(x=\langle x_1,\dots,x_n\rangle),\\ lh(x,y)&\Leftrightarrow y\in\omega\,\&\,x=\langle x_1,\dots,x_y\rangle,\\ q(x,y,z)&\Leftrightarrow y\in\omega\,\&\,x=\langle x_1,\dots,x_n\rangle\,\&\,z=x_y.\end{aligned}$$

By an argument of Moschovakis [1974], this ensures that all the arithmetical relations over ω are positive elementary on $\mathscr{V}$. In particular, this is the case for the following recursive ones:

$$\begin{aligned}F_1(x,y)&\Leftrightarrow x\in\omega\,\&\,y=2x,\\ F_2(x,y)&\Leftrightarrow x\in\omega\,\&\,y=2x+1,\\ F_3(x,y,z)&\Leftrightarrow\exists u\exists v(x+1=2^u(2v+1)\,\&\,z+1=2^u(2y+1)).\end{aligned}$$

Proposition 30.4. Let $\mathscr{V}$ be ω-acceptable and ϕ be primitive in $\{\mathscr{R}^*_1,\dots,\mathscr{R}^*_l\}\cup\mathscr{M}_1$. Then for all n the relation $\mathscr{P}_{\phi,n}$ of signature $(n+1,n+1)$ is positive elementary on $\mathscr{V}$, where

$$\mathscr{P}_{\phi,n}(\vec{x},S)\Leftrightarrow\phi(\vec{x},\tilde{S}),$$

$\tilde{S}$ standing for the unique $X\in\mathscr{F}$ to concentrate on M^n such that $X\cap\omega\times M^n=S\cap\omega\times M^n$.

Proof. By induction on the construction of ϕ.

1. Let $\phi=L_1$. Then

$$\begin{aligned}\mathscr{P}_{\phi,n}(s,x_1,\dots,x_n,S)&\Leftrightarrow L_1(s,x_1,\dots,x_n,\tilde{S})\Leftrightarrow\tilde{S}(2s,x_1,\dots,x_n)\\ &\Leftrightarrow S(2s,x_1,\dots,x_n)\Leftrightarrow\exists x(F_1(s,x)\,\&\,S(x,x_1,\dots,x_n));\end{aligned}$$

hence $\mathscr{P}_{\phi,n}$ is positive elementary on $\mathscr{V}$.

2. Let $\phi=\mathscr{R}^*_i$. Taking a $k\ge n_i,n$, one gets

$$\mathscr{P}_{\phi,n}(s,x_1,\dots,x_n,S)\Leftrightarrow\mathscr{R}^*_i(s,x_1,\dots,x_n,\tilde{S})\Leftrightarrow\forall x_{n+1}\dots\forall x_k\mathscr{R}_i(x_1,\dots,x_{n_i});$$

hence $\mathscr{P}_{\phi,n}$ is positive elementary on $\mathscr{V}$.

3. Let $\phi=\Psi_\exists$. Then $\mathscr{P}_{\phi,0}(s,S)\Leftrightarrow S(s)$ and

$$\begin{aligned}\mathscr{P}_{\phi,n}(s,x_1,\dots,x_n,S)&\Leftrightarrow\Psi_\exists(s,x_1,\dots,x_n,\tilde{S})\Leftrightarrow\exists y\tilde{S}(s,y,x_1,\dots,x_n)\\ &\Leftrightarrow\exists yS(s,y,x_1,\dots,x_{n-1})\text{ for }n>0.\end{aligned}$$

4. The cases $R_1,\Psi_0,\Psi_\&,\Psi_\vee,\Psi_\forall,\Psi_{i_0,\dots,i_m}$ are treated similarly.

5. Let $\mathscr{P}_{\phi,n},\mathscr{P}_{\Psi,n}$ be positive elementary on $\mathscr{V}$ for all n. Take a $k\ge n$ such

that whenever Y concentrates on M^k, then so do $\lambda t\vec{y}.\Psi(t,\vec{y},Y)$ and $\lambda s\vec{x}.\phi(s,\vec{x},Y)$; hence $\lambda t\vec{y}.\Psi(t,\vec{y},Y)=(\lambda_{k+1}t\vec{y}.\Psi(t,\vec{y},Y))^{\sim}$. It follows that

$$\begin{aligned}
&\mathscr{P}_{\phi\Psi,n}(s,x_1,\ldots,x_n,S)\Leftrightarrow\phi\Psi(s,x_1,\ldots,x_n,\tilde{S})\\
&\quad\Leftrightarrow\phi(s,x_1,\ldots,x_n,\lambda t\vec{y}.\Psi(t,\vec{y},\tilde{S}))\\
&\quad\Leftrightarrow\phi(s,x_1,\ldots,x_n,(\lambda_{k+1}t\vec{y}.\Psi(t,\vec{y},\tilde{S}))^{\sim})\\
&\quad\Leftrightarrow\phi(s,x_1,\ldots,x_n,(\lambda_{k+1}t\vec{y}.\mathscr{P}_{\Psi,k}(t,\vec{y},S\times M^{k-n}))^{\sim})\\
&\quad\Leftrightarrow\forall x_{n+1}\ldots\forall x_k\phi(s,x_1,\ldots,x_k,(\lambda_{k+1}t\vec{y}.\mathscr{P}_{\Psi,k}(t,\vec{y},S\times M^{k-n}))^{\sim})\\
&\quad\Leftrightarrow\forall x_{n+1}\ldots\forall x_k\mathscr{P}_{\phi,k}(s,x_1,\ldots,x_k,\lambda_{k+1}t\vec{y}.\mathscr{P}_{\Psi,k}(t,\vec{y},S\times M^{k-n})),
\end{aligned}$$

hence $\mathscr{P}_{\phi\Psi,n}$ is positive elementary on $\mathscr{V}$.

6. Let $\mathscr{P}_{\phi,n},\mathscr{P}_{\Psi,n}$ be positive elementary on $\mathscr{V}$. Then

$$\begin{aligned}
&\mathscr{P}_{(\phi,\Psi),n}(s,x_1,\ldots,x_n,S)\\
&\quad\Leftrightarrow(\phi,\Psi)(s,x_1,\ldots,x_n,\tilde{S})\\
&\quad\Leftrightarrow\exists x(F_1(x,s)\&\phi(x,x_1,\ldots,x_n,\tilde{S})\vee F_2(x,s)\&\Psi(x,x_1,\ldots,x_n,\tilde{S}))\\
&\quad\Leftrightarrow\exists x(F_1(x,s)\&\mathscr{P}_{\phi,n}(x,x_1,\ldots,x_n,S)\vee F_2(x,s)\&\mathscr{P}_{\Psi,n}(x,x_1,\ldots,x_n,S)),
\end{aligned}$$

hence $\mathscr{P}_{(\phi,\Psi),n}$ is positive elementary on $\mathscr{V}$.

7. Let $\mathscr{P}_{\phi,n}$ be positive elementary on $\mathscr{V}$. Then

$$\begin{aligned}
&\mathscr{P}_{\langle\phi\rangle,n}(s,x_1,\ldots,x_n,S)\\
&\quad\Leftrightarrow\langle\phi\rangle(s,x_1,\ldots,x_n,\tilde{S})\\
&\quad\Leftrightarrow\exists u\exists v(F_3(s,u,s)\&\phi(u,x_1,\ldots,x_n,\lambda t\vec{y}.F_3(s,t,v)\&\tilde{S}(v,\vec{y})))\\
&\quad\Leftrightarrow\exists u\exists v(F_3(s,u,s)\&\phi(u,x_1,\ldots,x_n,(\lambda_{n+1}t\vec{y}.F_3(s,t,v)\&S(v,\vec{y}))^{\sim}))\\
&\quad\Leftrightarrow\exists u\exists v(F_3(s,u,s)\&\mathscr{P}_{\phi,n}(u,x_1,\ldots,x_n,\lambda_{n+1}t\vec{y}.F_3(s,t,v)\&S(v,\vec{y}))),
\end{aligned}$$

hence $\mathscr{P}_{\langle\phi\rangle,n}$ is positive elementary on $\mathscr{V}$. The proof is complete.

Proposition 30.5. Let $\mathscr{V}$ be ω-acceptable and ϕ recursive in $\{R_1^*,\ldots,\mathscr{R}_i^*\}\cup\mathscr{M}_1$. Then $\mathscr{P}_{\phi,n}$ is positive inductive on $\mathscr{V}$ for all n.

Proof. The Normal form Theorem 9.3 gives $\phi=\bar{1}[\Psi]$ with a certain Ψ primitive in $\{\mathscr{R}_1^*,\ldots,\mathscr{R}_i^*\}\cup\mathscr{M}_1$. It suffices to show that $\mathscr{P}_{\phi,n}$ is positive inductive on $\mathscr{V}$ for all n not less than the type of Ψ. Indeed, the type of ϕ is not greater than such n; hence

$$\mathscr{P}_{\phi,m}(s,x_1,\ldots,x_m,S)\Leftrightarrow\forall x_{m+1}\ldots\forall x_n\mathscr{P}_{\phi,n}(s,x_1,\ldots,x_n,S\times M^{n-m})$$

for all $m\leq n$.

Take $\Sigma=[\Psi]$. If Y concentrates on M^n, then so do $\lambda s\vec{x}.\Psi(s,\vec{x},Y)$, $\lambda t\vec{y}.\Sigma(t,\vec{y},Y)$ since the types of Ψ,Σ are not greater than n. Using the equality $\Sigma=(I_1,\Psi\Sigma)$ and the fact that $\mathscr{P}_{\Psi,n}$ is positive elementary on $\mathscr{V}$ by 30.4, one gets

$$\begin{aligned}
&\mathscr{P}_{\Sigma,n}(s,x_1,\ldots,x_n,S)\Leftrightarrow\Sigma(s,x_1,\ldots,x_n,\tilde{S})\\
&\quad\Leftrightarrow\exists x(F_1(x,s)\&\tilde{S}(x,x_1,\ldots,x_n)\vee F_2(x,s)\\
&\qquad\&\Psi(x,x_1,\ldots,x_n,\lambda t\vec{y}.\Sigma(t,\vec{y},\tilde{S})))
\end{aligned}$$

$$\Leftrightarrow \exists x(F_1(x,s) \& S(x,x_1,\ldots,x_n) \vee F_2(x,s)$$
$$\& \Psi(x,x_1,\ldots,x_n,(\lambda_{n+1} t\vec{y}.\mathscr{P}_{\Sigma,n}(t,\vec{y},S))^{\sim}))$$
$$\Leftrightarrow \exists x(F_1(x,s) \& S(x,x_1,\ldots,x_n) \vee F_2(x,s)$$
$$\& \mathscr{P}_{\Psi,n}(x,x_1,\ldots,x_n,\lambda_{n+1} t\vec{y}.\mathscr{P}_{\Sigma,n}(t,\vec{y},S)))$$
$$\Leftrightarrow \mathscr{Q}(s,x_1,\ldots,x_n,S,\lambda_{n+1} t\vec{y}.\mathscr{P}_{\Sigma,n}(t,\vec{y},S))$$

with a certain relation $\mathscr{Q}$ positive elementary on $\mathscr{V}$.

Suppose that

(1) $\mathscr{P}(s,x_1,\ldots,x_n,S) \Leftrightarrow \mathscr{Q}(s,x_1,\ldots,x_n,S,\lambda_{n+1} t\vec{y}.\mathscr{P}(t,\vec{y},S))$.

Take the unique $\phi_1, \phi_2 \in \mathscr{F}_1$ such that for all Y the relations $\lambda s\vec{x}.\phi_1(s,\vec{x},Y)$, $\lambda s\vec{x}.\phi_2(s,\vec{x},Y)$ concentrate on M^n and

$$\phi_1(s,x_1,\ldots,x_n,Y) \Leftrightarrow Y(s,x_1,\ldots,x_n),$$
$$\phi_2(s,x_1,\ldots,x_n,Y) \Leftrightarrow \mathscr{P}(s,x_1,\ldots,x_n,Y \cap \omega \times M^n).$$

Then $\phi_2 = (\phi_1, \Psi\phi_2)$, hence $\Sigma\phi_1 \leq \phi_2$ by (££), which implies $\mathscr{P}_{\Sigma,n}(s,x_1,\ldots,x_n,S) \Rightarrow \mathscr{P}(s,x_1,\ldots,x_n,S)$ for all $s,x_1,\ldots,x_n,S$. Therefore, $\mathscr{P}_{\Sigma,n}$ is the least relation satisfying (1); hence it is positive inductive on $\mathscr{V}$. Finally, $\mathscr{P}_{\phi,n}$ is positive inductive on $\mathscr{V}$ since

$$\mathscr{P}_{\phi,n}(s,x_1,\ldots,x_n,S) \Leftrightarrow \exists y \exists z(F_1(s,y) \& F_2(y,z) \& \mathscr{P}_{\Sigma,n}(z,x_1,\ldots,x_n,S)).$$

The proof is complete.

The following Positive Inductiveness Theorem establishes the desired characterization of positive inductiveness.

Proposition 30.6. Let $\mathscr{V}$ be ω-acceptable. Then a second order relation $\mathscr{P}(\vec{x},\vec{S})$ is positive inductive on $\mathscr{V}$ iff $\mathscr{P}^*$ is recursive in $\{\mathscr{R}_1^*,\ldots,\mathscr{R}_l^*\} \cup \mathscr{M}_1$.

Proof. The 'only if'-part of the equivalence follows by 30.3.

Assume that $\mathscr{P}$ is of signature $(n,m_1,\ldots,m_k)$, monotonic in its relation arguments, and $\mathscr{P}^*$ is recursive in $\{\mathscr{R}_1^*,\ldots,\mathscr{R}_l^*\} \cup \mathscr{M}_1$. Fix a number $m \geq n, m_1,\ldots,m_k$. Then

$$\mathscr{P}(x_1,\ldots,x_n,S_1,\ldots,S_k)$$
$$\Leftrightarrow \mathscr{P}^*(0,x_1,\ldots,x_n,\lambda t\vec{y}.(S_1^*,\ldots,S_k^*)(t,\vec{y},O))$$
$$\Leftrightarrow \mathscr{P}^*(0,x_1,\ldots,x_n,(\lambda_{m+1} t\vec{y}.(S_1^*,\ldots,S_k^*)(t,\vec{y},O))^{\sim})$$
$$\Leftrightarrow \forall x_{n+1}\ldots\forall x_m \mathscr{P}_{\mathscr{P}^*,m}(0,x_1,\ldots,x_m,\lambda_{m+1} t\vec{y}.(S_1^*,\ldots,S_k^*)(t,\vec{y},O))$$
$$\Leftrightarrow \forall x_{n+1}\ldots\forall x_m \mathscr{P}_{\mathscr{P}^*,m}(0,x_1,\ldots,x_m,\lambda_{m+1} t\vec{y}.\mathscr{Q}(t,\vec{y},S_1,\ldots,S_k)),$$

where

$$\mathscr{Q}(t,y_1,\ldots,y_m,S_1,\ldots,S_k)$$
$$\Leftrightarrow \exists y(t+1 = 2^0(2y+1) \& S_1(y_1,\ldots,y_{m_1}) \vee \cdots \vee t+1 = 2^{k-2}(2y+1)$$
$$\& S_{k-1}(y_1,\ldots,y_{m_{k-1}}) \vee t+1 = 2^{k-1}(y+1) \& S_k(y_1,\ldots,y_{m_k})).$$

Therefore, $\mathscr{P}$ is positive inductive on $\mathscr{V}$ since $\mathscr{P}_{\mathscr{P}^*,m}$ is by 30.5 and positive inductiveness is preserved by substitutions and negation-free explicit definitions. The proof is complete.

Proposition 30.6 makes it possible to transfer results of the general IOS-theory to the theory of inductive relations: for example the Enumeration Theorem 9.18 and the First Recursion Theorem 9.13* for $\mathscr{S}_1$. The latter is a nontrivial generalization of the Positive Induction Completeness Theorem 6B.4 of Moschovakis [1974] (cf. the first Recursion Theorem in example 28.2).

The space $\mathscr{S}_1$ is equally suitable for treating more general notions of inductiveness introduced e.g. by adding a monotonic quantifier Q to the language $\mathscr{L}^{\mathscr{V}}$. One just adds corresponding elements Ψ_Q, Ψ_Q to $\mathscr{M}_1$.

It is worth mentioning finally that the abstract structures $\mathscr{V}$ considered in this chapter are not necessarily acceptable in the sense of Moschovakis [1974], though acceptability implies ω-acceptability and helps to simplify some formulations and proofs.

EXERCISES TO CHAPTER 30

Exercise 30.1. Show that the elements $\Psi_\vee, \Psi_\&$ of example 30.1 satisfy respectively conditions (1) of exercises 7.10, 7.14.

Example 30.2. Take a set $M, \omega \subseteq M$, fix an injective function $J: M^2 \to M$, then take $\mathscr{F} = \{X/X \subseteq M\}$, $X \leq Y$ iff $X \subseteq Y$, $\Pi(X, Y) = J(0, X) \cup J(1, Y)$, $L'(X) = \{y/X(\langle 0, y\rangle)\}$ and $R'(X) = \{y/X(\langle 1, y\rangle)\}$, writing $\langle x, y\rangle$ for $J(x, y)$. Construct a $(**)_0$-complete OS $\mathscr{S}'$ from the SCPS $\mathscr{S} = (\mathscr{F}, \Pi, L', R')$ by 19.9. Finally, $\mathscr{S}_1$ is the isomorphic copy of $\mathscr{S}'$ obtained by transforming mappings $\varphi': \mathscr{F} \to \mathscr{F}$ into second order relations $\{(x, Y)/\varphi'(Y)(x)\}$. Therefore, $\mathscr{F}_1 = \{\phi/\phi \subseteq M \times \mathscr{F} \,\&\, \phi \text{ is monotonic}\}$,

$$\phi\Psi(x, Y) \Leftrightarrow \phi(x, \lambda y.\Psi(y, Y)),$$
$$(\phi, \Psi)(x, Y) \Leftrightarrow \exists y(x = \langle 0, y\rangle \,\&\, \phi(y, Y) \vee x = \langle 1, y\rangle \,\&\, \Psi(y, Y)),$$
$$I_1(x, Y) \Leftrightarrow Y(x), L_1(x, Y) \Leftrightarrow Y(\langle 0, x\rangle) \text{ and } R_1(x, Y) \Leftrightarrow Y(\langle 1, x\rangle).$$

Exercise 30.2. Let $\mathscr{S}_1$ be the IOS of example 30.2 and let be $St: \mathscr{F}_1 \to \mathscr{F}_1$ such that

$$St(\phi)(x, Y) \Leftrightarrow \exists yz(x = \langle y, z\rangle \,\&\, \phi(z, \lambda u.\, Y(\langle y, u\rangle))).$$

Show that St is a t-operation satisfying the axiom $t\mu A_3$.

Hint. Take $\mathscr{L} = M\tilde{\ }$, where $\tilde{x}(y, Y) \Leftrightarrow Y(\langle x, y\rangle)$,

$$K_0(x, Y) \Leftrightarrow \exists yz(x = \langle y, \langle 0, z\rangle\rangle \,\&\, Y(\langle 0, \langle y, z\rangle\rangle)$$
$$\vee\, x = \langle y, \langle 1, z\rangle\rangle \,\&\, Y(\langle 1, \langle y, z\rangle\rangle)).$$
$$K_1(x, Y) \Leftrightarrow \exists yzu(x = \langle y, \langle z, u\rangle\rangle \,\&\, Y(\langle\langle y, z\rangle, u\rangle)),$$
$$K_2(x, Y) \Leftrightarrow \exists yzu(x = \langle\langle y, z\rangle, u\rangle \,\&\, Y(\langle y, \langle z, u\rangle\rangle)).$$

Use 10.18, 18.21.

Exercise 30.3. Show that the above operation St satisfies the assumptions of exercise 10.9 for appropriate $K_3 - K_6 \in \mathscr{F}_1$ expressible by $L_1, R_1, K_0^* - K_2^*$, multiplication, Π_1 and St, where

$$K_0^*(x, Y) \Leftrightarrow Y(\langle x, x\rangle),$$
$$K_1^*(x, Y) \Leftrightarrow \exists yz(x = \langle y, z\rangle \,\&\, Y(\langle z, y\rangle)),$$
$$K_2^*(x, Y) \Leftrightarrow \exists yz(x = \langle y, z\rangle \,\&\, Y(z)).$$

Hint. Cf. example 24.1.

Since $K_0 - K_2$ are also expressible by $K_0^* - K_2^*$, let us extend the notion of st-recursiveness so that ϕ is st-recursive in $\mathscr{B}_1 \subseteq \mathscr{F}_1$ iff

$$\phi \in cl(\{L_1, R_1, K_0^* - K_2^*\} \cup \mathscr{B}_1/\circ, \Pi_1, \langle\ \ \rangle, [\ \ ], St).$$

st-Recursiveness provides a simpler characterization of the inductive relations over acceptable structures, where $\mathscr{V}$ is *acceptable* iff it admits a *coding scheme* $\mathscr{C} = (\omega, \leq, \langle\ \ \rangle), \langle\ \ \rangle : \bigcup_n M^n \to M$, with relations ω, $\leq$, Seq, lh, q and their negations positive elementary on $\mathscr{V}$.

Take $\mathscr{M}_1 = \{\Psi_s, \Psi_\&, \Psi_\vee, \Psi_\exists, \Psi_\forall\}$, where

$$\Psi_s(x, Y) \Leftrightarrow \exists yz(x = \langle y, z\rangle \,\&\, y \in \omega \,\&\, Y(\langle y+1, z\rangle)),$$
$$\Psi_\& = L_1 \cap R_1, \quad \Psi_\vee = L_1 \cup R_1,$$
$$\Psi_\exists(x, Y) \Leftrightarrow \exists y Y(\langle y, x\rangle), \quad \Psi_\forall(x, Y) \Leftrightarrow \forall y Y(\langle y, x\rangle).$$

To each relation $\mathscr{P}$ of signature $(n, m_1, \ldots, m_k)$ assign a $\mathscr{P}^\sim \subseteq M \times \mathscr{F}$ such that

$$\mathscr{P}^\sim(x, Y) \Leftrightarrow \exists x_1 \ldots x_n(x = \langle x_1, \ldots, \langle x_{n-1}, x_n\rangle \cdots\rangle \,\&\, \mathscr{P}(x_1, \ldots, x_n, S_1, \ldots, S_k)).$$

where $S_i(z_1, \ldots, z_{m_i}) \Leftrightarrow Y(\langle i, \langle z_1, \ldots, \langle z_{m_i-1}, z_{m_i}\rangle \cdots\rangle\rangle)$. In particular, if $k = 0$, then

$$\mathscr{P}^\sim(x, Y) \Leftrightarrow \exists x_1 \ldots x_n(x = \langle x_1, \ldots, \langle x_{n-1}, x_n\rangle \cdots\rangle \,\&\, \mathscr{P}(x_1, \ldots, x_n)).$$

Exercise 30.4. Let $\mathscr{V} = (M, \mathscr{R}_1, \ldots, \mathscr{R}_l)$ be acceptable. Fix a coding scheme $\mathscr{C}$ positive elementary on $\mathscr{V}$, then take the space $\mathscr{S}_1$ of example 30.2 based on M and $J = \lambda xy.\langle x, y\rangle$. Show that a second order relation $\mathscr{P}$ is positive inductive on $\mathscr{V}$ iff $\mathscr{P}^\sim$ is st-recursive in $\{\mathscr{R}_1^\sim, \ldots, \mathscr{R}_l^\sim\} \cup \mathscr{M}_1$. In particular, $\mathscr{P} \subseteq M \times 2^M$ is positive inductive on $\mathscr{V}$ iff it is st-recursive in $\{\mathscr{R}_1^\sim, \ldots, \mathscr{R}_l^\sim\} \cup \mathscr{M}_1$.

Remark. To incorporate the original 'boldface' inductiveness of Moschovakis [1974] one should therefore substitute $\mathscr{L} \cup \mathscr{M}_1$ for $\mathscr{M}_1$ and turn to the 'boldface' version of IOS-theory referred to in the remarks to exercise 10.9.

PART G

Epilogue

Following Kreisel [1971], the main purposes of axiomatizing (fragments of) Recursion Theory are: (i) To help advance and better understand other parts of logic and mathematics, especially Ordinary Recursion Theory and its specific generalizations known as Generalized Recursion Theory, as well as Theoretical Computer Science; (ii) To analyse the concept of effective computability; (iii) Ultimately, to provide both independent foundations and genuine axiomatic development of Recursion Theory.

Axiomatic approaches to Recursion Theory have been suggested or studied by Wagner [1969], Strong [1968], Friedman [1971a], Moschovakis [1971], Moldestad [1977], Fenstad [1980], Skordev [1980], Ivanov [1980], Fitting [1981], Zashev [1983] and others. Especially elaborate and deep is the exploration of the so called computation theories in Fenstad's book. Our discussion will concentrate on the present approach and those of Skordev and Zashev which form a separate trend, Algebraic Recursion Theory (ART for short).

Although it would be overoptimistic to expect early contributions of ART to other parts of logic and mathematics, some encouraging examples can be pointed out.

Let $\mathscr{F}$ be a class of elements together with a notion of effective computability, the subclass $\mathscr{U}$ consisting of all the computable elements. Informally speaking, if there exist elements $\mathscr{B}_0$ and finitely many operations $\mathscr{B}'$ simple enough to be assumed initial and satisfying

(1) $$\mathscr{U} = cl(\mathscr{B}_0/\mathscr{B}'),$$

then the notion of effective computability concerned is *structurizable.* For instance, the notion of μ-recursiveness is structurized by Kleene's definition adduced in chapter 2. (To be more precise, by its modification for unary functions given in the same chapter.) As an application of ART one obtains from 24.3 or by the corresponding result of Skordev [1980] that Moschovakis' prime computability is structurizable, a by no means obvious fact. Normal forms for prime and search computable functions are also obtained.

Theorem 29.4 however does not imply that the Kleene-recursiveness in higher types is structurizable which would indeed be surprising if true. One can easily get (1) by taking in $\mathscr{B}_0$ the universal function σ from chapter 29 but this would not yield a structurization since σ is not at all simple.

In another application of ART, the First Recursion Theorems of Moschovakis for functionals and second order relations are improved in chapters 28 and 30, respectively.

Results established in chapters 23 and 26 suggest that the mathematical foundations of Computer Science provide another perspective field for conceptual and practical applications. ART is closely related to such topics as semantics of programming languages, structured and functional programming, proving program correctness etc.

There is little doubt that algebraization brings new coherence and insight to Recursion Theory. In chapter 9 we obtained new proofs of basic classical theorems, establishing a direct connection between results of Ordinary and Generalized Recursion Theories which turn out to be not mere analogues but particular instances of more general results of ART. The new concepts of ART also help to better understand classical ones. For example, the algebraic treatment of the transition property and the least fixed point operator respectively as translation and iteration in consecutive spaces is new both in Ordinary and Generalized Recursion Theories. The storing operation throws more light on the nature of certain technicalities of Generalized Recursion Theory, Inductive Definability Theory and Computer Science, contributing also to the 'lightface–boldface' division in the general theory itself. Splitting has also fully established its importance.

And now, ART and effective computability; we have referred to our starting point in chapter 2, presumably with a deeper knowledge. The motivation adduced in chapters 2, 12, 13 and the practical work carried out in parts E, F lead to the thesis that all natural concepts of effective computability can be formulated within suitable IOS, at worst having to use consecutive spaces. Therefore, the comparison with abstract IOS-notions of effective computability such as recursiveness, t-recursiveness and $\mathscr{B}'$-recursiveness provides an objective criterion for judging the naturalness of a proposed notion of effective computability and aids the choice between several such notions.

To make practical use of the above thesis one should be able to identify those mathematical notions which are notions of effective computability. (They need not be necessarily so called.) This can be done either by their intuitive content or by their effectivity properties such as transition, recursion and enumeration.

While Church's thesis and its various extensions identify the widest intuitively acceptable notion of effective computability in a given context, it could be claimed that narrower notions are in a sense more basic. For one can describe relative partial recursiveness in terms of relative μ-recursiveness but not vice versa; prime computability similarly embraces the wider notions of Friedman's computability and search computability. This underlines once again the fundamentality of the initial elements and operations of IOS which seem to form a minimal collection.

In addition to structurization, a principal problem of effective computability partly solved so far is the characterization of all those collections of operations $\mathscr{B}'$ and elements $\mathscr{B}_0$ which generate reasonable notions of effective computability $\mathscr{U} = cl(\mathscr{B}_0/\mathscr{B}')$. Such a notion should in particular satisfy the recursion property, which means that all unary mappings constructed by composing operations from $\mathscr{B}'$ and using constants from $\mathscr{B}_0$ have fixed points in $\mathscr{U}$ which are least with respect to a certain partial order. A related question. When does a given notion $\mathscr{U}$ have structurizable extension's which satisfy the recursion property? It would be interesting in connection with these problems to characterize alternatively the operations over an arbitrary IOS which are (equivalent to) t-operations.

Broadly speaking, it is to be hoped that ART will help one analyse effective computability as Group Theory helps to analyse symmetry.

We complete our discussion by outlining some possible directions for further work.

The first such direction deals with recursion theory on structures more general than IOS; developments of this kind were initiated by Zashev's dissertation devoted to general recursion theory on combinatory incomplete applicative systems. These are partially ordered sets with several constants called combinators, multiplication and either union or pairing operation satisfying certain weaker associative and distributive laws. Among the systems in question is a generalized OS, partially ordered set $\mathscr{F}$ with monotonic operations $\circ$, Π and constants I, L, R, $A_1 - A_4$, $D_1 - D_4$ such that $\varphi I = \varphi$, $(\varphi\psi)\chi = \varphi(\psi(\chi A_1))$, $\varphi(\psi\chi) = (\varphi(\psi A_2))(\chi A_3)$, $\varphi\psi = (\varphi A_4)(\psi A_3)$, $(\varphi, \psi)\chi = D_1(\varphi(\chi D_2), \psi(\chi D_3))$, $(\varphi\chi, \psi\chi) = ((\varphi, \psi)D_4)(\chi A_3)$, $L(\varphi, \psi) = \varphi$ and $R(\varphi, \psi) = \psi$. All the OS satisfy these axioms with $A_1 = \cdots = D_4 = I$. To get an example which is not an OS take an infinite set M such that $M^2 = M$, then take $\mathscr{F} = \{\varphi/\varphi \subseteq M\}$ and $\varphi\psi = \{s/\exists t \in \varphi((s, t) \in \psi)\}$. A further introduction to the subject can be found in Zashev [1983–198?].

In another development Skordev combinatory spaces are modified in Petrov and Skordev [1979] by considering categories instead of semigroups; the same can be done with IOS. So far, however, such generalizations have been less successful.

On the other hand, one may find even IOS too general a system and try to distinguish classes of models by means of additional axioms. For instance, 'multiple-valued' spaces are distinguished by a constant U such that $L, R \leq U$. (The element U itself need not necessarily be 'multiple-valued'; in the space of exercise 19.4 it is a single-valued function.) The availability of a second counter is expressed by making use of the constants W, W_1, W_2 in chapter 21.

A more ambitious task would be to axiomatize (parts of) particular recursion theories, turning to their more delicate and specific problems hitherto untouched by ART. Since consecutive spaces will be needed in some cases, the following remark is in order. The exposition of chapters 12–14 is not properly axiomatic but can be made so by introducing the notion of *higher* IOS which is an IOS $\mathscr{S}' = (\mathscr{F}', I', \Pi', L', R')$ with constants Id, $Ml \in \mathscr{F}'$ such that $I'^{\wedge} = I'$, $L'^{\wedge} = L'$, $R'^{\wedge} = R'$, $Id\varphi' = Id$, $Ml = Ml(L', R')$, $Ml(I', Id) = I'$, $Ml((L', L'R'), R'^2)$, $\langle \varphi' \rangle = (R'[(Ml(L', L'Id), Ml(R', R'Id))]\varphi')^{\vee}$, $Ml((L', L'R'), R'^2)$, $\langle \varphi'^{\wedge} \rangle = (R'[(Ml(L', L'Id), Ml(R', R'Id))]\varphi')^{\wedge}$, $[\varphi'^{\wedge}] = (R'[(Id, Ml)]\varphi')^{\wedge}$ and $\forall\theta'(\varphi'\theta' Id \leq \psi'\theta' Id) \Rightarrow \varphi' \leq \psi'$, $\varphi'^{\wedge}$ standing for $Ml(\varphi' Id, I')$. If $\mathscr{S}$, $\mathscr{S}'$ are consecutive IOS, then $\mathscr{S}'$ is a higher IOS and its subspace $\mathscr{S}'^{\wedge}$ based on $\{\varphi'^{\wedge}/\varphi' \in \mathscr{F}'\}$ is just the isomorphic copy $\bar{\mathscr{S}}$ of $\mathscr{S}$.

Conversely, let $\mathscr{S}'$ be a higher *IOS*. Assign to each $\varphi' \in \mathscr{F}'$ a mapping $\varphi'^{\smile} = \lambda\theta'^{\wedge}.(\varphi'\theta')^{\wedge}$ and take the isomorphic copy $\mathscr{S}'^{\smile}$ of $\mathscr{S}'$ based on $\{\varphi'^{\smile}/\varphi' \in \mathscr{F}'\}$. Then $\mathscr{S}'^{\wedge}, \mathscr{S}'^{\smile}$ are consecutive IOS.

The theory of consecutive spaces $\mathscr{S}$, $\mathscr{S}'$ of which the former is a pairing space and the latter an IOS (as in example 28.1) is similarly axiomatized by merely adding to the IOS-axioms the last axiom listed above with O' substituted for *Id*.

Another important direction for further work is the development of an axiomatic degree theory on IOS. (We regard it as a fascinating challenge of its own rather than a necessary part of the justification of the present approach.)

One way to introduce degrees in IOS is as follows. Consider the *adjunction operation* $\vee : \mathscr{F}^2 \to \mathscr{F}$ satisfying the 'light face' axioms $O \vee \varphi = \varphi \vee O = \varphi$, $(\varphi \vee \psi) \vee \chi = \varphi \vee (\psi \vee \chi)$, $L \vee R = L$, $\alpha(\varphi \vee \psi) = \alpha\varphi \vee \alpha\psi$, $(\varphi \vee \psi)\alpha = \varphi\alpha \vee \psi\alpha$ and $(\varphi L \vee R)(\alpha, \psi) = \varphi\alpha \vee \psi$ for $\alpha = L, R$, assuming also $\langle I \rangle = I$ and $\varphi = \varphi RL \Rightarrow \varphi = O$. One takes $\varphi \vee \psi = \sup\{\varphi, \psi \restriction (M \setminus \mathrm{Dom}\,\varphi)\}$ in example 4.7 and similarly in other spaces; compare with Moschovakis [1977].

Writing φ^+ for $\varphi L \vee R$, let $\varphi \leq_{\mathrm{RE}} \psi$ iff φ is recursive in ψ^+, $\varphi \leq_{\mathrm{R}} \psi$ iff $\varphi^+ \leq_{\mathrm{RE}} \psi$, $\varphi =_{\mathrm{R}} \psi$ iff $\varphi \leq_{\mathrm{R}} \psi$ & $\psi \leq_{\mathrm{R}} \varphi$,

$$D = \{a \subseteq \mathscr{F} / a \neq \varnothing \ \&\ \forall \varphi\psi \in a(\varphi =_{\mathrm{R}} \psi)\},$$

$a \leq_{\mathrm{RE}} b$ iff $\exists\varphi \in a \exists\psi \in b(\varphi \leq_{\mathrm{RE}} \psi)$, and $a \leq b$ iff $\exists\varphi \in a \exists\psi \in b(\varphi \leq_{\mathrm{R}} \psi)$. (Recursiveness could be replaced by other abstract notions of effective computability, e.g. t-recursiveness and $\mathscr{B}'$-recursiveness.) Now it can be shown that for every *degree a* in *D* there is a degree a', the *jump of a*, such that $a < a' \leq_{\mathrm{RE}} a$ and whenever $b \leq_{\mathrm{RE}} a$, then $b \leq a'$. The proposed object of study is the upper semilattice $(D, \leq)$ with or without the jump operator. It is certain however that in order to establish abstract analogues of the basic Degree Theory results one will need to assume further axioms and, perhaps, invent new techniques. This also seems to be an occasion for the concept of 'finite', much advocated by Kreisel, to duly enter.

Apart from Degree Theory, the adjunction operation can be employed in building up abstract analogues of the arithmetical hierarchy of Ordinary Recursion Theory; cf. Ivanov [1985] for some details on these hierarchies and degrees and their relationship.

An obvious question suggested by the abovementioned analogy between IOS and groups is whether the former support an interesting theory of a classically algebraic kind beyond the Semigroup Theory. And, if the answer is positive, then what is the benefit for Recursion Theory?

To summarize, while the present state of ART shows convincingly the feasibility of some algebraization of Recursion Theory, it will fall to future works both in the pure and applied theory to supply a more definite idea of how far can all this actually go.

References

Apt, K.R. (1981) Ten years of Hoare's logic, a survey, *ACM Transl. Progr. Lang. Syst.*, **3**, 431–483.

Backus, J. (1978) Can programming be liberated from the von Neumann style? A functional style and its algebra of programs, *Comm. ACM*, **21**, 613–641.

Barwise, J. (ed.) (1977) *Handbook of Mathematical Logic*, Amsterdam.

Böhm, C., Jacopini, G. (1966) Flow diagrams, Turing machines and languages with only two formation rules, *Comm. ACM*, **9**, 366–371.

Cutland, N. (1980) *Computability, an Introduction to Recursive Function Theory*, Cambridge.

de Bakker, J.W. (1971) Recursive procedures, *Math. Centre Tracts*, No. 24.

de Bakker, J.W., Scott, D. (1969) A theory of programs, IBM Seminar, Vienna.

Ershov, Yu.L. (1977) *Theory of Numberings*, Moscow (in Russian).

Feferman, S. (1977) Inductive schemata and recursively continuous functionals, in: *Logic Colloquium '76* (eds. R.O. Gandy, J.M.E. Hyland), Amsterdam, pp. 373–392.

Fenstad, J.E. (1980) *General Recursion Theory, an Axiomatic Approach*, Berlin.

Fitting, M. (1981) *Fundamentals of Generalized Recursion Theory*, Amsterdam.

Friedman, H. (1971) Algorithmic procedures, generalized Turing algorithms, and elementary recursion theories, in: *Logic Colloquium '69* (eds. R.O. Gandy, C.E.M. Yates), Amsterdam, pp. 361–389.

Friedman, H. (1971a) Axiomatic recursive function theory, in: *Logic Colloquium '69* (eds. R.O. Gandy, C.E.M. Yates), Amsterdam, pp. 113–137.

Georgieva, N.V. (1980) Normal form theorems for some recursive elements and mappings, *Compt. Rend. Acad. Bulg. Sci.*, **33**, 1577–1580 (in Russian).

Germano, G., Maggiolo-Schettini, A. (1976) Recursivity, sequence recursivity, stack recursivity and semantics of programs, in: *Lect. Notes in Comp. Sci.*, No. 45, pp. 52–64.

Greibach, S.A. (1975) Theory of program structures: schemes, semantics, verification, *Lect. Notes in Comp. Sci.*, No. 36.

Hoare, C.A.R. (1969) An axiomatic basis for computer programming. *Comm. ACM*, **12**, 576–580.

Ivanov, L.L. (1980) Iterative operative spaces, Dissertation, Sofia University (in Bulgarian).

Ivanov, L.L. (1980a) Iterative operative spaces, *Compt. Rend. Acad. Bulg. Sci.*, **33**, 735–738 (in Russian).

Ivanov, L.L. (1980b) Some examples of iterative operative spaces, *Compt. Rend. Acad. Bulg. Sci.*, **33**, 877–879 (in Russian).

Ivanov. L.L. (1983) Iterative operative spaces and the system of Scott and de Bakker, *Serdica Bulg. Math. Publ.*, **9**, 275–288.

Ivanov, L.L. (1984) Kleene-recursiveness and iterative operative spaces, in: *Proceedings of the 1980 Conference on Mathematical Logic held at Sofia* (eds. D.G. Skordev *et al.*), Sofia, pp. 47–62.

Ivanov, L.L. (1984a) First order axioms for the foundations of Recursion Theory in:

Extended Abstracts of Short Talks of the 1982 Summer Institute on Recursion Theory (ed. I. Kalantari), *Recursive Function Theory Newsletter*, 55–60.

Ivanov, L.L. (1985) Abstract hierarchies and degrees, Oslo University, Institute of Mathematics (June 1985), Preprint No. 5.

Ivanov, L.L. (1987) Distributive spaces, in: *Mathematical Logic and Applications: Proceedings of the 1986 Summer School and Conference held at Druzhba* (ed. D.G. Skordev), London (to appear).

Kechris, A.S., Moschovakis, Y.N. (1977) Recursion in higher types, in: *Handbook of Mathematical Logic* (ed. J. Barwise), Amsterdam, pp. 681–737.

Kleene, S.C. (1952) *Introduction to Metamathematics*, Amsterdam.

Kleene, S.C. (1955) Hierarchies of number-theoretic predicates, *Bull. Amer. Math. Soc.*, **61**, 193–213.

Kleene, S.C. (1959) Recursive functionals and quantifiers of finite types I, *Trans. Amer. Math. Soc.*, **91**, 1–52.

Kleene, S.C. (1963) Recursive functionals and quantifiers of finite types II, *Trans. Amer. Math. Soc.*, **108**, 106–142.

Kleene, S.C. (1978) Recursive functionals and quantifiers of finite types revisited I, in: *Generalized Recursion Theory II* (eds. J.E. Fenstad, R.O. Gandy, G.E. Sacks), Amsterdam, pp. 185–222.

Kleene, S.C. (1981) The theory of recursive functions, approaching its centennial, *Bull. Amer. Math. Soc.* **5**, 43–61.

Kolaitis, Ph.G. (1978) On recursion in $\underset{\sim}{E}$ and semi-Spector classes, in: *Lect. Notes in Math.*, No. 689, pp. 209–243.

Kreisel, G. (1971) Some reasons for generalizing Recursion Theory, in: *Logic Colloquium '69* (eds. R.O. Gandy, C.E.M. Yates), Amsterdam, pp. 139–198.

Maltsev, A.I. (1961) Constructive algebra I, *Usp. Mat. Nauk*, **16**, 3, 3–60 (in Russian).

Maltsev, A.I. (1965) *Algorithms and Recursive Functions*, Moscow (in Russian).

Manna, Z. (1974) *Mathematical Theory of Computation*, New York.

Mazurkievicz, A. (1971) Proving algorithms by tail functions, *Inform. and Control*, **18**, 220–226.

McCarthy, J. (1963) A basis for a mathematical theory of computation, in: *Computer Programming and Formal Systems* (eds. P. Braffort, D. Hirschberg), Amsterdam, pp. 33–70.

Moldestad, J. (1977) Computations in higher types, *Lect. Notes in Math.*, No. 574.

Moldestad, J., Stoltenberg-Hansen V., Tucker, J.V. (1981) Finite algorithmic procedures and computation theories, *Math. Scand.*, **46**, 77–94.

Moschovakis, Y.N. (1969) Abstract first order computability, *Trans. Amer. Math. Soc.*, **138**, 427–504.

Moschovakis, Y.N. (1971) Axioms for computation theories—first draft, in: *Logic Colloquium '69* (eds. R.O. Gandy, C.E.M. Yates), Amsterdam, pp. 199–255.

Moschovakis, Y.N. (1974) *Elementary Induction on Abstract Structures*, Amsterdam.

Moschovakis, Y.N. (1977) On the basic notions in the theory of induction, in: *Logic, Foundations of Mathematics and Computability Theory* (eds. R.E. Butts, J. Hintikka), Dordrecht, pp. 207–236.

Moschovakis, Y.N. (1984) Abstract recursion as a foundation for the theory of algorithms, in: *Lect. Notes in Math.*, No. 1104, pp. 289–364.

Myhill, J. (1961) Note on degrees of partial functions, *Proc. Amer. Math. Soc.*, **12**, 519–521.

Normann, D. (1980) Recursion on the countable functionals, *Lect. Notes in Math.*, No. 811.

Petrov, V.P., Skordev, D.G. (1979) Combinatory structures, *Serdica Bulg. Math. Publ.*, **5**, 128–148 (in Russian).

Platek, R.A. (1966) Foundations of recursion theory, Dissertation, Stanford University.

Rice, H.G. (1953) Classes of recursively enumerable sets and their decision problems, *Trans. Amer. Math. Soc.*, **74**, 358–366.

Rogers, H., Jr. (1967) *Theory of Recursive Functions and Effective Computability*, New York.

Scott, D. (1971) The lattice of flow diagrams, in: *Lect. Notes in Math.*, No. 188, pp. 311–366.

Shepherdson, J.C. (1975) Computation over abstract structures, in: *Logic Colloquium '73* (eds. H.E. Rose, J.C. Shepherdson), Amsterdam, pp. 445–513.

Shoenfield, J.R. (1967) *Mathematical Logic*, Reading, Massachusetts.

Skordev, D.G. (1963) Computable and μ-recursive operators, *Izv. Mat. Inst. BAN*, **7**, 5–43 (in Bulgarian, Russian summary).

Skordev, D.G. (1976) Recursion theory on iterative combinatory spaces, *Bull. Acad. Polon. Sci.*, **24**, 23–31.

Skordev, D.G. (1979) The First Recursion Theorem for iterative combinatory spaces, *Z. Math. Logik Grundl. Math.*, **25**, 69–77.

Skordev, D.G. (1980) *Combinatory Spaces and Recursiveness in them*, Sofia (in Russian, English summary).

Skordev, D.G. (1980a) Semicombinatory spaces, *Compt. Rend. Acad. Bulg. Sci.*, **33**, 739–742 (in Russian).

Skordev, D.G. (1982) An algebraic treatment of flow diagrams and its application to Generalized Recursion Theory, *Banach Center Publ.*, No. 9, pp. 277–287.

Skordev, D.G. (1982a) Applications of abstract recursion theory for studying the capability of functional programming systems, in: *Mathematical Theory and Practice of Softwear Systems* (ed. A.P. Ershov) Novosibirsk, pp. 7–16 (in Russian).

Skordev, D.G. (1983) A reduction of polyadic recursive programs to monadic ones, in: *Symposium on Mathematical Foundations of Computer Science, Seminarbericht 52, Humboldt University Berlin*, pp. 124–132.

Skordev, D.G. (1984) First Recursion Theorem for iterative semicombinatory spaces, in: *Proceedings of the 1980 Conference on Mathematical Logic held at Sofia* (eds. D.G. Skordev *et al.*), Sofia, pp. 89–111 (in Russian).

Soare, R.I. (1986) *Recursively Enumerable Sets and Degrees: The Study of Computable Functions and Computably Generated Sets*, Berlin.

Soskov, I.N. (1983) Computability in algebraic systems, *Compt. Rend. Acad. Bulg. Sci.*, **36**, 301–304 (in Russian).

Strong, H.R., Jr. (1968) Algebraically generalized recursive function theory, *IBM J. Res. Devel.*, **12**, 465–475.

Tarski, A. (1955) A lattice theoretical fixpoint theorem and its applications, *Pacific J. Math.*, **5**, 289–309.

Wagner, E.G. (1969) Uniformly reflexive structures: On the nature of Gödelization and relative computability, *Trans. Amer. Math. Soc.*, **144**, 1–41.

Zashev, J.A. (1983) Recursion theory in partially ordered combinatory models, Dissertation, Sofia University (in Bulgarian).

Zashev, J.A. (1984) Basic recursion theory in partially ordered models of some fragments of Combinatory Logic, *Compt. Rend. Acad. Bulg. Sci.*, **37**, 561–564.

Zashev, J.A., B-combinatory algebras, *Serdica Bulg. Math. Publ.* (to appear).

Zashev, J.A. Recursion theory in B-combinatory algebras, *Serdica Bulg. Math. Publ.* (to appear).

List of symbols

Subject Index

Mathematics and its Applications

Series Editor: G. M. BELL, Professor of Mathematics, King's College (KQC), University of London

Artmann, B.	**The Concept of Number***
Balcerzyk, S. & Joszefiak, T.	**Commutative Rings***
Balcerzyk, S. & Joszefiak, T.	**Noetherian and Krull Rings***
Baldock, G.R. & Bridgeman, T.	**Mathematical Theory of Wave Motion**
Ball, M.A.	**Mathematics in the Social and Life Sciences: Theories, Models and Methods**
de Barra, G.	**Measure Theory and Integration**
Bell, G.M. and Lavis, D.A.	**Co-operative Phenomena in Lattice Models Vols. I & II***
Berkshire, F.H.	**Mountain and Lee Waves**
Berry, J.S., Burghes, D.N., Huntley, I.D., James, D.J.G. & Moscardini, A.O.	**Teaching and Applying Mathematical Modelling**
Burghes, D.N. & Borrie, M.	**Modelling with Differential Equations**
Burghes, D.N. & Downs, A.M.	**Modern Introduction to Classical Mechanics and Control**
Burghes, D.N. & Graham, A.	**Introduction to Control Theory, including Optimal Control**
Burghes, D.N., Huntley, I. & McDonald, J.	**Applying Mathematics**
Burghes, D.N. & Wood, A.D.	**Mathematical Models in the Social, Management and Life Sciences**
Butkovskiy, A.G.	**Green's Functions and Transfer Functions Handbook**
Butkovskiy, A.G.	**Structural Theory of Distributed Systems**
Cao, Z-Q., Kim, K.H. & Roush, F.W.	**Incline Algebra and Applications**
Chorlton, F.	**Textbook of Dynamics, 2nd Edition**
Chorlton, F.	**Vector and Tensor Methods**
Crapper, G.D.	**Introduction to Water Waves**
Cross, M. & Moscardini, A.O.	**Learning the Art of Mathematical Modelling**
Cullen, M.R.	**Linear Models in Biology**
Dunning-Davies, J.	**Mathematical Methods for Mathematicians, Physical Scientists and Engineers**
Eason, G., Coles, C.W. & Gettinby, G.	**Mathematics and Statistics for the Bio-sciences**
Exton, H.	**Handbook of Hypergeometric Integrals**
Exton, H.	**Multiple Hypergeometric Functions and Applications**
Exton, H.	***q*-Hypergeometric Functions and Applications**
Faux, I.D. & Pratt, M.J.	**Computational Geometry for Design and Manufacture**
Firby, P.A. & Gardiner, C.F.	**Surface Topology**
Gardiner, C.F.	**Modern Algebra**
Gardiner, C.F.	**Algebraic Structures: with Applications**
Gasson, P.C.	**Geometry of Spatial Forms**
Goodbody, A.M.	**Cartesian Tensors**
Goult, R.J.	**Applied Linear Algebra**
Graham, A.	**Kronecker Products and Matrix Calculus: with Applications**
Graham, A.	**Matrix Theory and Applications for Engineers and Mathematicians**
Griffel, D.H.	**Applied Functional Analysis**
Griffel, D.H.	**Linear Algebra***
Hanyga, A.	**Mathematical Theory of Non-linear Elasticity**
Harris, D.J.	**Mathematics for Business, Management and Economics**
Hoksins, R.F.	**Generalised Functions**
Hoskins, R.F.	**Standard and Non-standard Analysis***
Hunter, S.C.	**Mechanics of Continuous Media, 2nd (Revised) Edition**
Huntley, I. & Johnson, R.M.	**Linear and Nonlinear Differential Equations**
Jaswon, M.A. & Rose, M.A.	**Crystal Symmetry: The Theory of Colour Crystallography**
Johnson, R.M.	**Theory and Applications of Linear Differential and Difference Equations**
Kim, K.H. & Roush, F.W.	**Applied Abstract Algebra**
Kosinski, W.	**Field Singularities and Wave Analysis in Continuum Mechanics**
Krishnamurthy, V.	**Combinatorics: Theory and Applications**
Lindfield, G. & Penny, J.E.T.	**Microcomputers in Numerical Analysis**
Lord, E.A. & Wilson, C.B.	**The Mathematical Description of Shape and Form**
Marichev, O.I.	**Integral Transforms of Higher Transcendental Functions**
Massey, B.S.	**Measures in Science and Engineering**
Meek, B.L. & Fairthorne, S.	**Using Computers**
Mikolas, M.	**Real Function and Orchogonal Series**
Moore, R.	**Computational Functional Analysis**
Müller-Pfeiffer, E.	**Spectral Theory of Ordinary Differential Operators**
Murphy, J.A. & McShane, B.	**Computation in Numerical Analysis***
Nonweiller, T.R.F.	**Computational Mathematics: An Introduction to Numerical Approximation**
Ogden, R.W.	**Non-linear Elastic Deformations**
Oldknow, A. & Smith, D.	**Learning Mathematics with Micros**
O'Neill, M.E. & Chorlton, F.	**Ideal and Incompressible Fluid Dynamics**
O'Neill, M.E. & Chorlton, F.	**Viscous and Compressible Fluid Dynamics***

Page, S. G.	**Mathematics: A Second Start**
Rankin, R.A.	**Modular Forms**
Ratschek, H. & Rokne, J.	**Computer Methods for the Range of Functions**
Scorer, R.S.	**Environmental Aerodynamics**
Smith, D.K.	**Network Optimisation Practice: A Computational Guide**
Srivastava, H.M. & Karlsson, P.W.	**Multiple Gaussian Hypergeometric Series**
Srivastava, H.M. & Manocha, H.L.	**A Treatise on Generating Functions**
Shivamoggi, B.K.	**Stability of Parallel Gas Flows***
Stirling, D.S.G.	**Mathematical Analysis***
Sweet, M.V.	**Algebra, Geometry and Trigonometry in Science, Engineering and Mathematics**
Temperley, H.N.V. & Trevena, D.H.	**Liquids and Their Properties**
Temperley, H.N.V.	**Graph Theory and Applications**
Thom, R.	**Mathematical Models of Morphogenesis**
Toth, G.	**Harmonic and Minimal Maps**
Townend, M. S.	**Mathematics in Sport**
Twizell, E.H.	**Computational Methods for Partial Differential Equations**
Wheeler, R.F.	**Rethinking Mathematical Concepts**
Willmore, T.J.	**Total Curvature in Riemannian Geometry**
Willmore, T.J. & Hitchin, N.	**Global Riemannian Geometry**
Wojtynski, W.	**Lie Groups and Lie Algebras***

Statistics and Operational Research

Editor: B. W. CONOLLY, Professor of Operational Research, Queen Mary College, University of London

Beaumont, G.P.	**Introductory Applied Probability**
Beaumont, G.P.	**Probability and Random Variables***
Conolly, B.W.	**Techniques in Operational Research: Vol. 1, Queueing Systems***
Conolly, B.W.	**Techniques in Operational Research: Vol. 2, Models, Search, Randomization**
Conolly, B.W.	**Lecture Notes in Queueing Systems**
French, S.	**Sequencing and Scheduling: Mathematics of the Job Shop**
French, S.	**Decision Theory: An Introduction to the Mathematics of Rationality**
Griffiths, P. & Hill, I.D.	**Applied Statistics Algorithms**
Hartley, R.	**Linear and Non-linear Programming**
Jolliffe, F.R.	**Survey Design and Analysis**
Jones, A.J.	**Game Theory**
Kemp, K.W.	**Dice, Data and Decisions: Introductory Statistics**
Oliveira-Pinto, F.	**Simulation Concepts in Mathematical Modelling***
Oliveira-Pinto, F. & Conolly, B.W.	**Applicable Mathematics of Non-physical Phenomena**
Schendel, U.	**Introduction to Numerical Methods for Parallel Computers**
Stoodley, K.D.C.	**Applied and Computational Statistics: A First Course**
Stoodley, K.D.C., Lewis, T. & Stainton, C.L.S.	**Applied Statistical Techniques**
Thomas, L.C.	**Games, Theory and Applications**
Whitehead, J.R.	**The Design and Analysis of Sequential Clinical Trials**

**In preparation*